# Lecture Notes in Mathematics

1989

**Editors:**
J.-M. Morel, Cachan
F. Takens, Groningen
B. Teissier, Paris

T0184485

**FONDAZIONE CIME**
ROBERTO CONTI
CENTRO INTERNAZIONALE MATEMATICO ESTIVO
INTERNATIONAL MATHEMATICAL SUMMER CENTER

C.I.M.E. means Centro Internazionale Matematico Estivo, that is, International Mathematical Summer Center. Conceived in the early fifties, it was born in 1954 and made welcome by the world mathematical community where it remains in good health and spirit. Many mathematicians from all over the world have been involved in a way or another in C.I.M.E.'s activities during the past years.

So they already know what the C.I.M.E. is all about. For the benefit of future potential users and co-operators the main purposes and the functioning of the Centre may be summarized as follows: every year, during the summer, Sessions (three or four as a rule) on different themes from pure and applied mathematics are offered by application to mathematicians from all countries. Each session is generally based on three or four main courses (24–30 hours over a period of 6-8 working days) held from specialists of international renown, plus a certain number of seminars.

A C.I.M.E. Session, therefore, is neither a Symposium, nor just a School, but maybe a blend of both. The aim is that of bringing to the attention of younger researchers the origins, later developments, and perspectives of some branch of live mathematics.

The topics of the courses are generally of international resonance and the participation of the courses cover the expertise of different countries and continents. Such combination, gave an excellent opportunity to young participants to be acquainted with the most advance research in the topics of the courses and the possibility of an interchange with the world famous specialists. The full immersion atmosphere of the courses and the daily exchange among participants are a first building brick in the edifice of international collaboration in mathematical research.

C.I.M.E. Director
Pietro ZECCA
Dipartimento di Energetica "S. Stecco"
Università di Firenze
Via S. Marta, 3
50139 Florence
Italy
e-mail: zecca@unifi.it

C.I.M.E. Secretary
Elvira MASCOLO
Dipartimento di Matematica
Università di Firenze
viale G.B. Morgagni 67/A
50134 Florence
Italy
e-mail: mascolo@math.unifi.it

For more information see CIME's homepage: http://www.cime.unifi.it

CIME activity is carried out with the collaboration and financial support of:

– INdAM (Istituto Nazionale di Alta Matematica)

Immanuel M. Bomze · Vladimir Demyanov
Roger Fletcher · Tamás Terlaky

# Nonlinear Optimization

Lectures given at the
C.I.M.E. Summer School
held in Cetraro, Italy
July 1–7, 2007

Editors:
Gianni Di Pillo
Fabio Schoen

 Springer

FONDAZIONE
CIME
ROBERTO CONTI

*Editors*
Gianni Di Pillo
Sapienza, Università di Roma
Dipto. Informatica e Sistemistica
Via Ariosto, 25
00185 Rome
Italy
dipillo@dis.uniroma1.it

Fabio Schoen
Università degli Studi di Firenze
Dipto. Sistemi e Informatica
Via di Santa Marta, 3
50139 Firenze
Italy
fabio.schoen@unifi.it

*Authors: see List of Contributors*

ISSN: 0075-8434          e-ISSN: 1617-9692
ISBN: 978-3-642-11338-3     e-ISBN: 978-3-642-11339-0
DOI: 10.1007/978-3-642-11339-0
Springer Heidelberg Dordrecht London New York

Library of Congress Control Number: 2010923242

Mathematics Subject Classification (2000): 90C30; 90C26; 90C25; 90C55

*Cover images*: stills images from video "labirinto" - antonellabussanich@yahoo.fr, © VG Bild-Kunst, Bonn 2010

*Cover design*: WMXDesign GmbH

Printed on acid-free paper

springer.com

# Preface

This volume collects the expanded notes of four series of lectures given on the occasion of the CIME course on *Nonlinear Optimization* held in Cetraro, Italy, from July 1 to 7, 2007.

The Nonlinear Optimization problem of main concern here is the problem of determining a vector of *decision variables* $x \in \mathbb{R}^n$ that minimizes (maximizes) an *objective function* $f(\cdot): \mathbb{R}^n \to \mathbb{R}$, when $x$ is restricted to belong to some *feasible set* $\mathcal{F} \subseteq \mathbb{R}^n$, usually described by a set of *equality and inequality constraints*: $\mathcal{F} = \{x \in \mathbb{R}^n : h(x) = 0, h(\cdot): \mathbb{R}^n \to \mathbb{R}^m; g(x) \leq 0, g(\cdot): \mathbb{R}^n \to \mathbb{R}^p\}$; of course it is intended that at least one of the functions $f, h, g$ is nonlinear. Although the problem can be stated in very simple terms, its solution may result very difficult due to the analytical properties of the functions involved and/or to the number $n, m, p$ of variables and constraints. On the other hand, the problem has been recognized to be of main relevance in engineering, economics, and other applied sciences, so that a great lot of effort has been devoted to develop methods and algorithms able to solve the problem even in its more difficult and large instances.

The lectures have been given by eminent scholars, who contributed to a great extent to the development of Nonlinear Optimization theory, methods and algorithms. Namely, they are:

- Professor Immanuel M. BOMZE, University of Vienna, Austria
- Professor Vladimir F. DEMYANOV, St. Petersburg State University, Russia
- Professor Roger FLETCHER, University of Dundee, UK
- Professor Tamás TERLAKY, McMaster University, Hamilton, Ontario, Canada (now at Lehigh University, Bethlehem, PA - USA).

The lectures given by Roger Fletcher deal with a basic framework for treating the Nonlinear Optimization problem in the smooth case, that is the Sequential Quadratic Programming (SQP) approach. The SQP approach can be considered as an extension to constrained problems of Newton's method for unconstrained minimization. Indeed, the underlying idea of the SQP approach is that of applying Newton's method to solve the nonlinear equations given by the first order necessary conditions for optimality. In order to fully develop the idea, the required optimality conditions for the constrained problem are recalled. Then the basic SQP method is introduced and some issues of the method are discussed: in particular the requirement of avoiding the evaluation of second order derivatives, and the occurrence of infeasibility in solving the QP subproblems. However the basic SQP method turns out to be only locally convergent, even if with a superlinear convergence rate. Therefore the need arises of some globalization strategy, that retains the good convergence rate. In particular, two classes of globalization strategies are considered, the first one using some merit function, the second one resorting to some filter method. Filter methods are of main concern in the context of the course, since they have been introduced and developed by Roger Fletcher himself. A last section of the lecture notes deals with the practical problem of interfacing a model of the nonlinear programming problem with a code for its solution. Different modelling languages are mentioned, and a short introduction to AMPL is provided.

The lectures given by Tamás Terlaky, whose chapter is co-authored by Imre Pólik, focus on the Interior Point Methods (IPM), that arose as the main novelty in linear optimization in the eighties of the last century. The interesting point is that the IPM, originally developed for linear optimization, is deeply rooted in nonlinear optimization, unlike the simplex method used until before. It was just the broadening of the horizon from linear to nonlinear, that allowed to describe for the first time an algorithm for linear optimization not only with polynomial complexity but also with competitive performances. The lecture notes first review the IPM for linear optimization, by introducing the self dual-model into which every linear optimization problem can be embedded; the basic notion of central path is defined, and its existence and convergence are analyzed; it is shown that, by a rounding procedure on the central path, a solution of the problem can be found in a polynomial number of arithmetic operations. On these bases, a general scheme of IP algorithms for linear optimization is provided, and several implementation issues are considered. Then, the more general problem of conic optimization is addressed, relying on the fact that most of theoretical results and algorithmic considerations valid for the linear case carry over to the conic case with only minor modifications. Moreover conic optimization represents a step in the pathway from linear to nonlinear optimization. The interest in conic optimization is motivated by important applications, like robust linear optimization, eigenvalue optimization, relaxing of binary variables. In particular, two special classes of conic optimization problems are considered, namely second order conic optimization and semidefinite optimization, and for each class a well suited IPM is described. Finally, the interior point approach is extended to nonlinear optimization, by employing the key of a reformulation of the nonlinear optimization problem as a nonlinear complementarity problem. In this way a central path can be defined also for the nonlinear case, even if its existence and convergence require stronger assumptions than in the linear or conic cases, and complexity results hold only in the convex case. The analytical and algorithmic analysis of IPM is complemented by an overview of existing software implementations, pointing out that some of them are available in leading commercial packages. A challenging list of open questions, concerning mainly algorithmic issues, concludes these lecture notes.

The methods mentioned before are able to find only local solutions of the Nonlinear Optimization problem. In his lectures Immanuel Bomze considers the much more difficult problem of Global Optimization, that is the problem of finding global, rather than local, solutions. In order to fully explain how a gap in difficulty of the problem arises, he makes reference to the simplest nonlinear optimization problem, that is quadratic programming, minimizing a quadratic objective function under linear constraints. If the quadratic objective function is nonconvex, this problem may have so many local non global solutions that any enumerative strategy is not viable. A particular feature of the nonconvex quadratic programming is that necessary and sufficient global optimality conditions can be stated, which not only provide a certificate of optimality for a current tentative solution, but also an improving feasible point if the conditions are not satisfied. These conditions rely on the notion of copositivity, which is central in algorithmic developments. Moreover, additional optimality conditions can be stated in terms of nonsmooth analysis, thus establishing a link with the contents of the lectures by Vladimir Demyanov introduced below. A particular instance of a quadratic programming problem is the so-called Standard Quadratic Programming Problem (StQP), where the feasible set is the unitary simplex. StQP is used to illustrate the basic techniques available for searching global solutions; among these, the well known branch-and-bound approach borrowed from combinatorial optimization. Again, StQP is used to illustrate approaches by which the problem may be in some way reformulated, relaxed or approximated in order to obtain a good proxy of its exact global solution. Finally, a section deals with detecting copositivity, a problem known to be in general NP-hard.

In the Nonlinear Optimization problems considered up to now, the functions $f$, $h$, $g$, are assumed to be smooth, that is at least continuously differentiable. In his lectures, Vladimir Demyanov faces the much more difficult case of nonsmooth optimization. The smooth case can be characterized as the "kingdom of gradient", due to the main role played by the notion of gradient in establishing optimality conditions and in detecting improving feasible solutions. Therefore, a first challenge, when moving outside of that kingdom, is to provide analytical notions able to perform, in some way, the same role. To this aim, different definitions of differentiability and of set-valued subdifferential are introduced, where each element of the subdifferential is, in some sense, a generalized gradient. On these bases, it is possible to establish optimality conditions for nonsmooth optimization problems, not only when the decision variable belongs to the usual $\mathbb{R}^n$ finite dimensional space, but also when it belongs to a more general metric or normed space. More in particular, first the case of unconstrained optimization problems, and then the case of constrained optimization problems are considered. It is remarkable the fact that a nonsmooth constrained optimization problem can always be transformed into a nonsmooth unconstrained optimization problem by resorting to an exact nondifferentiable penalty functions that accounts for the constraints. Therefore, an amazing feature of nonsmooth optimization is that, in principle, the presence of constraints does not add analytical difficulties with respect to the unconstrained case, as it happens if the same exact penalty approach is adopted in smooth optimization.

The course took place in the wonderful location of San Michele Hotel in Cetraro and was attended by 34 researchers from 9 different countries. The course was organized in 6 days of lectures, with each lecturer presenting his course material in 5 parts. The course was indeed successful for its scientific interest and for the friendly environment - this was greatly facilitated by the beauty of the course location and by the professional and warm atmosphere created by the organizers and by all of the staff of Hotel San Michele.

We are very grateful with CIME for the opportunity given of organizing this event and for the financial as well as logistic support; we would like to thank in particular CIME Director, prof. Pietro Zecca, for his continuous encouragement and friendly support before, during and after the School; we also would like to thank Irene Benedetti for her help and participation during the School, and all of the staff of CIME, who made a great effort for the success of this course. In particular we would like to thank Elvira Mascolo, CIME Scientific Secretary, for her precious work in all parts of the organization of the School, and Francesco Mugelli who maintained the web site.

Gianni Di Pillo and Fabio Schoen

# Contents

# Contributors

**Immanuel M. Bomze** Department of Statistics and Operations Research, University of Vienna, 1010 Wien, Austria, immanuel.bomze@univie.ac.at

**Vladimir F. Demyanov** Applied Mathematics Dept., St.-Petersburg State University, Staryi Peterhof, St.-Petersburg, 198504 Russia, vfd@ad9503.spb.edu

**Roger Fletcher** Department of Mathematics, University of Dundee, Dundee DD1 4HN, fletcher@maths.dundee.ac.uk

**Imre Pólik** Department of Industrial and Systems Engineering, Lehigh University, 200 West Packer Avenue, 18015-1582, Bethlehem, PA, USA, imre@polik.net

**Tamás Terlaky** Department of Industrial and Systems Engineering, Lehigh University, 200 West Packer Avenue, 18015-1582, Bethlehem, PA, USA, terlaky@lehigh.edu

# Global Optimization: A Quadratic Programming Perspective

Immanuel M. Bomze

*Special thanks to Gianni Di Pillo and Fabio Schoen,*
*for their continuous support and valuable remarks*

## Introduction

Global optimization is a highly active research field in the intersection of continuous and combinatorial optimization (a basic web search delivers over a million hits for this phrase and for its British cousin, *Global Optimisation*). A variety of methods have been devised to deal with this problem class, which – borrowing biological taxonomy terminology in a very superficial way – may be divided roughly into the two domains of *exact/rigorous methods* and *heuristics*, the difference between them probably being that you can prove less theorems in the latter domain. Breaking the domain of exact methods into two phyla of *deterministic methods* and *stochastic methods*, we may have the following further taxonomy of the deterministic phylum:

$$\text{exhaustive methods} \begin{cases} \text{passive/direct, streamlined enumeration} \\ \text{homotopy, trajectory methods} \end{cases}$$

$$\text{methods using global structure} \begin{cases} \text{smoothing, filling, parameter continuation} \\ \text{hierarchical funnel, difference-of-convex} \end{cases}$$

I.M. Bomze (✉)
Department of Statistics and Operations Research, University of Vienna,
1010 Wien, Austria
e-mail: immanuel.bomze@univie.ac.at

G. Di Pillo et al. (eds.), *Nonlinear Optimization*, Lecture Notes in
Mathematics 1989, DOI 10.1007/978-3-642-11339-0_1,
© Springer-Verlag Berlin Heidelberg 2010

$$\text{iterative improvement methods} \begin{cases} \text{escape, tunneling, deflation, aux.functions} \\ \text{successive approximation, minorants} \end{cases}$$

implicit enumeration methods: branch & bound

In these notes, we will focus on a problem class which serves as an application model for some of the above techniques, but which mathematically nevertheless is of surprisingly simple structure – basically the next step after Linear Optimization, namely Quadratic Optimization. Despite the fact that curvature of the objective function is constant and that constraints are linear, quadratic problems exhibit all basic difficulties you may encounter in global optimization: a multitude of inefficient local solutions; global solutions with a very narrow domain of attraction for local solvers; and instances where you encounter very early the optimal solution, but where you find a certificate for global optimality of this solutions, or even only a satisfying rigorous bound very late – a case of particular nuisance in applications.

The contents of these notes are organized as follows: Section 1 deals with local and global optimality conditions in the quadratic world. Due to the constant curvature of the objective function, conditions for both local and global optimality can be formulated in a compact way using the notion of copositivity. It turns out that this class also allows for closing the gap between necessary and sufficient conditions in most cases. $\varepsilon$-subdifferential calculus is used to analyse these conditions in a more general framework, going beyond Quadratic Optimization to the quite general theory of difference-of-convex (d.c.) optimization. To emphasize how close continuous global optimization is tied to discrete problems, we investigate a particular class of quadratic problems, the so-called Standard Quadratic Problems (StQPs) which simply consist of extremizing a quadratic form of the standard simplex – and yet form an NP hard problem class with immediate applications in combinatorial optimization. We continue the study of StPQs in Section 2, to exemplify some basic global optimization techniques like determining escape directions and rigorous bounds, as well as the basic steps in branch-and-bound. Section 3 is devoted to different approaches to global quadratic optimization, namely relaxation and approximation, but also exact reformulation. As an example for the latter, we discuss an emerging branch of optimization which receives rapidly increasing interest in contemporary scientific community: copositive optimization. Again applied to StQPs, the copositive reformulation means that a global quadratic optimization problem is rewritten as a linear programming problem over a convex cone of matrices, thereby completely avoiding the problem of inefficient local solutions. The hardness of the problem is completely shifted to sheer feasibility, and this new aspect opens up a variety of different methods to approach the global solution of the original problem (the StQP in our case). The cone of copositive matrices is known since the mid-fifties of the last century, however, algorithmic approaches to detect whether or not a given matrix satisfies this condition, are much more recent. The last Section 4 is devoted to these aspects, and also discusses some complexity issues. A by now well-established technology for conic optimization which gained momentum since practical implementations of interior-point methods were available, is Semidefinite Programming (SDP), where the matrix cone is that of positive-semidefinite matrices. Since checking positive-semidefiniteness of a given matrix is relatively easy, SDPs can be solved to any prescribed accuracy in polynomial time. Section 4 also describes how SDP-based bounds arising from approximating copositivity via SDP technology can be reinterpreted in the decomposition context of the earlier Section 2.

In the sequel, we will employ the following notation: $^\top$ stands for transposition of a (column) vector in $n$-dimensional Euclidean space $\mathbb{R}^n$; for two such vectors $\{x, y\} \subset \mathbb{R}^n$, we denote by $x \leq y$ the fact that $x_i \leq y_i$ for all $i$. The letters $o$, $O$, $0$ stand for the zero vector, matrix, or number, respectively, all of appropriate dimension. The positive orthant is denoted by $\mathbb{R}^n_+ = \{x \in \mathbb{R}^n : o \leq x\}$, the $n \times n$ identity matrix by $I_n$, with $i$-th column $e_i$ (the $i$th standard basis vector). $e = \sum_i e_i \in \mathbb{R}^n$ is the all-ones vector and $E_n = ee^\top$ the all-ones $n \times n$ matrix. For a finite set $A$, we denote its cardinality by $|A|$. If $v$ is a vector in $\mathbb{R}^n$, we denote by $\operatorname{Diag} v$ the diagonal $n \times n$ matrix $D$ with $d_{ii} = v_i$, for $i = 1, \ldots, n$. Conversely, for an $n \times n$ matrix $B$, $\operatorname{diag} B = [b_{ii}]_i \in \mathbb{R}^n$ denotes the $n$-dimensional vector formed by the diagonal elements of $B$. Finally, we abbreviate by $S \succeq O$ the fact that a symmetric $n \times n$ matrix $S$ is positive-semidefinite (psd), and by $N \geq O$ the fact that $N$ has no negative entries.

# 1 Global Optimization of Simplest Structure: Quadratic Optimization

Let $Q = Q^\top$ be a symmetric $n \times n$ matrix, $A$ an $m \times n$ matrix and $b \in \mathbb{R}^m$. The feasible set of a quadratic optimization problem (QP) is a polyhedron which can be described as the intersection of finitely many half-spaces:

$$M = \{x \in \mathbb{R}^n : Ax \leq b\}.$$

Hence let us consider here

$$\min\left\{f(x) = \tfrac{1}{2}x^\top Q x + c^\top x : x \in M\right\}, \qquad (1.1)$$

as the basic model of a QP. To conveniently formulate optimality conditions, we frequently employ the *Lagrangian function*

$$L(x; u) = f(x) + u^\top(Ax - b), \quad u \in \mathbb{R}^m_+$$

being the vector of (non-negative) Lagrange multipliers for the (inequality) constraints. In our case, the gradient of $L$ with respect to $x$ at a point $(\overline{x}; u)$ is affine (i.e., a linear map plus a constant vector) and reads $\nabla_x L(\overline{x}; u) = Q\overline{x} + c + A^\top u$.

If $Q$ is positive-semidefinite ($x^\top Q x \geq 0$ for all $x \in \mathbb{R}^n$), then (1.1) is (relatively) easy: the basin(s) of attraction of the global solution(s) are universal, and virtually all local solvers will deliver it. Note that the first-order optimality conditions

$$\nabla_x L(\overline{x}; u) = o \quad \text{and} \quad u^\top (A\overline{x} - b) = 0, \quad (\overline{x}, u) \in M \times \mathbb{R}^m_+ \qquad (1.2)$$

are in this case necessary and sufficient for global optimality, and can be recast into a *Linear Complementarity Problem* which may be solved by complementary pivoting methods like Lemke's algorithm [24].

However, if $Q$ is indefinite, then even with a few simple constraints, problem (1.1) is really difficult. As a running example, we will consider the problem to determine the farthest point from a point $c$, in the hypercube $M = [-1, 1]^n$:

**Example 1.** Let $Q = -I_n$, and $A = [I_n | - I_n]^\top$ with $b = [e^\top | - e^\top]^\top$. If $c_i \in [-1, 1]$, then it is easily seen that all $y \in \mathbb{R}^n$ with $y_i \in \{-1, c_i, 1\}$, all $n$, are KKT points, and that all $x \in \{-1, 1\}^n$ (the vertices of the hypercube) are (local) solutions. If $c = o$, all vertices are evidently global solutions. However, if we consider $c \neq o$, this renders a unique global solution, while now all other $(2^n - 1)$ local solutions are inefficient; even more drastically, $(3^n - 1)$ KKT points, i.e., solutions of (1.2) are inefficient. We slightly simplify calculations by restricting ourselves to $c = -\mu e$ where $0 < \mu < 1$. The unique global solution then is the positive vertex $x^* = e$ of $M$.

## 1.1 Local Optimality Conditions in QPs

The first-order KKT conditions (1.2) help us to single out finitely many candidates ($3^n$ in Example 1) for being optimal solutions. Note that the complementary slackness conditions – in our case $u^\top (A\overline{x} - b) = 0$ – at $(\overline{x}, u) \in M \times \mathbb{R}^m_+$ always mean $L(\overline{x}; u) = f(\overline{x})$ in terms of the Lagrangian, while primal-dual feasibility always implies $L(x; u) \leq f(x)$, by construction of $L$, for all $(x, u) \in M \times \mathbb{R}^m_+$.

Now, to remove $3^n - 2^n$ candidates in Example 1 above, we have to employ second-order optimality conditions, using constant curvature of the objective and/or the Lagrangian: both functions have the same Hessian matrix $D^2_x L(x; u) = D^2 f(x) = Q$ for all $(x, u) \in M \times \mathbb{R}^m_+$.

The local view of $M$ from $\overline{x}$ is captured by the tangent cone, which due to linearity of constraints coincides with the cone of feasible directions at $\overline{x}$,

$$\Gamma(\overline{x}) = \{v \in \mathbb{R}^n : \overline{x} + tv \in M \text{ for all small enough } t > 0\}$$
$$= \mathbb{R}_+(M - \overline{x})$$
$$= \{v \in \mathbb{R}^n : (Av)_i \leq 0 \text{ if } (A\overline{x})_i = b_i, \quad \text{all } i \in \{1, \ldots, m\}\}. \quad (1.3)$$

If $\overline{x}$ is a local solution, then a decrease along the feasible direction $v$ is impossible, and this follows if $v^\top Q v \geq 0$ for all feasible directions $v \in \Gamma(\overline{x})$:

$$f(x) - f(\overline{x}) \geq L(x; u) - L(\overline{x}; u)$$
$$= v^\top \nabla_x L(\overline{x}; u) + \tfrac{1}{2} v^\top Q v$$
$$= v^\top o + \tfrac{1}{2} v^\top Q v \geq 0.$$

However, this condition is too strong: no locality is involved at all ! Hence we have to repeat the argument directly, with $f$ replacing $L$. To account for locality, we also put $x = \overline{x} + tv$ with $t > 0$ small. Note that the KKT condition (1.2) implies the weak first-order ascent condition $v^\top \nabla f(\overline{x}) \geq 0$ for the increment function $\theta_v(t) = f(\overline{x} + tv) - f(\overline{x})$ and

$$f(x) - f(\overline{x}) = \theta_v(t) = tv^\top \nabla f(\overline{x}) + \frac{t^2}{2} v^\top Q v > 0, \quad (1.4)$$

if $t > 0$ small and $v^\top \nabla f(\overline{x}) > 0$, even if $v^\top Q v < 0$: *strict first-order ascent directions may be negative curvature directions.* Clearly, the sign of $v^\top Q v$ determines curvature of the univariate function $\theta_v$ which is convex if and only if $v^\top Q v \geq 0$, and strictly concave otherwise. In the latter case, the condition $\theta_v(t) \geq 0$ for all $t \in [0, \overline{t}]$ is equivalent to $\theta_v(\overline{t}) \geq 0$, as always $\theta_v(0) = 0$ holds.

Thus we concentrate on the *reduced tangent cone*

$$\Gamma_{\text{red}}(\overline{x}) = \{v \in \Gamma(\overline{x}) : v^\top \nabla f(\overline{x}) = 0\} \quad (1.5)$$

and stipulate only $v^\top Q v \geq 0$ for all $v \in \Gamma_{\text{red}}(\overline{x})$.

**Theorem 1 (2nd order local optimality condition)** [20, 23, 48]:

*A KKT point $\overline{x}$ (i.e., satisfying (1.2) for some $u \in \mathbb{R}_+^m$) is a local solution to (1.1) if and only if*

$$v^\top Q v \geq 0 \quad \text{for all } v \in \Gamma_{\text{red}}(\overline{x}), \quad (1.6)$$

*i.e., if $Q$ is $\Gamma_{\text{red}}(\overline{x})$-copositive. If (1.6) is violated, then $v \in \Gamma_{\text{red}}(\overline{x})$ with $v^\top Q v < 0$ is a strictly improving feasible direction.*

Copositivity conditions of the form (1.6) will be central also later on in these notes. Here it may suffice to notice that this condition is clearly satisfied if $Q$ is positive-semidefinite, an aspect of the well-known result that in convex problems, KKT points are (global) solutions. Generally, it is NP-hard to check copositivity even if the cone $\Gamma_{\mathrm{red}}(\overline{x})$ is very simple. This is not surprising in view of the result that checking local optimality of KKT points is NP-hard [52].

**Example 1, continued.** Consider $\overline{x} = -e$. Then as $\nabla f(\overline{x}) = -\overline{x} + c = (1 - \mu)e$ and $Av = [v^\top | - v^\top]$, the set of all feasible directions at $\overline{x}$ coincides with $\mathbb{R}^n_+$, and we have $\Gamma_{\mathrm{red}}(\overline{x}) = \{0\}$. So (1.6) is satisfied trivially, despite of negative-definiteness of $Q = -I_n$, and $\overline{x}$ is a local solution (the same is of course true for $x^*$ where $\nabla f(x^*) = -(1 + \mu)e$). However, if, e.g., $y$ is a KKT point with, say, $y_i = -\mu$, then $v = e_i$ is a feasible direction as $y + tv \in M$ for all small $|t|$. Further $\nabla f(y)^\top v = 0$, so $v \in \Gamma_{\mathrm{red}}(y)$ with $v^\top Q v = -1 < 0$, showing that $y$ is no local solution. Indeed, $f(y + tv) < f(y)$ for all small $|t|$. This way, we got rid of $3^n - 2^n$ inefficient solution candidates.

## 1.2  *Extremal Increments and Global Optimality*

We have seen that negative curvature along strict first-order ascent directions,

$$v^\top \nabla f(\overline{x}) > 0 \quad \text{while} \quad v^\top Q v < 0$$

pose no problems for checking local optimality. For global optimality of a KKT point $\overline{x}$, we need the *extremal increment* for non-binding constraints at $\overline{x}$, as follows: denote by $s = b - A\overline{x}$ the vector of slack variables, and by

$$J(\overline{x}) = \{0\} \cup \{i \in \{1, \ldots, m\} : s_i > 0\} \tag{1.7}$$

To have a consistent terminology, we also can view as $J(\overline{x})$ as the set of non-binding constraints if we add an auxiliary non-binding constraint of the form $0 < 1$ by enriching $(A, b)$ with a 0-th row to

$$\bar{A} = \begin{pmatrix} a_0^\top \\ A \end{pmatrix} = \begin{pmatrix} o^\top \\ a_1^\top \\ \vdots \\ a_m^\top \end{pmatrix}, \quad \bar{b} = \begin{pmatrix} b_0 \\ b \end{pmatrix} = \begin{pmatrix} 1 \\ b_1 \\ \vdots \\ b_m \end{pmatrix},$$

and put $\bar{s} = \bar{b} - \bar{A}\overline{x} \geq o$ as well as $J(\overline{x}) = \{i \in \{0, \ldots, m\} : \bar{s}_i > 0\}$. The 0-th slack and the corresponding constraint will be needed for dealing with unbounded feasible directions. However, if $v$ is a bounded feasible direction, then there is an $i \in J(\overline{x})$ such that $a_i^\top (\overline{x} + tv) > b_i$ for some $t > 0$, and the *maximal feasible stepsize* in direction of $v$

$$\bar{t}_v = \min\left\{\bar{s}_i/(\bar{a}_i^\top v) : i \in J(\bar{x}),\ \bar{a}_i^\top v > 0\right\}$$

is finite. Note that feasibility of a direction $v \in \mathbb{R}^n$ is fully characterized by the property $a_i^\top v \le 0$ for all $i$ with $s_i = 0$, i.e., for all $i$ in the complement of $J(\bar{x})$. If in addition, $a_i^\top v \le 0$ for all $i \in J(\bar{x})$, i.e., $Av \le o$ but $v \ne o$, then we have an unbounded feasible direction with $\bar{t}_v = +\infty$ by the usual default rules, consistently with the property that $\bar{x} + tv \in M$ for all $t > 0$ in this case. In the opposite case $\bar{t}_v = s_i/(\bar{a}_i^\top v) < +\infty$, we have ($i \ne 0$ and) that the $i$-th constraint is the first non-binding constraint which becomes binding when travelling from $\bar{x}$ along the ray given by $v$: then $\bar{x} + \bar{t}_v v \in M$, but $\bar{x} + tv \notin M$ for all $t > \bar{t}_v$.

By consequence, the feasible polyhedron $M$ is decomposed into a union of polytopes $M_i(\bar{x}) = \{x = \bar{x} + tv \in M : 0 \le t \le \bar{t}_v = s_i/(\bar{a}_i^\top v)\}$ across $i \in J(\bar{x}) \setminus \{0\}$, and $M_0(\bar{x}) = \{x \in \mathbb{R}^m : Ax \le A\bar{x}\}$, the (possibly trivial) unbounded polyhedral part of $M$.

To be more precise, we need the $(m+1) \times n$-matrices $D_i = \bar{s}\,a_i^\top - \bar{s}_i\bar{A}$, rank-one updates of $\bar{A}$ with zeroes in the $i$-th row, to define the polyhedral cones

$$\Gamma_i = \{v \in \mathbb{R}^n : D_i v \ge o\},\quad i \in J(\bar{x}). \tag{1.8}$$

Then $M_i(\bar{x}) = M \cap (\Gamma_i + \bar{x})$ for all $i \in J(\bar{x})$, as is easily checked by elementary calculations. Similarly, we have $\bigcup_{i \in J(\bar{x})} \Gamma_i = \Gamma(\bar{x})$ from (1.3). By this decomposition, the purely local view of $M$ from $\bar{x}$ is appropriately "globalized", focussing on the first non-binding constraint which becomes binding along a certain direction.

After these preparations dealing with the feasible set only, we turn to the objective function. Clearly, the extremal increment is simply $\theta_v(\bar{t}_v)$ – see (1.4) – and for $v^\top Q v < 0$ we have

$$f(x) - f(\bar{x}) = \theta_v(t) \ge 0 \quad \text{for all } x = \bar{x} + tv \in M, \text{i.e., } t \in [0, \bar{t}_v],$$

if and only if $\theta_v(\bar{t}_v) \ge 0$. In the sequel, we will also need the gradient $\bar{g} = \nabla f(\bar{x}) = Q\bar{x} + c$, and the following symmetric rank-two updates of $Q$:

$$Q_i = a_i\,\bar{g}^\top + \bar{g}\,a_i^\top + \bar{s}_i Q,\quad i \in J(\bar{x}). \tag{1.9}$$

Note that the theorem below also applies to QPs for which the Frank/Wolfe theorem applies, i.e., where the objective function $f$ is bounded from below over an unbounded polyhedron $M$ (see [55] for a recent survey and extension on this topic).

**Theorem 2 (Global optimality condition for QP)** [3]:

*A KKT point $\bar{x}$ (i.e., satisfying (1.2) for some $u \in \mathbb{R}_+^m$) is a global solution to (1.1) if and only if $Q_i$ are $\Gamma_i$- copositive for all $i \in J(\bar{x})$:*

$$v^\top Q_i v \ge 0 \quad \text{if} \quad D_i v \ge o; \quad \text{for all } i \in J(\bar{x}). \tag{1.10}$$

*If this is violated, i.e., if $v^\top Q_i v < 0$ and $D_i v \geq o$ for some $i \in J(\bar{x}) \setminus \{0\}$, then*

$$\tilde{x} = \bar{x} + \bar{t}v \quad \text{is an improving feasible point if} \quad \bar{t} = \bar{s}_i/(\bar{a}_i^\top v),$$

*whereas $v^\top Q_0 v < 0$ for some $v$ with $D_0 v \geq o$ if and only if (1.1) is unbounded.*

*Proof.* We distinguish cases according to the sign of $v^\top Q v$ along a feasible direction $v$. The KKT condition (1.2) means always $v^\top(Q\bar{x} + c) = v^\top \bar{g} \geq 0$, so by (1.4) we are fine if $v^\top Q v \geq 0$; if $\bar{x} + tv \in M$ for all $t \geq 0$, convexity of $\theta_v$ must necessarily hold, unless the QP is unbounded. This case is dealt with the index $i = 0$: then $D_0 = -A$ and $Q_0 = Q$, and $v$ is an unbounded feasible direction $v$ if and only if $D_0 v = -Av \geq o$, whereas convexity (boundedness) along $v$ means $v^\top Q_0 v = v^\top Q v \geq 0$. So it remains to deal with the concave case whenever the maximal feasible stepsize is finite: $\bar{t}_v = \bar{s}_i/(\bar{a}_i^\top v)$ for some $i \in J(\bar{x}) \setminus \{0\}$ and $v^\top Q v < 0$. Hence $v \in \Gamma_i$, since the definition of $\bar{t}_v$ implies $\bar{s}_i a_j^\top v \leq \frac{\bar{s}_j}{a_i^\top v}$ for all $j \in J(\bar{x})$, i.e., $D_i v \geq o$ (note that this includes also $a_j^\top v \leq 0$ if $j \notin J(\bar{x})$ for feasibility of the direction $v$). Now we check the sign of the extremal increment $\theta_v(\bar{t}_v)$, to test for global optimality:

$$
\begin{aligned}
f(\bar{x} + \bar{t}_v v) - f(\bar{x}) &= \tfrac{\bar{t}_v^2}{2} v^\top Q v + \bar{t}_v\, v^\top \bar{g} \\
&= \tfrac{\bar{t}_v^2}{2} \left[ v^\top Q v + \tfrac{2}{\bar{t}_v} v^\top \bar{g} \right] \\
&= \tfrac{\bar{t}_v^2}{2} \left[ v^\top Q v + \tfrac{2}{\bar{s}_i} (v^\top \bar{g})(a_i^\top v) \right] \\
&= \tfrac{\bar{t}_v^2}{2\bar{s}_i} v^\top \left[ \bar{s}_i Q + a_i\, \bar{g}^\top + \bar{g}\, a_i^\top \right] v \\
&= \tfrac{\bar{t}_v^2}{2\bar{s}_i} v^\top Q_i v \geq 0.
\end{aligned}
$$

Hence we have reduced global optimality to condition (1.10).

Comparing Theorems 1 and 2, we see that the effort of checking local versus global optimality is not that different: at most $m$ copositivity checks instead of merely one.

Further, from definitions (1.5) and (1.9) we directly see that (1.10) implies (1.6). Finally, if $Q$ is positive-semidefinite and $\bar{x}$ is a KKT point, then $v \in \Gamma_i$ for $i \in J(\bar{x}) \setminus \{0\}$ implies $v^\top Q_i v \geq 2(a_i^\top v)(\bar{g}^\top v) \geq 0$, since $a_i^\top v > 0$ and $\bar{g}^\top v \geq 0$ by (1.2); note that $Q_0 = Q$ anyhow so that (1.10) is automatically satisfied for $i = 0$ in the convex case. Hence Theorem 2 also implies that for convex QPs, KKT points solve the problem globally.

**Example 1, continued.** Now we will get rid of the remaining $2^n - 1$ non-global local solutions. For simplicity of notation, we only deal with $\bar{x} = -e$ instead of all local solutions different from $x^* = e$, but the general case is as easy to handle. Now, $J(\bar{x}) = \{0, 1, \ldots, n\}$ and $s_i = 2$ for $i \in \{1, \ldots, n\}$. Since $M$ is bounded, $\Gamma_0 = \{o\}$ and we can safely ignore $i = 0$ in the sequel. For $1 \leq i \leq n$, we have $\Gamma_i = \{v \in \mathbb{R}_+^n : 2v_i \geq 2v_j, 1 \leq j \leq n\}$ which contains $v = e_i$ but also $v^* = e$. As before, $\bar{g} = (1 - \mu)e$, and hence $Q_i = (1 - \mu) \left[ e_i e^\top + e e_i^\top \right] + 2Q$, so that $v^\top Q_i v = 2[(1 - \mu) v_i (e^\top v) - v^\top v]$ which gives $v^\top Q_i v = -2\mu < 0$ for $v = e_i$ and even $(v^*)^\top Q_i(v^*) = -2n\mu < 0$. So either $\tilde{x} = -e + 2e_i$ or (even better) $x^* = -e + 2v^*$ emerges as an improving feasible point, the latter as the best one. The picture changes completely for $x^*$ where $J(x^*) = \{0\} \cup \{n+1, \ldots, 2n\}$ with the same slack values $s_i = 2$ but now $\Gamma_i^* = \{v \in -\mathbb{R}_+^n : 2v_{i-n} \le 2v_j, \, 1 \le j \le n\}$. The gradient is $g^* = -(1 + \mu)e$ and $a_i = -e_{i-n}$ for $i \in \{n+1, \ldots, 2n\}$, so that $Q_i^* = (1 + \mu) \left[ e_{i-n} e^\top + e e_{i-n}^\top \right] + 2Q$ with

$$v^\top Q_i^* v = 2 \sum_{j=1}^n \left[ (1 + \mu)v_{i-n} - v_j \right] v_j \ge 2 \sum_{j=1}^n \left[ (1 + \mu)v_j - v_j \right] v_j = 2\mu v^\top v \ge 0$$

for all $v \in \Gamma_i^*$ and all $i \in \{n+1, \ldots, 2n\}$, establishing global optimality of $x^*$.

Theorem 2 was first proved for the concave case [27] using $\varepsilon$-subdifferential calculus, and later on (but published earlier) for the indefinite case along the arguments above [3]. The following sections will synthesize both approaches.

## 1.3  Global Optimality and $\varepsilon$-Subdifferential Calculus

Given an extended-valued convex function $h : \mathbb{R}^n \to \mathbb{R} \cup \{+\infty\}$ and $\varepsilon \ge 0$, the $\varepsilon$-subdifferential of $h$ at $\overline{x}$ with $h(\overline{x}) < +\infty$ is defined as

$$\partial_\varepsilon h(\overline{x}) = \{y \in \mathbb{R}^n : h(x) - h(\overline{x}) \ge y^\top (x - \overline{x}) - \varepsilon \text{ for all } x \in \mathbb{R}^n\}. \quad (1.11)$$

Note that for $\varepsilon > 0$ always $\partial_\varepsilon h(\overline{x}) \ne \emptyset$; indeed, any normal $[y^\top | - 1]^\top$ of a hyperplane in $\mathbb{R}^{n+1}$ separating the point $[\overline{x}, h(\overline{x}) - \varepsilon] \in \mathbb{R}^{n+1}$ from the convex set

$$\text{epi } h = \{[x, t] \in \mathbb{R}^{n+1} : h(x) \le t\}$$

satisfies

$$y^\top x - h(x) = [y^\top | - 1] \begin{pmatrix} x \\ h(x) \end{pmatrix} \le [y^\top | - 1] \begin{pmatrix} \overline{x} \\ h(\overline{x}) - \varepsilon \end{pmatrix} = y^\top \overline{x} - h(\overline{x}) + \varepsilon$$

for all $x \in \mathbb{R}^n$ with $h(x) < +\infty$, since by definition $[x, h(x)] \in \text{epi } h$. Similarly one can show $\partial_0 h(\overline{x}) \ne \emptyset$, if $h$ takes only real values and therefore is continuous on $\mathbb{R}^n$ so that epi $h$ is closed and admits a supporting hyperplane at $[\overline{x}, h(\overline{x})]$.

Consider the general d.c. (difference-of-convex) optimization problem, where $h : \mathbb{R}^n \to \mathbb{R}$ is a real-valued convex function and $g : \mathbb{R}^n \to \mathbb{R} \cup \{+\infty\}$ is an extended-valued convex function:

$$\min \{g(x) - h(x) : x \in \mathbb{R}^n\}. \quad (1.12)$$

Then we have the following characterization of global optimality:

**Theorem 3 (Optimality condition for d.c. optimization)** [39]:
*Suppose that $\overline{x}$ satisfies $g(\overline{x}) < +\infty$. Then $\overline{x}$ is a global solution to (1.12) if and only if*

$$\partial_\varepsilon h(\overline{x}) \subseteq \partial_\varepsilon g(\overline{x}) \quad \text{for all } \varepsilon > 0. \tag{1.13}$$

*Proof.* Suppose that $\overline{x}$ is a global minimizer of $g - h$ and pick $y \in \partial_\varepsilon h(\overline{x})$; then for any $x \in \mathbb{R}^n$

$$g(x) - g(\overline{x}) \geq h(x) - h(\overline{x}) \geq y^\top(x - \overline{x}) - \varepsilon,$$

so that also $y \in \partial_\varepsilon g(\overline{x})$ holds. To show the converse, assume that there is a $\hat{x} \in \mathbb{R}^n$ such that

$$\delta = \tfrac{1}{2}[g(\overline{x}) - g(\hat{x}) - h(\overline{x}) + h(\hat{x})] > 0,$$

pick $y \in \partial_\delta h(\hat{x})$ and put

$$\varepsilon = h(\overline{x}) - h(\hat{x}) - y^\top(\overline{x} - \hat{x}) + \delta > 0.$$

This enables parallel transport from $(\hat{x}, \delta)$ to $(\overline{x}, \varepsilon)$:

$$\begin{aligned}
y^\top(x - \overline{x}) - \varepsilon &= y^\top(x - \hat{x}) + y^\top(\hat{x} - \overline{x}) - h(\overline{x}) + h(\hat{x}) - y^\top(\hat{x} - \overline{x}) - \delta \\
&= y^\top(x - \hat{x}) - \delta - h(\overline{x}) + h(\hat{x}) \\
&\leq h(x) - h(\hat{x}) - h(\overline{x}) + h(\hat{x}) = h(x) - h(\overline{x})
\end{aligned}$$

for all $x \in \mathbb{R}^n$, so that $y \in \partial_\varepsilon h(\overline{x})$. But then, by (1.13), also $y \in \partial_\varepsilon g(\overline{x})$, so that in particular for $x = \hat{x}$, we get $y^\top(\hat{x} - \overline{x}) - \varepsilon \leq g(\hat{x}) - g(\overline{x})$. This yields

$$\begin{aligned}
2\delta = g(\overline{x}) - g(\hat{x}) - [h(\overline{x}) - h(\hat{x})] \\
\leq \varepsilon - y^\top(\hat{x} - \overline{x}) - [h(\overline{x}) - h(\hat{x})] \\
= \delta + y^\top(\hat{x} - \overline{x}) - y^\top(\hat{x} - \overline{x}) = \delta < 2\delta,
\end{aligned}$$

a contradiction, proving that $\overline{x}$ is indeed globally optimal.

For a closed convex set $M \subseteq \mathbb{R}^n$, define the characteristic function

$$\Psi_M(x) = \begin{cases} 0, & \text{if } x \in M, \\ +\infty, & \text{else.} \end{cases}$$

Next we consider a *convex maximization* problem over $M$

$$\max\{h(x) : x \in M\}. \tag{1.14}$$

**Theorem 4 (Optimality condition for convex maximization)** [39]:
*Let $M$ be a closed convex set and $h : M \to \mathbb{R}$ be a real-valued convex function. Then $\bar{x} \in M$ is a global solution to (1.14) if and only if*

$$\partial_\varepsilon h(\bar{x}) \subseteq N_\varepsilon(M, \bar{x}) \quad \text{for all } \varepsilon > 0, \tag{1.15}$$

*where $N_\varepsilon(M, \bar{x})$ denotes the set of $\varepsilon$-normal directions to $M$ at $\bar{x}$,*

$$N_\varepsilon(M, \bar{x}) = \{y \in \mathbb{R}^n : y^\top(x - \bar{x}) \le \varepsilon \text{ for all } x \in M\}. \tag{1.16}$$

*Proof.* It is easily seen that $\max\{h(x) : x \in M\} = \max\{h(x) - g(x) : x \in \mathbb{R}^n\}$, for $g(x) = \Psi_M(x)$. Likewise, we have $\partial_\varepsilon g(\bar{x}) = N_\varepsilon(M, \bar{x})$ as defined in (1.16). Now the assertion follows from the general d.c. result in Theorem 3.

Sometimes it may be more convenient to rephrase the inclusion condition in (1.15) into inequality relations of support functionals. Recall that given a convex set $Y \subseteq \mathbb{R}^n$, the support functional of $Y$ is defined as

$$\sigma_Y(v) = \sup\{v^\top y : y \in Y\}.$$

Now both $S(\varepsilon) = \partial_\varepsilon h(\bar{x})$ and $N(\varepsilon) = N_\varepsilon(M, \bar{x})$ are convex sets, so that the inclusion $S(\varepsilon) \subseteq N(\varepsilon)$ holds if and only

$$\sigma_{S(\varepsilon)}(v) \le \sigma_{N(\varepsilon)}(v) \quad \text{for all directions } v \in \mathbb{R}^n \setminus \{o\}. \tag{1.17}$$

In Section 1.5 below, we will follow this approach for QPs.

## 1.4 Local Optimality and $\varepsilon$-Subdifferential Calculus

Also a local optimality condition can be formulated by means of $\varepsilon$- subdifferential calculus; remember that we deal with the d.c. problem (1.12), where $h : \mathbb{R}^n \to \mathbb{R}$ is a real-valued convex function and $g : \mathbb{R}^n \to \mathbb{R} \cup \{+\infty\}$ is an extended-valued convex function:

**Theorem 5 (Dür's local optimality condition, d.c. case)** [31]:
*Suppose that $\bar{x}$ satisfies $g(\bar{x}) < +\infty$. Then $\bar{x}$ is a local solution to (1.12) if for some $\delta > 0$*

$$\partial_\varepsilon h(\bar{x}) \subseteq \partial_\varepsilon g(\bar{x}) \quad \text{if } 0 < \varepsilon < \delta. \tag{1.18}$$

*Proof.* Arguing by contradiction, suppose there is a sequence $x_k \to \overline{x}$ as $k \to \infty$ such that

$$g(x_k) - h(x_k) < g(\overline{x}) - h(\overline{x}) \quad \text{for all } k \in \mathcal{N}. \tag{1.19}$$

Choose $y_k \in \partial_0 h(x_k)$ and put

$$\varepsilon_k = h(\overline{x}) - h(x_k) + y_k^\top (x_k - \overline{x}) \geq 0.$$

Now by definition of $\partial_0 h(x_k)$, we have $y_k^\top (\overline{x} - x_k) \leq h(\overline{x}) - h(x_k)$, but also

$$y_k^\top (x_k - \overline{x}) = y_k^\top (2x_k - \overline{x} - x_k) \leq h(2x_k - \overline{x}) - h(x_k),$$

and as both $h(x_k) - h(\overline{x})$ and $h(2x_k - \overline{x}) - h(x_k)$ tend to zero as $k \to \infty$ by continuity of the convex function $h$ (observe that $\|(2x_k - \overline{x}) - \overline{x}\| \to 0$) we obtain

$$|y_k^\top (x_k - \overline{x})| \leq \max\{h(\overline{x}) - h(x_k), h(2x_k - \overline{x}) - h(x_k)\} \to 0 \quad \text{as } k \to \infty,$$

entailing $\varepsilon_k \to 0$ as $k \to \infty$, so that $\varepsilon_k < \delta$ for all $k$ sufficiently large. For these $k$ we know by assumption of the theorem $\partial_{\varepsilon_k} h(\overline{x}) \subseteq \partial_{\varepsilon_k} g(\overline{x})$ and also by construction, using parallel transport again, i.e., establishing

$$\begin{aligned} h(x) - h(\overline{x}) &= h(x) - h(x_k) + h(x_k) - h(\overline{x}) \\ &\geq y_k^\top (x - x_k) - \varepsilon_k + y_k^\top (x_k - \overline{x}) = y_k^\top (x - \overline{x}) - \varepsilon_k \end{aligned}$$

for all $x \in \mathbb{R}^n$, that $y_k \in \partial_{\varepsilon_k} h(\overline{x}) \subseteq \partial_{\varepsilon_k} g(\overline{x})$, which is absurd as (1.19) implies

$$g(x_k) - g(\overline{x}) < h(x_k) - h(\overline{x}) = y_k^\top (x_k - \overline{x}) - \varepsilon_k.$$

Hence the result.

So define $G(\overline{x}) = \{\varepsilon > 0 : \partial_\varepsilon h(\overline{x}) \not\subseteq \partial_\varepsilon g(\overline{x})\}$ and $\delta(\overline{x}) = \inf G(\overline{x})$.

$$\delta(\overline{x}) = \infty \quad \Longleftrightarrow \quad \overline{x} \text{ is a global minimizer of } g - h\,;$$

$$\delta(\overline{x}) > 0 \quad \Longrightarrow \quad \overline{x} \text{ is a local minimizer of } g - h.$$

Suppose $\overline{x}$ is a local, nonglobal solution.
Then an upper bound for this parameter $\delta(\overline{x})$ can be found if we know an improving feasible solution, which suggests that searching for $\varepsilon$ beyond $\delta(\overline{x})$ may be a good strategy for escaping from inefficient solutions:

**Theorem 6 (Upper bound for Dür's delta)** [32]:
*If $x \in \mathbb{R}^n$ satisfies $g(x) - h(x) < g(\overline{x}) - h(\overline{x}) < +\infty$ and $y \in \partial_0 h(x)$, then*

$$\varepsilon = h(x) - h(\overline{x}) - y^\top (\overline{x} - x) \geq \delta(\overline{x}).$$

**Exercise:** Prove Theorem 6 by establishing $y \in \partial_\varepsilon h(\overline{x}) \setminus \partial_\varepsilon g(\overline{x})$.

Of course, there is a again a variant for convex constrained maximization:

**Theorem 7 (Local optimality for convex maximization)** [31]:

*Let $M$ be a closed convex set and $h : M \to \mathbb{R}$ be a real-valued convex function. Then $\overline{x} \in M$ is a local solution to (1.14) if for some $\delta > 0$*

$$\partial_\varepsilon h(\overline{x}) \subseteq N_\varepsilon(M, \overline{x}) \quad \text{if} \quad 0 < \varepsilon < \delta. \tag{1.20}$$

*Proof.* Follows as in Theorem 4 above.

Condition (1.20) above is sufficient but not necessary, even over a polytope $M$:

**Example 2.** Denote by $t_+ = \max\{t, 0\}$ and let $h(x) = [(x - 1)_+]^2$ and $M = [0, 3]$. Then condition (1.20) is violated at the local maximizer $\overline{x} = 0$ of $h$ over $M$. Indeed, $h(x) - h(\overline{x}) - y(x - \overline{x}) = h(x) - yx \geq -\varepsilon$ for all $x \in \mathbb{R}$ implies $y \geq 0$ (consider $x \to -\infty$) but also $|\frac{y}{2} + 1| \leq \sqrt{1 + \varepsilon}$ (consider $x = 1 + \frac{y}{2} \geq 1$). On the other, hand, every $y \in [0, 2(\sqrt{1+\varepsilon} - 1)]$ satisfies $y \in S(\varepsilon) = \partial_\varepsilon h(\overline{x})$, as can easily be verified in a straightforward manner. Further, obviously $N_\varepsilon(\overline{x}; M) = \,]-\infty, \frac{\varepsilon}{3}]$. Since $\frac{\varepsilon}{3} < 2(\sqrt{1+\varepsilon} - 1)$ if and only if $\varepsilon < 24$, we have $G(\overline{x}) = \,]0, 24[$ and $\delta(\overline{x}) = 0$. A similar example is provided in [31] with a local minimizer of d.c. function $g - h$ where $g$ is finite everywhere (rather than being equal to $\Psi_M$). Note however, if we employ the quadratic function $q(x) = (x-1)^2$ instead of the function $h$ in our example above, we obtain $\partial_\varepsilon q(\overline{x}) = [-2 - 2\sqrt{\varepsilon}, -2 + 2\sqrt{\varepsilon}] \subset \,]-\infty, \frac{\varepsilon}{3}] = N_\varepsilon(\overline{x}; M)$ whenever $\varepsilon \notin \tilde{G}(\overline{x}) = \,]12 - \sqrt{108}, 12 + \sqrt{108}[$, so that $\delta(\overline{x}) = 12 - \sqrt{108} > 0$. Theorem 8 below shows that this is no coincidence: for (indefinite or convex) *quadratic* maximization over a polyhedron, local optimality of $\overline{x}$ is equivalent to $\delta(\overline{x}) > 0$.

Whether $G(\overline{x})$ an interval (convex) or not, remained an open question for a considerable period. However, our running example will answer this in the negative:

**Example 1, continued.** We again turn to the local, non-global solution $\overline{x} = -e$, considering the maximization problem $h(x) = \frac{1}{2}x^\top x + \mu e^\top x$ instead of $\frac{1}{2}\|x + \mu e\|^2$ (the difference is the constant $\frac{\mu^2 n}{2}$), over $M = [-1, 1]^n$. Now, $y \in S(\varepsilon) = \partial_\varepsilon h(\overline{x})$ if and only if

$$\min\{\varphi_y(x) : x \in \mathbb{R}^n\} \geq n(\tfrac{1}{2} - \mu) + e^\top y - \varepsilon, \quad \text{where} \quad \varphi_y(x) = \tfrac{1}{2}x^\top x + (\mu e - y)^\top x.$$

Since the minimizer of $\varphi_y$ is obviously $x_y = y - \mu e$ with the value $\varphi_y(x_y) = -\frac{1}{2}\|y - \mu e\|^2$, we thus get $y \in S(\varepsilon)$ if and only if $-\frac{1}{2}\|y - \mu e\|^2 \geq n(\frac{1}{2} - \mu) + e^\top y - \varepsilon$, which after elementary calculations yields the ball representation of the subdifferential

$$S(\varepsilon) = \left\{ y \in \mathbb{R}^n : \|y + (1 - \mu)e\| \leq \sqrt{2\varepsilon} \right\}.$$

Later, we will need to solve the linear trust region problem $\max\left\{e_J^\top y : y \in S(\varepsilon)\right\}$ with $e_J = \sum_{j \in J} e_j \in \mathbb{R}^n$ for any subset $J \subseteq \{1, \ldots, n\}$ with the usual convention $e_\emptyset = o$. For $J \neq \emptyset$, the solution is of course $y_J = m_\varepsilon + \sqrt{\frac{2\varepsilon}{|J|}}\, e_J$ where $m_\varepsilon = (\mu - 1)e \le o$ is the center of the ball $S(\varepsilon)$. The optimal objective of this trust region problem is $e_J^\top y_J = (\mu - 1)e_J^\top e + \sqrt{2\varepsilon|J|}$, so that we arrive at

$$\max\left\{e_J^\top y : y \in S(\varepsilon)\right\} = \sqrt{2\varepsilon|J|} - (1 - \mu)|J|, \qquad (1.21)$$

which trivially also holds if $J = \emptyset$. The reason for considering (1.21) lies in the structure of $N_\varepsilon(\overline{x}; M)$: indeed, the vertices of $M$ are of the form $2e_J - e$, so that

$$N_\varepsilon(\overline{x}; M) = \left\{y \in \mathbb{R}^n : \max\left\{y^\top(x - \overline{x}) : x \in M\right\} \le \varepsilon\right\}$$

$$= \left\{y \in \mathbb{R}^n : \max\left\{y^\top x : x \in \operatorname{ext} M\right\} \le y^\top \overline{x} + \varepsilon\right\}$$

$$= \left\{y \in \mathbb{R}^n : \max\left\{y^\top(2e_J - e) : J \subseteq \{1, \ldots, n\}\right\} \le y^\top \overline{x} + \varepsilon\right\}$$

$$= \left\{y \in \mathbb{R}^n : y^\top e_J \le \tfrac{\varepsilon}{2} \text{ for all } J \subseteq \{1, \ldots, n\}\right\}. \qquad (1.22)$$

Now (1.21) and (1.22) imply that the inclusion $S(\varepsilon) \subseteq N_\varepsilon(\overline{x}; M)$ holds if and only if

$$\sqrt{2\varepsilon|J|} - (1 - \mu)|J| \le \frac{\varepsilon}{2} \text{ for all } J \subseteq \{1, \ldots, n\}. \qquad (1.23)$$

Putting $q = \sqrt{|J|} \in [0, \sqrt{n}]$, we see that (1.23) holds whenever $(1 - \mu)q^2 - \sqrt{2\varepsilon}q + \frac{\varepsilon}{2} \ge 0$, which means that

$$q \notin I_{\mu,\varepsilon} = \sqrt{\frac{\varepsilon}{2(1-\mu)^2}}\, \left]1 - \sqrt{\mu}, 1 + \sqrt{\mu}\right[.$$

So (1.23) means $\left\{0, 1, \sqrt{2}, \ldots, \sqrt{n}\right\} \cap I_{\mu,\varepsilon} = \emptyset$. In particular, $I_{\mu,\varepsilon}$ should not contain 1. But since $\sup I_{\mu,\varepsilon} \searrow 0$ as $\varepsilon \searrow 0$, this means that Dür's delta $\delta(\overline{x})$ must satisfy $\sup I_{\mu,\varepsilon} \le 1$ for all $\varepsilon < \delta(\overline{x})$, which amounts to

$$\delta(\overline{x}) = 2\left(\tfrac{1-\mu}{1+\sqrt{\mu}}\right)^2.$$

Furthermore, by the same reasoning, we obtain

$$G(\overline{x}) = \left]\delta(\overline{x}), 2\left(\tfrac{1-\mu}{1-\sqrt{\mu}}\right)^2\right[ \bigcup \left]4\left(\tfrac{1-\mu}{1+\sqrt{\mu}}\right)^2, 4\left(\tfrac{1-\mu}{1-\sqrt{\mu}}\right)^2\right[ \bigcup \cdots,$$

so that for suitable values of $\mu$, the set $G(\overline{x})$ cannot be connected. A general formula for $G(\overline{x})$ is

$$G(\overline{x}) = \bigcup_{k=1}^{n} \, ] \, 2k \left( \tfrac{1-\mu}{1+\sqrt{\mu}} \right)^2, 2k \left( \tfrac{1-\mu}{1-\sqrt{\mu}} \right)^2 [ \, ,$$

which shows that $G(\overline{x})$ can have up to $n$ connected components.

## 1.5 $\varepsilon$-Subdifferential Optimality Conditions in QPs

We start with an explicit characterization of $\varepsilon$-subdifferentials and their support functionals in the quadratic world, as in [11, 27]:

**Proposition 1 (subdifferentials for convex quadratic functions)** [26]:
*If $q(x) = \tfrac{1}{2}x^\top P x + d^\top x$ with positive-definite $P$, and $\varepsilon \geq 0$, then for any $\overline{x} \in \mathbb{R}^n$*

$$\partial_\varepsilon q(\overline{x}) = S(\varepsilon) = \left\{ \nabla q(\overline{x}) + P u : u^\top P u \leq 2\varepsilon \right\}, \qquad (1.24)$$

*and*

$$\sigma_{S(\varepsilon)}(v) = v^\top \nabla q(\overline{x}) + \sqrt{2\varepsilon \, v^\top P v}. \qquad (1.25)$$

*Proof.* Let $u$ and $y$ be such that $P u = y - \nabla q(\overline{x}) = y - P\overline{x} - d$. Then we get

$$q(x) - q(\overline{x}) - y^\top (x - \overline{x})$$
$$= \tfrac{1}{2}(x^\top P x - \overline{x}^\top P \overline{x}) + (d - y)^\top (x - \overline{x})$$
$$= \tfrac{1}{2}x^\top P x - x^\top P \overline{x} + \tfrac{1}{2}\overline{x}^\top P \overline{x} + (x - \overline{x})^\top P \overline{x} + (d - y)^\top (x - \overline{x})$$
$$= \tfrac{1}{2}(x - \overline{x})^\top P(x - \overline{x}) + (x - \overline{x})^\top P \overline{x} - (P\overline{x} + P u)^\top (x - \overline{x})$$
$$= \tfrac{1}{2}(x - \overline{x})^\top P(x - \overline{x}) - u^\top P(x - \overline{x}) + \tfrac{1}{2}u^\top P u - \tfrac{1}{2}u^\top P u$$
$$= \tfrac{1}{2}(x - \overline{x} - u)^\top P(x - \overline{x} - u) - \tfrac{1}{2}u^\top P u \quad \text{for all } x \in \mathbb{R}^n.$$

Now, if $y \in \partial_\varepsilon q(\overline{x})$, put $x = \overline{x} + u$ with $u = P^{-1}[y - \nabla q(\overline{x})]$ and arrive at $\tfrac{1}{2}u^\top P u \leq \varepsilon$ from above, as well as $y = \nabla q(\overline{x}) + P u$, hence $y \in S(\varepsilon)$. Conversely, for any $y \in S(\varepsilon)$ as defined in (1.24), we get by $(x - \overline{x} - u)^\top P(x - \overline{x} - u) \geq 0$ the required inequality $q(x) - q(\overline{x}) - y^\top (x - \overline{x}) \geq -\varepsilon$ for arbitrary $x \in \mathbb{R}^n$, which shows $y \in \partial_\varepsilon q(\overline{x})$. Hence (1.24) is established. Next, denote the inner product induced by the positive-definite matrix $P$ with $\langle u, v \rangle_P = u^\top P v$, and the corresponding norm by $\|u\|_P = \sqrt{u^\top P u}$. Then the Cauchy-Schwarz-Bunyakovsky inequality

$$|\langle u, v \rangle_P| \leq \|u\|_P \|v\|_P$$

holds. Hence we have, by (1.24), for all $v \neq o$,

$$
\begin{aligned}
\sigma_{S(\varepsilon)}(v) &= v^{\top} \nabla q(\overline{x}) + \sup \left\{ v^{\top} Pu : u^{\top} Pu \le 2\varepsilon \right\} \\
&= v^{\top} \nabla q(\overline{x}) + \sup \left\{ \langle v, u \rangle_P : \|u\|_P \le \sqrt{2\varepsilon} \right\} \\
&\le v^{\top} \nabla q(\overline{x}) + \sup \left\{ \|v\|_P \|u\|_P : \|u\|_P \le \sqrt{2\varepsilon} \right\} \\
&\le v^{\top} \nabla q(\overline{x}) + \sqrt{2\varepsilon} \|v\|_P,
\end{aligned}
$$

with equality for $u = \frac{\sqrt{2\varepsilon}}{\|v\|_P} v \in S(\varepsilon)$. This proves (1.25).

A slightly more refined argument shows that (1.24) and (1.25) also hold if $P$ is merely positive-semidefinite. Then $P^{-1}$ has to be replaced with the Moore-Penrose generalized inverse $P^{\dagger}$, see [26]. Anyhow, we here only need to discuss the case of $P$ nonsingular, as detailed below.

Next, if $M = \{x \in \mathbb{R}^n : Ax \le b\}$ is a polyhedron, consider any unbounded feasible direction $v \in \Gamma_0$, i.e., $Av \le o$ (if such $v$ exists). Then $x = \overline{x} + tv \in M$ for all $t > 0$, so that $v^{\top} y = \frac{1}{t} y^{\top}(x - \overline{x}) \le \frac{\varepsilon}{t} \searrow 0$ as $t \nearrow \infty$ for all $y \in N(\varepsilon) = N_\varepsilon(\overline{x}; M)$. This shows $\sigma_{N(\varepsilon)}(v) = 0$ for all $v \in \Gamma_0(\overline{x})$. If $v \in \Gamma(\overline{x}) \setminus \Gamma_0$ is a feasible direction at $\overline{x}$ with finite maximal feasible stepsize $\overline{t}_v$, we by the same reasoning get $v^{\top} y \le \frac{\varepsilon}{t}$ for all $t \in \,]0, \overline{t}_v]$ for all $y \in N(\varepsilon)$, so that $\sigma_{N(\varepsilon)}(v) \le \frac{\varepsilon}{\overline{t}_v}$. This upper bound is indeed tight, as shown in [26]:

$$
\sigma_{N(\varepsilon)}(v) = \varepsilon z(v) \text{ with } z(v) = \begin{cases} \max \left\{ \frac{a_i^{\top} v}{\overline{s}_i} : i \in J(\overline{x}) \right\}, & \text{if } v \in \Gamma(\overline{x}), \\ +\infty, & \text{otherwise,} \end{cases} \tag{1.26}
$$

with $\Gamma(\overline{x})$ the tangent cone as in (1.3) and $J(\overline{x})$ the set of non-binding constraints at $\overline{x}$ as in (1.7).

Suppose for the moment that we have to minimize a strictly concave quadratic function $f = -h$, putting $Q = -P$ and $\overline{g} = -\nabla h(\overline{x})$. Put $\eta = \sqrt{\varepsilon} > 0$. Then (1.15) and therefore (1.17) holds if and only if

$$
f_v(\eta) = \eta^2 z(v) - \eta \sqrt{2 v^{\top} P v} + v^{\top} \overline{g} \ge 0 \quad \text{for all } \eta > 0 \text{ and all } v \in \Gamma(\overline{x}).
$$

Since $z(v) \ge 0$ always holds, the function $f_v$ is convex. Now, if in particular $z(v) = 0$ (which is equivalent to $Av \le o$), $f_v$ is affine, and hence nonnegative if and only if

$$
\begin{cases} f_v(0) = \quad v^{\top} \overline{g} \quad \ge 0 & \text{— this is the KKT condition (1.2) — and} \\ f_v'(0) = -\sqrt{2 v^{\top} P v} \ge 0 & \text{— this is equivalent to } v^{\top} P v = 0. \end{cases}
$$

In the strictly convex case $z(v) > 0$, the function $f_v$ attains its minimum value at $\eta^* = \frac{\sqrt{2 v^{\top} P v}}{2 z(v)} > 0$, so $f_v$ nonnegative if and only if

$$f_v(\eta^*) = v^\top \bar{g} - \frac{1}{2z(v)}\, v^\top P v \geq 0,$$

which can be rephrased as

$$2v^\top \bar{g}\, z(v) - v^\top P v \geq 0,$$

which is exactly

$$v^\top Q_i v \geq 0 \quad \text{if } D_i v \geq o.$$

To enforce the same argument for indefinite rather than negative-definite $Q$, one has to resort directly to the d.c. case treated in Theorem 3. So put $Q = P_+ - P_-$ with $P_\pm$ positive-definite (this we can enforce by adding the same positive multiple of $I_n$ to both, if necessary), and define $g(x) = \frac{1}{2}x^\top P_+ x + \Psi_M(x)$ whereas $h(x) = \frac{1}{2}x^\top P_- x - c^\top x$ as above.

Then we need some subdifferential calculus:

**Proposition 2 (addition of subdifferentials)** [40]:

*If $q : \mathbb{R}^n \to \mathbb{R}$ is continuously differentiable and convex, and if $r : \mathbb{R}^n \to \mathbb{R} \cup \{+\infty\}$ is convex, then for any $\bar{x}$ with $r(\bar{x}) \in \mathbb{R}$*

$$\partial_\varepsilon r(\bar{x}) + \nabla q(\bar{x}) \subseteq \partial_\varepsilon (q + r)(\bar{x}) \tag{1.27}$$

*holds and moreover*

$$\partial_\varepsilon (q + r)(\bar{x}) = \bigcup_{\delta \in [0,\varepsilon]} [\partial_\delta q(\bar{x}) + \partial_{\varepsilon-\delta} r(\bar{x})] \quad \text{for all } \varepsilon > 0. \tag{1.28}$$

*Proof.* Suppose $y \in \partial_\varepsilon r(\bar{x})$. For any $x \in \mathbb{R}^n$, we then get by convexity of $q$

$$[y + \nabla q(\bar{x})]^\top (x - \bar{x}) \leq y^\top (x - \bar{x}) + (x - \bar{x})^\top \nabla q(\bar{x})$$
$$\leq r(x) - r(\bar{x}) + \varepsilon + q(x) - q(\bar{x}).$$

Hence (1.27) results. The inclusion $\supseteq$ in (1.28) follows by elementary addition of the relations in (1.11), whereas the reverse inclusion $\subseteq$ in (1.28) is established, e.g., in [40, Thm.3.1.1]. Observe that by the smoothness assumption, we have $\partial_0 q(\bar{x}) = \{\nabla q(\bar{x})\}$.

**Exercise:** Show that the assumptions of Proposition 2 imply assumption [40, (3.1.2)] which are sufficient for [40, Thm.3.1.1].

For $q(x) = r(x) = x^\top x$ and any $\varepsilon > 0$ we have $\partial_\varepsilon r(\bar{x}) = \{y \in \mathbb{R}^n : y^\top y \leq 4\varepsilon\}$ at $\bar{x} = o$, so evidently $\partial_\varepsilon (2r)(\bar{x}) = 2\partial_{\varepsilon/2} r(\bar{x}) \not\subseteq \partial_\varepsilon r(\bar{x}) = \partial_\varepsilon r(\bar{x}) + \nabla q(\bar{x})$. Hence inclusion (1.27) in Proposition 2 is strict in general.

Next we need some more elementary relations, namely

$$\sigma_{A+B} = \sigma_A + \sigma_B \quad \text{and} \quad \sigma_Y = \sup_{i \in I} \sigma_{Y(i)} \quad \text{for } Y = \bigcup_{i \in I} Y(i). \tag{1.29}$$

**Proposition 3 (subdifferentials for constrained quadratic case):**
*Let $g(x) = \frac{1}{2}x^\top P_+ x + \Psi_M(x)$, $\overline{x} \in M$ and $\varepsilon > 0$, and denote $T(\varepsilon) = \partial_\varepsilon g(\overline{x})$. Then for any $v \in \mathbb{R}^n$ we have*

$$\sigma_{T(\varepsilon)}(v) = v^\top P_+ \overline{x} + \begin{cases} \sqrt{2\varepsilon\, v^\top P_+ v}, & \text{if} \quad \varepsilon < \frac{v^\top P_+ v}{2z^2(v)}, \\[2mm] \frac{v^\top P_+ v}{2z(v)} + \varepsilon z(v), & \text{if} \quad \varepsilon \geq \frac{v^\top P_+ v}{2z^2(v)}. \end{cases} \tag{1.30}$$

*Proof.* Put $q(x) = \frac{1}{2}x^\top P_+ x$ and $r(x) = \Psi_M(x)$. Then $g = q + r$, and from (1.28), (1.29), (1.25) and (1.26) we get

$$\sigma_{T(\varepsilon)}(v) = \sup\left\{ v^\top P_+ \overline{x} + \sqrt{2\delta\, v^\top P_+ v} + \frac{\varepsilon - \delta}{z(v)} : 0 \leq \delta \leq \varepsilon \right\}. \tag{1.31}$$

Next we calculate the sup in (1.31). Putting $\rho = \sqrt{\delta}$, we arrive at the strictly concave quadratic function

$$\psi_v(\rho) = v^\top P_+ \overline{x} + \rho\sqrt{2v^\top P_+ v} + \frac{\varepsilon - \rho^2}{z(v)}, \quad \rho \in [0, \sqrt{\varepsilon}],$$

with zero derivative at $\rho = \bar{\rho}_v = \sqrt{\frac{v^\top P_+ v}{2z^2(v)}}$, where

$$\psi_v(\bar{\rho}_v) = v^\top P_+ \overline{x} + \frac{v^\top P_+ v}{z(v)} + \left(\varepsilon - \frac{v^\top P_+ v}{2z^2(v)}\right) z(v) = v^\top P_+ \overline{x} + \frac{v^\top P_+ v}{2z(v)} + \varepsilon z(v).$$

However, if $\bar{\rho}_v > \sqrt{\varepsilon}$, then $\psi_v(\rho) \leq \psi_v(\sqrt{\varepsilon}) = v^\top P_+ \overline{x} + \sqrt{2\varepsilon\, v^\top P_+ v} + 0$ for all $\rho \in [0, \sqrt{\varepsilon}]$, whence (1.30) follows.

The remainder follows the same lines as for convex quadratic maximization: again, we put $S(\varepsilon) = \partial_\varepsilon h(\overline{x})$ where $h(x) = \frac{1}{2}x^\top P_- x - c^\top x$ and $\eta = \sqrt{\varepsilon}$, to discuss the condition

$$f_v(\eta) = \sigma_{T(\eta^2)}(v) - \sigma_{S(\eta^2)}(v) \geq 0 \quad \text{for all} \quad \eta > 0, \tag{1.32}$$

to arrive at the desired inclusion (1.13). Now, from (1.30) and $\bar{g} = Q\overline{x} + c = P_+\overline{x} - P_-\overline{x} + c = P_+\overline{x} - \nabla h(\overline{x})$ we get via (1.25)

$$f_v(\eta) = v^\top \bar{g} - \eta\sqrt{2v^\top P_- v} + \begin{cases} \eta\sqrt{2v^\top P_+ v}, & \text{if } \eta < \eta_+(v), \\[2mm] \frac{v^\top P_+ v}{2z(v)} + z(v)\eta^2, & \text{if } \eta \geq \eta_+(v), \end{cases} \tag{1.33}$$

with $\eta_+(v) = \sqrt{\frac{v^\top P_+ v}{2z^2(v)}}$.

**Proposition 4 (subdifferentials and optimality in indefinite QPs):**
*Let $\bar{x}$ be a KKT point of (1.1) and assume $Q = P_+ - P_-$ with $P_\pm$ positive-definite. Suppose $v \in \Gamma(\bar{x})$. Then the function $f_v$ from (1.33) takes no negative values over the positive reals if and only if (1.10) holds.*

*Proof.* The function $f_v$ is piecewise quadratic and continuously differentiable at $\eta = \eta_+(v)$ because $2z(v)\eta_+(v) = \sqrt{2\,v^\top P_+ v}$. For the affine part of $f_v$ over the interval $[0, \eta_+(v)]$ we need to worry only about the values at the end points. However, $f_v(0) = v^\top \bar{g} \geq 0$ as $\bar{x}$ is a KKT point and $v$ is a feasible direction. Further, if $v^\top Q v \geq 0$, we get $\sqrt{v^\top P_- v} \leq \sqrt{v^\top P_+ v}$, and the slope is nonnegative. If the slope of $f_v$ on $[0, \eta_+(v)]$ is negative, i.e., $v^\top Q v < 0$, then by continuous differentiability of $f_v$ over $\mathbb{R}_+$, the global minimizer of $f_v$ must be larger than $\eta_+(v)$ and we have only to investigate the quadratic part over $[\eta_+(v), +\infty[$. So $v^\top Q v < 0$ entails that the minimum of the convex function $f_v$ is attained at $\eta = \eta_-(v) = \sqrt{\frac{v^\top P_- v}{2z^2(v)}} > \sqrt{\frac{v^\top P_+ v}{2z^2(v)}} = \eta_+(v)$. However, the minimal value is exactly

$$f_v(\eta_-(v)) = v^\top \bar{g} - \frac{v^\top P_- v}{z(v)} + \frac{v^\top P_- v + v^\top P_+ v}{2z(v)}$$
$$= v^\top \bar{g} + \frac{v^\top P_+ v - v^\top P_- v}{2z(v)}$$
$$= v^\top \bar{g} + \frac{v^\top Q v}{2z(v)},$$

and this expression is not negative if and only if $v^\top Q v + 2z(v)\bar{g}^\top v \geq 0$.

By means of this analysis we are now in a position to establish necessity of condition (1.18) in the quadratic world. Note that this result complements previous ones for piecewise affine objective functions [38].

**Theorem 8 (Dür's condition is an exact criterion for QPs):**
*Let $f(x) = \frac{1}{2}x^\top Q x + c^\top x$ be a (possibly indefinite) quadratic function and $M$ be a polyhedron. Decompose $Q = P_+ - P_-$ with $P_\pm$ positive-definite, and consider $h(x) = \frac{1}{2}x^\top P_- x - c^\top x$ as well as $g(x) = \frac{1}{2}x^\top P_+ x + \Psi_M(x)$. Then $\bar{x} \in M$ is a local solution to (1.1) if and only if for some $\delta > 0$*

$$\partial_\varepsilon h(\bar{x}) \subseteq \partial_\varepsilon g(\bar{x}) \quad if \quad 0 < \varepsilon < \delta. \tag{1.34}$$

*Proof.* In view of Theorem 5 we only have to prove necessity. So assume that $\bar{x}$ is a local solution to (1.1). Then by linearity of constraints, $\bar{x}$ is also a KKT point, and therefore $v^\top \bar{g} \geq 0$ for all $v \in \Gamma(\bar{x})$. We again consider the function $f_v$ from (1.32) and (1.33). Fix a positive $\nu > 0$ to be determined later. By positive homogeneity, we may and do restrict ourselves to directions $v \in \Gamma(\bar{x})$ with $\|v\| = \nu$, to establish $f_v(\eta) \geq 0$ for a certain $\eta > 0$, which would then imply the inclusion $S(\eta^2) \subseteq T(\eta^2)$. Now if $v \in \Gamma_{\text{red}}(\bar{x}) = \{v \in \Gamma(\bar{x}) : v^\top \bar{g} = 0\}$, we get $v^\top Q v \geq 0$ by Theorem 1. So Lemma 1 in [3] guarantees the existence of a (possibly small) $\nu > 0$ such that

$$v^\top \bar{g} + \tfrac{1}{2} v^\top Q v \geq 0 \quad \text{for all } v \in \Gamma(\overline{x}) \text{ with } \|v\| = \nu. \tag{1.35}$$

Of course we know that $v \notin \Gamma_{\text{red}}(\overline{x})$ for any direction $v \in \Gamma(\overline{x})$ with $v^\top Q v < 0$, i.e., that $v^\top \bar{g} > 0$, and thus $f_v(\eta) > 0$ for all $\eta > 0$ small enough, by the properties discussed in the proof of Proposition 4 above. However, we have to establish some $\widehat{\eta} > 0$ such that $f_v(\eta) \geq 0$ for all $\eta \in [0, \widehat{\eta}]$ *and all* $v \in \Gamma(\overline{x})$ *simultaneously*. To this end, we first investigate the (positive) zeroes of the affine parts of $f_v$, namely $\eta = \dfrac{v^\top \bar{g}}{\sqrt{2 v^\top P_- v} - \sqrt{2 v^\top P_+ v}}$ in the case $v^\top Q v < 0$. Now, if

$$\tau = \tfrac{1}{\sqrt{8}} \min \left\{ \sqrt{v^\top P_- v} + \sqrt{v^\top P_+ v} : \|v\| = \nu \right\} > 0,$$

we derive from (1.35)

$$\frac{v^\top \bar{g}}{\sqrt{2 v^\top P_- v} - \sqrt{2 v^\top P_+ v}} = \frac{v^\top \bar{g} + \tfrac{1}{2} v^\top Q v}{\sqrt{2 v^\top P_- v} - \sqrt{2 v^\top P_+ v}} - \frac{\tfrac{1}{2} v^\top Q v}{\sqrt{2 v^\top P_- v} - \sqrt{2 v^\top P_+ v}}$$

$$\geq 0 + \frac{\tfrac{1}{2}(v^\top P_- v - v^\top P_+ v)}{\sqrt{2 v^\top P_- v} - \sqrt{2 v^\top P_+ v}}$$

$$\geq \tfrac{1}{\sqrt{8}} \left( \sqrt{v^\top P_- v} + \sqrt{v^\top P_+ v} \right)$$

$$\geq \tau. \tag{1.36}$$

Next we turn our attention to $\eta_+(v) = \sqrt{\dfrac{v^\top P_+ v}{2 z^2(v)}}$. By polyhedrality of $M$, we may pick an $\omega > 0$ such that $\{v \in \Gamma(\overline{x}) : \|v\| \leq \omega\} \subseteq M$. Then define

$$\bar{\eta} = \omega \min \left\{ \sqrt{\frac{v^\top P_+ v}{2}} : \|v\| = 1 \right\} > 0$$

and observe $\overline{x} + \omega v \in M$ for all $v \in \Gamma(\overline{x})$, so that $1/z(v) = \sup\{t : \overline{x} + tv \in M\} \geq \omega$. Hence we have $\eta_+(v) \geq \bar{\eta} > 0$ for all $v \in \Gamma(\overline{x})$. Finally, we see that for $\widehat{\eta} = \min\{\tau, \bar{\eta}\} > 0$, we have $f_v(\eta) \geq 0$ for all $v \in \Gamma(\overline{x})$ and all $\eta \leq \widehat{\eta}$ by construction. Indeed, if $v^\top Q v \geq 0$, this follows by $\eta \leq \bar{\eta} \leq \eta_+(v)$ and $f_v(0) \geq 0$, while for $v^\top Q v < 0$ we know by (1.36) that $\eta \leq \tau$ cannot exceed the smallest positive zero of $f_v$ (if it exists). Hence $\widehat{\eta} \leq \delta(\overline{x})$ as defined in Section 1.4, and we are done.

**Exercise:** Consider a KKT point $\overline{x}$ of (1.1) and infer from $\delta(\overline{x}) > 0$ as defined after Theorem 5 the condition that $Q$ be $\Gamma_{\text{red}}(\overline{x})$-copositive, thus deducing Theorem 1 from Theorem 8.

## 1.6  Standard Quadratic Problems (StQPs)

In the preceding sections, we saw that checking copositivity yields – in the positive case – a certificate for global optimality of a KKT point in a QP, and – in the negative – an improving feasible point. Hence we have to deal with the following question which is (almost) a homogeneous decision problem:

$$\text{is} \quad \min\left\{v^\top Q v : Dv \geq o\right\} < 0 \ ? \tag{1.37}$$

If all extremal rays of the polyhedral cone $\{Dv \geq o\}$ are known, this question reduces to

$$\text{is} \quad \min\left\{x^\top Q x : x \in \mathbb{R}_+^n\right\} < 0 \ ? \tag{1.38}$$

In the general case, one can decompose (1.37) into several problems of the form (1.38) [25]. Without loss of generality, we may add the normalization constraint $e^\top x = \sum_i x_i = 1$ to (1.38), to arrive at the decision form of a so-called *Standard QP:* is $\alpha_Q < 0$, where

$$\alpha_Q = \min\left\{x^\top Q x : x \in \Delta\right\} \tag{1.39}$$

with $\Delta = \left\{x \in \mathbb{R}_+^n : e^\top x = 1\right\}$ the **standard simplex** in $\mathbb{R}^n$. When $Q$ is indefinite, we can have – even in the generic case of finitely many local solutions – a multitude of inefficient solutions, up to $\sim \frac{2^n}{1.25\sqrt{n}}$ (see Theorem 12 below), which is less than $2^n$ as over the box in Example 1, but still a good indicator of NP-hardness of the problem.

We now proceed to an equivalence result which establishes exactness of penalization w.r.t. the constraint $e^\top x = 1$:

Consider the StQP $\min\left\{g(x) = x^\top Q x : x \in \Delta\right\}$ and the QP over the positive orthant

$$\min\left\{h(p) = \tfrac{1}{2}p^\top Q p - e^\top p : p \in \mathbb{R}_+^n\right\}. \tag{1.40}$$

Further, let us assume that the objective is strictly convex along rays starting from the origin, which amounts to strict $\mathbb{R}_+^n$-copositivity of $Q$:

$$p^\top Q p > 0 \quad \text{for all} \quad p \in \mathbb{R}_+^n \setminus \{o\}.$$

### Theorem 9 (StQPs are QPs over the positive orthant) [6]:

*Assume that $Q$ is strictly $\mathbb{R}_+^n$-copositive. This implies that problem (1.40) is bounded from below, and therefore has a solution. Further, local and global solutions to (1.39) and (1.40) are related as follows:*

*(a) If $\bar{x} \in \Delta$ is a local minimizer of $g(x) = x^\top Q x$ on $\Delta$, then $\bar{p} = \frac{1}{g(\bar{x})}\bar{x}$ is a local minimizer of $h(p) = \tfrac{1}{2}p^\top Q p - e^\top p$ on $\mathbb{R}_+^n$.*

*(b) If $\bar{p} \geq o$ is a local minimizer of $h(p)$, then $\bar{x} = \frac{1}{e^\top \bar{p}}\bar{p}$ is a local minimizer of $g(x) = x^\top Q x$ on $\Delta$.*

(c) *The objective values in cases* (a) *and* (b) *are related by* $\frac{1}{g(\bar{x})} = -2h(\bar{p})$.

*Hence global solutions correspond to each other.*

*Proof.* First note that $\alpha_Q = \min\{x^\top Q x : x \in \Delta\} > 0$ implies strict convexity of the function $\phi_x(t) = h(tx) = \frac{x^\top Q x}{2} t^2 - t$ as $t \geq 0$ for any fixed $x \in \Delta$. Now $\phi_x$ attains its minimal value at $t_x = \frac{1}{x^\top Q x}$, namely $\phi_x(t) \geq \phi_x(t_x) = -\frac{1}{2x^\top Q x} \geq -\frac{1}{2\alpha_Q}$, and the latter is the minimum value of $h(p)$ over $\mathbb{R}^n_+$. This way, we proved the existence of a (global) solution to (1.40) even without invoking the general Frank/Wolfe theorem. The assertion on the local solutions is a bit more involved, but follows essentially the same lines:

(a) For arbitrary $p \in \mathbb{R}^n_+ \setminus \{o\}$ let $\Phi(p) = \frac{1}{e^\top p} p$. Then $\Phi$ maps the domain $\mathbb{R}^n_+ \setminus \{o\}$ continuously on $\Delta$, and hence $V = \Phi^{-1}(U)$ is a neighbourhood of $\bar{p} = \frac{1}{g(\bar{x})} \bar{x}$ if $U$ is a neighbourhood of $\bar{x}$ in $\Delta$ satisfying $g(x) \geq g(\bar{x})$ for all $x \in U$. We now claim that $h(p) \geq h(\bar{p})$ for all $p \in V$. Indeed, using $p^\top Q p > 0$ and some algebra, it is easy to derive from $[p^\top Q p - e^\top p]^2 \geq 0$ the inequality

$$h(p) \geq -\frac{(e^\top p)^2}{2 p^\top Q p} \quad \text{for all } p \in \mathbb{R}^n_+ \setminus \{o\}. \tag{1.41}$$

Now $g(\bar{x}) \leq g(\Phi(p)) = \frac{1}{(e^\top p)^2} g(p) = (-\frac{1}{2})[-\frac{(e^\top p)^2}{2 p^\top Q p}]^{-1}$ so that by (1.41)

$$-\frac{1}{2g(\bar{x})} \leq h(p)$$

for all $p \in V$. What remains to show is the assertion in (c). But this is immediate from $e^\top \bar{x} = 1$.

(b) If $\bar{p}$ is a local solution of (1.40), then necessarily also the KKT conditions are satisfied due to linearity of the constraints. Hence there is a vector $r \geq o$ such that $r^\top \bar{p} = 0$ and $\nabla h(\bar{p}) = Q\bar{p} - e = r$, which in turn entails $g(\bar{p}) = \bar{p}^\top Q \bar{p} = e^\top \bar{p}$ and $h(\bar{p}) = -\frac{1}{2} e^\top \bar{p}$. On the other hand, by definition of $\bar{x}$ we now get

$$g(\bar{x}) = \frac{1}{(e^\top \bar{p})^2} g(\bar{p}) = \frac{1}{e^\top \bar{p}},$$

and thus again the relation in (c) is established. Next define $\Psi(x) = \frac{1}{g(x)} x$ which maps $\Delta$ continuously into $\mathbb{R}^n_+$. Hence $\Psi(\bar{x}) = (e^\top \bar{p})\bar{x} = \bar{p}$ implies that $U = \Psi^{-1}(V)$ is a neighbourhood of $\bar{x}$ in $\Delta$ provided that $V$ is one of $\bar{p}$ in $\mathbb{R}^n_+$ such that $h(p) \geq h(\bar{p})$ for all $p \in V$. Consequently,

$$-\frac{1}{2g(\bar{x})} = h(\bar{p}) \leq h(\Psi(x)) = \frac{1}{2[g(x)]^2} g(x) - \frac{1}{g(x)} e^\top x = -\frac{1}{2g(x)}$$

for all $x \in U$, which shows (b). The remaining assertion in (c) is immediate.

The central role of StQPs within the class of all QPs is not only motivated by copositivity conditions for optimality in the latter class. Moreover, there are quite many applications where StQPs either arise as auxiliary (sub-)problems, or where they can be used directly. This is the topic of the next section.

## 1.7 Some Applications of StQPs

Consider again the general QP over a bounded polyhedron

$$\min\left\{f(x) = \tfrac{1}{2}x^\top Qx + c^\top x : x \in M\right\},$$

where $M = \operatorname{conv}\{v_1,\dots,v_k\} = V(\Delta) \subset \mathbb{R}^n$ is now a polytope with

$$V = [v_1,\dots,v_k] \text{ an } n \times k\text{-matrix and } \Delta \subset \mathbb{R}^k.$$

Then for the $k \times k$ matrix $\hat{Q} = \tfrac{1}{2}\left(V^\top QV + e^\top V^\top c + c^\top Ve\right)$

$$f(x) = y^\top \hat{Q}y \quad \text{whenever} \quad x = Vy \in M, \text{ i.e., } y \in \Delta.$$

Thus every QP over a polytope is equivalent to the StQP $\min\left\{y^\top \hat{Q}y : y \in \Delta\right\}$. This approach is of course only practical if the vertices $V$ are known and $k$ is not too large. This is the case of QPs over the $\ell^1$ ball [14] where $V = [I_n|-I_n]$ and $\Delta \subset \mathbb{R}^{2n}$.

However, even for general QPs, we can use StQP as a relaxation without using all vertices. To this end, we represent $M = \left\{x \in \mathbb{R}^n_+ : Ax = b\right\}$ in standard form used in LPs rather than in the previously used inequality form. Here $A$ is an $m \times n$ matrix and $b \in \mathbb{R}^m$.

**Theorem 10 (Inclusion of $M$ in a simplex − StQP relaxation)** [15]:
*Suppose $M \neq \{o\}$ is bounded. Then there is $z \in \mathbb{R}^m$ with $A^\top z \geq e$ and $b^\top z > 0$. Put $D_z = b^\top z \left[\operatorname{Diag}(A^\top z)\right]^{-1}$, as well as*

$$\hat{Q}(z) = \tfrac{1}{2}\left(D_z QD_z + ec^\top D_z + D_z ce^\top\right).$$

*Then*

$$\alpha_{\hat{Q}(z)} \leq \min\left\{\tfrac{1}{2}x^\top Qx + c^\top x : x \in M\right\}.$$

*Proof.* Consider the polyhedron

$$R = \{z \in \mathbb{R}^m : A^\top z \geq e\} \tag{1.42}$$

Since $M$ is bounded, the function $e^\top y$ attains its maximum on $M$. Thus, by LP duality, $R$ is nonempty as it coincides with the feasible set of the dual LP. Also, since $M$ is bounded and different from $\{o\}$, we must have $b \neq o$. Let $z \in R$ and put $\pi^z = b^\top z$ as well as $p^z = A^\top z$. Hence we obtain from $A^\top z \geq e$ the relation $\pi^z = b^\top z = y^\top A^\top z \geq y^\top e > 0$ for all $y \in M$, since $Ay = b \neq o$ implies $y \neq o$. Further, for all $z \in R$ we have $p^z \geq e$ as well, and $M$ is contained in the simplex $S^z = \{y \in \mathbb{R}^n_+ : \frac{1}{\pi^z}(p^z)^\top y = 1\} = D_z^{-1}(\Delta)$. Moreover, for any $y \in S^z$ there is $x = D_z y \in \Delta$ with $f(y) = x^\top \hat{Q}(z)x \geq \alpha_{\hat{Q}(z)}$, which establishes the result.

We can refine this approach if $Q$ is negative-semidefinite, i.e., for concave minimization QPs. Then

$$h(z) = \min_i f\left(\frac{\pi(z)}{p_i(z)} e_i\right) = \alpha_{\hat{Q}(z)} \leq \alpha^* = \min\{f(x) : x \in M\}.$$

The best upper bound $h^* = \max\{h(z) : A^\top z \geq e\}$ is hard to find. However, put $f_i(t) = f(te_i)$ which is a concave function in one variable $t$. Then we can solve the easy problem

$$\bar{h} = \min_i \sup_{t \geq 0} f_i(t) \geq h^*.$$

Now check whether also $\bar{h} \leq \alpha^*$, i.e., put $[L_i, R_i] = \{t : f_i(t) \geq \bar{h}\}$ and check

$$L_i \leq \frac{\pi(z)}{p_i(z)} \leq R_i, \text{ all } i, \text{ some } z \text{ with } A^\top z \geq e.$$

This is a linear feasibility problem for $z$.

We now proceed to illustrate one of the most prominent direct applications of StQPs in combinatorial optimization, the StQP formulation of the Maximum Clique Problem due to Motzkin and Straus.

Consider an undirected graph $\mathcal{G} = (\mathcal{V}, \mathcal{E})$ with $|\mathcal{V}| = n$ vertices. A clique $\mathcal{S} \subseteq \mathcal{V}$ is a subset of vertices inducing a complete subgraph. $\mathcal{S}$ is called *maximal* if $\mathcal{S}$ is not contained in a larger clique, and a clique $\mathcal{S}^*$ is a *maximum clique* if

$$|\mathcal{S}^*| = \max\{|\mathcal{T}| : \mathcal{T} \text{ is a clique in } \mathcal{G}\}.$$

Finding the *clique number* $\omega(\mathcal{G}) = |\mathcal{S}^*|$ is an NP-complete combinatorial optimization problem, which can be formulated as continuous optimization problem, namely an StQP:

To this end, we employ the *adjacency matrix* $A_\mathcal{G}$ of a graph $\mathcal{G}$ defined as the indicator function of $\mathcal{E}$ over $\mathcal{V} \times \mathcal{V}$:

$$[A_\mathcal{G}]_{ij} = \begin{cases} 1, \text{ if } \{i,j\} \in \mathcal{E}, \\ 0, \text{ else.} \end{cases}$$

**Theorem 11 (Motzkin/Straus formulation of Maximum-Clique)** [51]: *For $Q = Q_\mathcal{G} = E - A_\mathcal{G}$, with $E = ee^\top$ the all-ones matrix, we get*

$$\frac{1}{\omega(\mathcal{G})} = \alpha_Q.$$

Recently an interesting converse has been established in the context of dominant sets and clustering, where $q_{ij}$ are interpreted as edge weights [58]. A different, even more recent approach for solving general StQPs has been taken by [62], who first define a convexity graph based on the instance $Q$, and then try to find maximal cliques in this graph, as the support $\sigma(\overline{x}) = \{i : \overline{x}_i > 0\} = J(\overline{x}) \setminus \{0\}$ (the non-binding constraints set) of the global solution $\overline{x}$ must be a clique. Given the support is $\sigma(\overline{x}) = \overline{J}$ is fixed, the solution $\overline{x}$ is easily obtained by solving the linear system arising from the KKT conditions. See also Theorem 12 below on how supports of coexisting local solutions are related.

## 2 Some Basic Techniques, Illustrated by StQPs

Quite generally, the interplay of local search and escape steps in global optimization can be algorithmically sketched as follows:

1. Start local search with some $x^0$;
2. check if result $x$ is a local solution;
    if not, restart with random perturbation of $x$;
    call the escape procedure, denote resulting point by $\tilde{x}$;
3. repeat from step 1., replacing $x^0$ with $\tilde{x}$.

   If there are only finitely many objective *values* for local solutions (which is true for any QP), and if the local search produces a (strict) local solution with probability one for the choice of starting point (true for the method (2.3) and StQPs, if no principal minor of $Q$ is zero), this algorithm stops after finitely many repetitions.

   Of further interest in global optimization is the question of coexistence of local solutions. For StQPs, an important property is established in the following

**Theorem 12 (Antichain property of local StQP solutions)** [5]:
*For $x \in \Delta$ define the support as*

$$\sigma(x) = \{i : x_i > 0\}.$$

*Let $x$ and $z$ be local solutions of StQP with supports $\sigma(x)$ and $\sigma(z)$. Suppose there are only finitely many local solutions to (1.39). Then neither*

$$\sigma(x) \subseteq \sigma(z) \quad nor \quad \sigma(x) \supseteq \sigma(z).$$

*Further, then there cannot be more than $\binom{n}{\lfloor \frac{n}{2} \rfloor}$ local solutions to (1.39).*

Hence, asymptotically, i.e., for large $n$, there are at most $2^n\sqrt{2/(\pi n)} \approx 2^n/(1.2533141373155\sqrt{n})$ local solutions of an StQP in variables, if there are only finitely many of them. This follows from Stirling's approximation formula for the factorials in $\binom{n}{\lfloor\frac{n}{2}\rfloor}$.

## 2.1  Local Search and Escape Directions

We start by a local search procedure which is called *replicator dynamics (RD)*. To this end, we switch to the maximization form

$$\beta_Q = \max\{x^\top Q x : x \in \Delta\} \tag{2.1}$$

of an StQP. By adding a nonnegative multiple of $E$ to $Q$ (which changes $x^\top Q x$ by only a constant on $\Delta$), we may and do assume in the sequel that

$$q_{ii} > 0 \quad \text{and} \quad q_{ij} \geq 0 \quad \text{for all } i, j. \tag{2.2}$$

Condition (2.2) is necessary and sufficient for $Qx \geq 0$ and $x^\top Q x > 0$ for all $x \in \Delta$. Now consider the following iteration process (a dynamical system in discrete time $t$) on $\Delta$:

$$x_i(t+1) = x_i(t) \frac{[Qx(t)]_i}{x(t)^\top Q x(t)}, \quad 1 \leq i \leq n. \tag{2.3}$$

This method is reminiscent of the power method for finding the largest (in magnitude) eigenvalue of a square matrix, with the unit Euclidean sphere replaced by the unit simplex. The RD method has a long history in mathematical biology, and it connects three different fields: optimization, evolutionary games, and qualitative analysis of dynamical systems; see [8] and references therein. It arises in population genetics under the name *selection equations* where it is used to model time evolution of haploid genotypes, with $Q$ being the (symmetric) fitness matrix, and $x_i(t)$ representing the relative frequency of allele $i$ in the population at time $t$ (see, e.g., [34], [50, Chapter III]).

Since it also serves to model replicating entities in a much more general context, it is often called *replicator dynamics* nowadays. The continuous-time version of the RD method is known to be a gradient system with respect to Shahshahani geometry; see [41]. This suggests that the method may be useful for local optimization. In fact, (2.3) has the remarkable property that, under assumption (2.2), the generated sequence of iterates $\{x(t)\}$ converges [47] (i.e., has a unique cluster point) and, given that we start in ri$\Delta$, the limit $\bar{x}$ is a KKT point of (1.39). This contrasts with other interior-point methods for solving (1.39), for which additional assumptions on $Q$ are required to prove convergence of the generated iterates; see [19, 35, 49, 63, 64]. Additionally, the objective values $f(x(t))$ increase with $t$, $\|x(t) - \bar{x}\| = \mathcal{O}(1/\sqrt{t})$, and convergence rate is linear if and only if strict complementarity holds at $\bar{x}$. In [9], the RD method (2.3) was applied to solve medium-sized test problems from portfolio selection and was shown to be superior in performance to classical feasible ascent methods using exact line search, including Rosen's gradient projection method and Wolfe's reduced gradient method.

**Theorem 13 (Monotonicity and convergence of RD)** [18, 34, 47]:
*If $Q$ satisfies (2.2), then any non-stationary trajectory $x(t)$ under (2.3)*
- *yields strictly increasing values $x(t)^{\top}Qx(t)$ and*
- *converges to a fixed point $\bar{x}$;*
- *further, with probability one (if the starting points $x(0)$ are distributed according to a law absolutely continuous w.r.t. Lebesgue measure) $\bar{x}$ is a strict local maximizer of $x^{\top}Qx$, provided no principal minor of $Q$ vanishes.*

**Exercise:** For symmetric $n\times n$ matrices $N \geq O$ and $x \in \mathbb{R}^n_+ \setminus \{o\}$ we have

$$\left(\frac{x^{\top}Nx}{x^{\top}x}\right)^m \leq \frac{x^{\top}N^mx}{x^{\top}x}$$

(this can be proven via induction on $n$, but is of no concern here). Using this inequality, establish monotonicity of $x(t)^{\top}Qx(t)$ along (2.3)-trajectories $x(t)$, as in Theorem 13. Hint: put $m = 3$ and $n_{ij} = \sqrt{x_i}q_{ij}\sqrt{x_j}$ in above inequality.

There are also variants of the replicator dynamics involving monotonic transformations of the quantities $[Qx(t)]_i$, among them the exponential function $\exp(\theta[Qx(t)]_i)$ for a parameter $\theta > 0$. For empirical experience see [59], for some rudimentary theory in the spirit of Theorem 13 see [9], and for further theoretical analysis [66]. In that paper, the fixed-step iteration of (2.3) is generalized as follows.

For an (2.3)-iteration, put $x' = x(t+1)$, $x = x(t)$, and $g = \nabla f(x) = 2Qx$. Then

$$x' - x = v^{(1)} = \frac{1}{x^{\top}g}(\text{Diag } x)\left[g - (x^{\top}g)e\right]$$

is an ascent direction which can be used for (an inexact) line search, e.g., by a limited maximization rule à la Armijo. A similar affine-scaling type approach was proposed by [35] with

$$v^{(2)} = (\text{Diag } x)^2\left[g - \frac{x^{\top}(\text{Diag } x)g}{x^{\top}x}e\right].$$

One common generalization of both approaches leads to

$$v^{(\gamma)} = (\text{Diag } x)^{\gamma}\left[g - \frac{x^{\top}(\text{Diag } x)^{\gamma-1}g}{\sum_i x_i^{\gamma}}e\right].$$

This iteration also converges globally with sublinear rate, and good large-scale experience for relatively large problems ($n = 1000$) is reported in [66] for $1 \leq \gamma < 2.4$.

For problems at an even larger scale ($n = 10^4$), another variant with a scaled projected reduced gradient (SPRG) is proposed by [65] which truncates negative entries in $[g - (x^\top g)e]$ to zero, i.e. puts $r_i = [(Qx)_i - x^\top Qx]_+$ and uses the search direction

$$v^+ = r - (e^\top r)x.$$

**Theorem 14 (SPRG ascent direction)** [8,65]:
$v^+$ is a feasible ascent direction for the problem (2.1), with $v^+ = o$ if and only if $x$ is a KKT point.

*Proof.* If $x_i = 0$, then $v_i^+ = r_i \geq 0$. Further, $e^\top v = e^\top r(1 - e^\top x) = 0$. Hence $v^+$ is a feasible direction. Moreover, put $s_j = [(Qx)_i - x^\top Qx]$ so that $r_j = [s_j]_+$. Then

$$
\begin{aligned}
(v^+)^\top Qx &= \sum_{i,j} x_i q_{ij}[r_j - (e^\top r)x_j] \\
&= \sum_{i,j} r_j q_{ji} x_i - (e^\top r)x^\top Qx \\
&= \sum_j r_j[(Qx)_j - x^\top Qx] \\
&= \sum_j s_j[s_j]_+ \geq 0,
\end{aligned}
$$

with equality if all $s_j \leq 0$, which just means that $x$ is a KKT point of (2.1).

If we now base an (inexact Armijo type) line search upon $v^+$, this yields a globally convergent iteration scheme with sublinear rate. Preliminary experiments in [65] suggest that SPRG is sometimes more accurate than MINOS, and robust also if strict complementarity fails.

Of course, escaping from inefficient local solutions is, generally speaking, hard. By applying the global optimality conditions by copositivity from Theorem 2 to the special case of StQP, we arrive at the so-called *genetic engineering via negative fitness* (GENF) procedure. Returning to the original biomathematical interpretation of (2.3), we determine *truly unfit alleles* (meaning coordinates) $i$ via fitness *minimization* rather than maximization, bearing in mind some alleles will go extinct in a non-global solution $x$ (else the antichain property from Theorem 12 will guarantee global optimality of $x$ due to $\sigma(x) = \{1,\ldots,n\}$). But *not all* extinct alleles $i \notin \sigma(x)$, i.e. $x_i = 0$, will remain zero in a global solution $x^*$. So to determine the *truly unfit*, we rather look at fitness *minimizers*, i.e., replace $Q$ with

$$\overline{Q} = [\gamma_{\sigma(x)} - q_{ij}]_{i,j \notin \sigma(x)}$$

with $\gamma_{\sigma(x)} \geq \max_{i,j \notin \sigma(x)} q_{ij}$ (which is the largest fitness of extinct alleles), and perform a local search for $\overline{Q}$ on the reduced simplex, to obtain a local solution

$z$ to this auxiliary problem. Then $\tau = \sigma(z) \backslash \sigma(x)$ can be seen as the set of truly unfit alleles. As it turns out, this rather heuristic approach indeed gives the construction of the improving feasible point as in the general case described in Theorem 2.

**Theorem 15 (GENF – escape directions for StQP)** [18]:
*Suppose $\bar{x}$ is local solution to the master problem*

$$\beta_Q = \max \left\{ x^\top Q x : x \in \Delta \right\}. \tag{2.4}$$

*Pick a set $\tau$ disjoint to the surviving allele set $\sigma(\bar{x})$ by 'negative genetic engineering', i.e., $\tau = \sigma(z) \backslash \sigma(x)$ where $z$ is a local solution to the auxiliary problem $\max \left\{ z^\top \bar{Q} z : z \in \bar{\Delta} \right\}$ as above. Denote by $m = |\tau|$.*
*For all $(s,t) \in \sigma(\bar{x}) \times \tau$ replace $q_{si}$ with $q_{ti}$ and remove all other $j \in \tau \backslash \{t\}$. For the resulting matrix $Q_{t \to s}$ consider the reduced problem on $\tilde{\Delta} \subset \mathbb{R}^{n-m}$*

$$\beta_{Q_{t \to s}} = \max \left\{ z^\top Q_{t \to s} z : z \in \tilde{\Delta} \right\}. \tag{2.5}$$

*Then $\bar{x}$ is a global solution to (2.4) if and only*

$$\max \left\{ \beta_{Q_{t \to s}} : (s,t) \in \sigma(x) \times \tau \right\} \le \bar{x}^\top Q \bar{x},$$

*i.e., if the optimal values of all the reduced problems (2.5) do not exceed the current best value in the master problem.*
*In the negative, enrich any (global) solution $z$ to the reduced problem (2.5) by suitable zeroes to an improving feasible point $\tilde{x} \in \mathbb{R}^n$.*

With this formulation, the hardness aspect of finding an improving feasible point is the requirement to find the global solution for the reduced problem (2.5), although this may be considerably smaller. However, in practice a good local solution to (2.5) may be enough to escape from the first local solutions found in the overall local search. Still, a satisfactory implementation of the GENF algorithm has yet to be done.

## 2.2 *Bounding and Additive Decompositions*

Let us return to a minimization StQP

$$\alpha_Q = \min \left\{ x^\top Q x : x \in \Delta \right\}.$$

The most elementary lower bound for $\alpha_Q$ is

$$\alpha_Q^0 = \min_{i,j} q_{ij}, \tag{2.6}$$

which is exact whenever the minimum entry of $Q$ is attained at the diagonal, and vice versa:

$$\alpha_Q^0 \le \alpha_Q \le \min_i q_{ii} = \min_i e_i^\top Q e_i. \tag{2.7}$$

In case $\alpha_Q^0 = \min_{i,j} q_{ij} < \min_i q_{ii}$ there is a refinement:

$$\alpha_Q^{\text{ref}} = \alpha_Q^0 + \left[ \sum_i (q_{ii} - \alpha_Q^0)^{-1} \right]^{-1} \tag{2.8}$$

(with the usual conventions for $1/0 = \infty$, $t + \infty = \infty$, and $1/\infty = 0$) is strictly improving, i.e., $\alpha_Q^0 < \alpha_Q^{\text{ref}} \le \alpha_Q$, and exactness equality $\alpha_Q^{\text{ref}} = \alpha_Q$ holds in some instances:

**Exercise:** Show that for $Q = \text{Diag } d$ with $d_i > 0$, all $i$, then

$$\alpha_Q = \left[ \sum_i d_i^{-1} \right]^{-1} = \alpha_Q^{\text{ref}} > \alpha_Q^0 = 0.$$

We proceed to discuss *Lagrangian dual bounds* for StQPs. Recall the Lagrange function

$$L(x; \nu, u) = x^\top Q x + \nu(1 - e^\top x) - u^\top x,$$

where $u \in \mathbb{R}_+^n$ and $\nu \in \mathbb{R}$. Now unless $Q$ is positive-semidefinite, the straightforward bound is useless:

$$\Theta(\nu, u) = \inf \{ L(x; \nu, u) : x \in \mathbb{R}^n \} = -\infty.$$

On the other hand, without any convexity assumptions on $Q$, we also may have a perfect bound:

**Theorem 16 (Lagrangian dual bounds for StQPs)** [16]:
*If $\alpha_Q^0 > 0$, then the duality gap is either infinity or zero, depending on relaxation:*

*for $\bar{\Theta}(\nu) = \inf \{ x^\top Q x + \nu(1 - e^\top x) : x \in \mathbb{R}_+^n \}$, we get*

$$\bar{\Theta}(\nu) = \begin{cases} \nu - \frac{1}{4\alpha_Q} \nu^2, & \text{if } \nu \ge 0, \\ \nu & \text{else.} \end{cases} \tag{2.9}$$

*Hence*

$$\max \{ \bar{\Theta}(\nu) : \nu \in \mathbb{R} \} = \alpha_Q.$$

Nevertheless also this bound is useless since the dual function involves the unknown parameter $\alpha_Q$.

The situation does not change if we employ Semi-Lagrangian bounds for StQPs; recall that a *semi-Lagrangian relaxation* (SLR) emerges if we split the (only) equality constraint $e^\top x = 1$ into two inequalities $e^\top x \leq 1$ and $e^\top x \geq 1$, and relax only the latter.

**Theorem 17 (Zero duality gap for SLR of StQPs):**
*For $\mu \geq 0$, define*

$$\tilde{\Theta}(\mu) = \inf \left\{ x^\top Q x + \mu(1 - e^\top x) : x \in \mathbb{R}^n_+, e^\top x \leq 1 \right\}. \qquad (2.10)$$

*Then*

$$\tilde{\Theta}(\mu) = \mu - \frac{1}{4\alpha_Q}\mu^2 \quad \text{for all } \mu \geq 0,$$

*and thus*

$$\max \left\{ \tilde{\Theta}(\mu) : \mu \geq 0 \right\} = \alpha_Q.$$

So even to evaluate $\tilde{\Theta}(\mu)$ at *any* $\mu$, we need to know $\alpha_Q$.

**Exercise:** By scaling a vector $x \in \mathbb{R}^n_+$ with $e^\top x \leq 1$ appropriately to a vector $y \in \Delta$, show that the semi-Lagrangian function $\tilde{\Theta}$ of an StQP as defined in (2.10) equals

$$\tilde{\Theta}(\mu) = \min_{t \in [0,1]} \min_{y \in \Delta} \left\{ t^2 y^\top Q y + \mu(1 - t) \right\} = \min_{t \in [0,1]} [t^2 \alpha_Q + \mu(1 - t)], \ \mu \geq 0,$$

and use this relation to establish Theorem 17. Remark: formula (2.9) in Theorem 16 is much harder to prove (e.g., via Theorem 9), whereas the implication for the zero duality gap there follows easily.

*Convex underestimation bounds* are more interesting. To this end, choose any positive-semidefinite matrix $S$ such that $x^\top S x \leq x^\top Q x$ for all $x \in \Delta$. Then

$$\alpha_S = \min_{x \in \Delta} x^\top S x \leq \alpha_Q,$$

i.e., $\alpha_S$ is a valid bound. Since $x^\top S x$ is convex, $\alpha_S$ is cheap, i.e., can be determinated to any desired accuracy in polynomial time.

**Theorem 18 (Best convex underestimation bound for StQPs) [16]:**
*The best such underestimation bound is given by*

$$\alpha_Q^{\text{conv}} := \max \left\{ \alpha_S : S \succeq O, \ s_{ij} \leq q_{ij} \text{ all } i, j, \text{ diag } S = \text{diag } Q \right\}. \qquad (2.11)$$

To obtain a solution to (2.11), one may employ *semidefinite programming* (SDP) via exact Shor relaxation [1]. We will discuss a similar approach in more detail below.

Continuing the idea above, one is lead quite naturally to the approach of *d.c. decomposition* (D.C.D.) bounds; first use subadditivity to obtain a lower bound for $\alpha_Q$ as follows: for any symmetric $n \times n$ matrix $S$, the relation $Q = S + (Q - S)$ implies

$$\alpha_Q \geq \alpha_S + \alpha_{Q-S}.$$

Recall that $\alpha_S$ is cheap if $S \succeq O$. Now if $-T$ is positive-semidefinite either, then $\alpha_T = \min_i t_{ii}$ is even cheaper, since we just have to find the smallest of $n$ given numbers.

Hence denote the set of all D.C.D.s of $Q$ by

$$\mathcal{P}^Q = \{S \succeq O : S - Q \succeq O\}. \tag{2.12}$$

Again, the best such D.C.D. bound can be obtained by SDP techniques:

$$\alpha_Q^{\mathrm{dcd}} := \max\left\{\alpha_S + \alpha_{Q-S} : S \in \mathcal{P}^Q\right\}. \tag{2.13}$$

Nevertheless, there is a dominating bound first considered in [1] which employs the same machinery. See Section 4.4 below.


## 2.3  Branch-and-Bound: Principles and Special Cases

Here, we concentrate on a key technology for global optimization, which is primarily used in combinatorial optimization, but has its justification as well in the continuous domain: branch-and-bound. As a motivating example, we consider a D.C.D. based branch-and-bound for StQPs.

The problem tree is generated as follows: at the root node, we start with some D.C.D. $Q = S + T$ with

$$\alpha_Q \geq \alpha_S + \alpha_T,$$

where $S$ and $-T$ are positive-semidefinite, so that above estimate is cheap. Note that this D.C.D. is not necessarily the one realizing (2.13).

At an internal node of the problem tree, characterized by a subsimplex $X = \mathrm{conv}\,(v_1, \ldots, v_n)$, we repeat this to obtain

$$\gamma_Q(X) = \min\{x^\top S x : x \in X\} + \min\left\{v_i^\top T v_i : i \in \{1, \ldots, n\}\right\}.$$

If $V = [v_1, \ldots, v_n]$, for any $x \in X$ there is some $y \in \Delta$ such that $x^\top S x = y^\top (V^\top S V) y$, thus

$$\min\{x^\top S x : x \in X\} = \alpha_{V^\top S V},$$

is automatically achieved by any local search as also $V^\top S V \succeq O$.

Hence

$$\gamma_Q(X) = \alpha_{V^\top SV} + \min\left\{v_i^\top T v_i : i \in \{1, \ldots, n\}\right\}$$

is the desired *local lower bound*.

For bounding from above, we employ a *local upper bound* $\delta(X)$ with the property that

$$\delta(X_k) \to (x^*)^\top Q(x^*) \quad \text{whenever} \quad \bigcap_{k=1}^\infty X_k = \{x^*\}.$$

For instance, any feasible point $x \in X$ yields such a bound $\delta(X) = x^\top Q x$. A better alternative would be to put $\delta(X)$ equal to the objective result of a local search for $V^\top Q V$. As always, the final choice of method depends on the balance between quality and effort: the more we invest in high quality local search, the more cutting power we gain, and hence less subproblems are generated. The *global upper bound* is then the smallest of the $\delta(X)$ values obtained so far.

The next important ingredient is the *branching rule*. Contrasting to combinatorial optimization where branching is canonically done by fixing a binary variable at either of its values zero or one, in continuous optimization there are a lot of choices. Here, we can choose a subsimplex $X$ with smallest $\gamma(X)$ for fast improvement, or with smallest $\delta(X)$ for early good intermediate results (depth-first or width-first traversal of the problem tree), then fathom the node labeled by $X$ if $\gamma(X)$ exceeds global upper bound: any global solution $x^* \notin X$. Else we bisect $X$ along longest edge (e.g., by halving it), to obtain two children of node $X$.

**Theorem 19 (Convergence of branch-and-bound)** [8]:
*If the problem tree is generated as sketched above, the branch-and-bound algorithm converges.*

*Proof.* To enforce the general convergence result of the branch-and-bound algorithm [42], one has to establish an asymptotically vanishing gap: $\delta(X_k) - \gamma(X_k) \to 0$ as $k \to \infty$, if $(X_k)$ is an exhaustive sequence of subsimplices, i.e., $\bigcap_{k=1}^\infty X_k = \{x^*\}$. In fact, in the StQP case we have $\gamma(X_k) \to (x^*)^\top Q(x^*)$ as $k \to \infty$. The result follows by assumption $\delta(X_k) \to (x^*)^\top Q(x^*)$ as $k \to \infty$.

# 3 Reformulation, Relaxation, Approximation

Again by the class of StQPs, we study three basic strategies to deal with NP-hard nonlinear problems. Reformulations find alternative representations of given problems with the aim to employ different, more suitable methods, while relaxations and approximations are used to find a good proxy of the exact solution.

## 3.1  Quartic and Unconstrained Reformulations

First we want to get rid of the constraint $x \geq o$ in an StQP. To this end, we use Parrilo's trick $x = T(y)$ with $x_i = y_i^2 \geq 0$, to arrive at *quartic formulation* of StQPs with a ball constraint. Indeed the partial inverse of $T$ is given by $y = T^{-1}(x)$ with $y_i = +\sqrt{x_i}$, and $y^\top y = 1$ means $e^\top x = 1$, i.e. Euclidean unit sphere coincides with $T^{-1}\Delta$.

To put down the quartic formulation in a compact way, we abbreviate $Y = \mathrm{diag}\, y$. Then $\alpha_Q = \min \left\{ x^\top Q x : x \in \Delta \right\}$ coincides with

$$\min \left\{ f_q(y) = \tfrac{1}{2} y^\top Y Q Y y : y^\top y = 1 \right\} \tag{3.1}$$

which is a homogeneous problem equivalent to the StQP:

**Theorem 20 (Quartic formulation of StQPs)** [17]:

$\bar{x} \in \Delta$ *is local/global solution to the StQP (1.39) if and only if*

$\bar{y} = T^{-1}(\bar{x})$ *is local/global solution to the quartic problem (3.1).*

Unfortunately, the above equivalence result does *not* hold for KKT points without any further assumptions. So we have to resort to *second-order necessary conditions* (SOC) for local optimality of $\bar{y}$ in the quartic problem:

$$z^\top \left[ 2\bar{Y}Q\bar{Y} + \mathrm{diag}\{Q\bar{Y}\bar{y}\} + \bar{\mu}I \right] z \geq 0 \quad \text{for all } z \in \mathbb{R}^n : \ z^\top \bar{x} = 0, \tag{3.2}$$

where $\bar{\mu}$ is the Lagrange multiplier of $y^\top y = 1$ in $\bar{y}$.

A similar SOC for local optimality of $\bar{y}$ in the StQP is $\Gamma(\bar{x})$-copositivity of $Q$ (stronger than condition (1.6), of course):

$$z^\top Q z \geq 0 \quad \text{for all } z \in \mathbb{R}^n : e^\top z = 0 \text{ and } z_i = 0 \text{ if } \bar{x}_i = 0. \tag{3.3}$$

**Theorem 21 (SOC in quartic and StQP are equivalent)** [17]:

$\bar{x} \in \Delta$ *is a KKT point for StQP satisfying (3.3) if and only if*

$\bar{y} = T^{-1}(\bar{x})$ *is a KKT point for (3.1) satisfying (3.2).*

*Further, if $\bar{y}$ is a local solution to (3.1), then necessarily $\bar{\mu} = -2f_q(\bar{y})$.*

Similar to Theorem 9, we follow an exact penalty approach to the quartic problem, and thus can discard the ball constraint. To this end, we employ the *multiplier function*

$$\mu(y) = -2f_q(y) = -y^\top Y Q Y y.$$

For the penalty parameter $\varepsilon > 0$, we define the merit function

$$P(y; \varepsilon) = f_q(y) + \tfrac{1}{\varepsilon}(\|y\|^2 - 1)^2 + \mu(y)(\|y\|^2 - 1)$$

$$= f_q(y) + \frac{1}{\varepsilon}(\|y\|^2 - 1)^2 - 2f_q(y)(\|y\|^2 - 1)$$

$$= f_q(y)(3 - 2\|y\|^2) + \frac{1}{\varepsilon}(\|y\|^2 - 1)^2.$$

Pick any $y_0 \in \mathbb{R}^n$, $\Theta \in (0,1)$ and $\gamma > 1$ and let $\varepsilon \le \varepsilon_Q$, where

$$\varepsilon_Q = \min\left\{ \tfrac{1}{2}\|Q\| \tfrac{(1+3\Theta^4)}{(\Theta^2 - 1)^2}, \ \tfrac{1}{2\|Q\|}\tfrac{(\gamma^2 - 1)^2}{\gamma^4}, \ \tfrac{2\Theta^2}{3\|Q\|\gamma^4} \right\}.$$

Then the level set $\mathcal{L}_0 = \{y \in \mathbb{R}^n : P(y; \varepsilon) \le P(y_0; \varepsilon)\}$ is compact.

**Theorem 22 (Equivalence of exact quartic penalization)** [17]:

*Pick any $\varepsilon \le \varepsilon_Q$. Let $\bar{y} \in \mathcal{L}_0$ be a critical point of $P(y; \varepsilon)$: $\nabla_y P(y; \varepsilon) = o$. If $\bar{y}$ satisfies $D_y^2 P(y; \varepsilon) \succeq O$ (the standard unconstrained SOC), then $\bar{y}$ satisfies (3.2) with $\bar{x} = T(\bar{y}) \in \Delta$, and therefore $\bar{x}$ is a KKT point for the StQP fulfilling (3.3).*

## 3.2  Convex Conic Reformulations

Recall that a linear optimization problem (LP) usually involves $n$ variables organized in a vector $x \in \mathbb{R}^n$, so that the linear objective function can be written as $c^\top x$, and $m$ linear equality constraints as $Ax = b$ while nonnegativity constraints read $x \ge o$ which in fact means $\min_i x_i \ge 0$.

To ensure feasibility, sometimes a *barrier function* $\beta(x) = -\sum_i \log(x_i) \nearrow \infty$ if $x_i \searrow 0$ is incorporated into the objective function, rendering a nonlinear optimization problem

$$\min\left\{ c^\top x + \gamma\beta(x) : Ax = b \right\} \tag{3.4}$$

with a parameter $\gamma > 0$. Interior point methods usually start with some $\gamma = \gamma_0$, solve (3.4) approximately, then decrease $\gamma$ and iterate.

Passing now to *semidefinite optimization problems* (SDP), the approach is quite similar, but instead of vectors we now arrange variables in a symmetric matrix $X = X^\top$, and consider (additional) constraints of the form $X \succeq O$, which means $\lambda_{\min}(X) \ge 0$.

Again, there is a logarithmic barrier function

$$\beta(X) = -\log \det X = -\sum_i \log \lambda_i(X) \nearrow \infty \quad \text{if} \quad \lambda_{\min}(X) \searrow 0,$$

which prevents us from leaving the feasible set during interior-point procedures, and again we consider a linear objective function and $m$ linear constraints:

$$\min\left\{ \langle C, X \rangle : \langle A_i, X \rangle = b_i \ (i = 1..m), \ X \succeq O \right\},$$

where $\langle C, X \rangle = \mathrm{trace}\,(CX) = \sum_{i,j} C_{ij} X_{ij}$ is now a substitute for the previously used $c^\top x$.

Observe that any LP can be written as

$$\min \left\{ \langle C, X \rangle : \langle A_i, X \rangle = b_i \,(i = 1..m),\; X \geq O \right\}.$$

Proceeding to the more general form of *conic linear optimization problems*, let $\mathcal{K}$ be a convex cone of $X$ matrices. A *conic linear program* is of the form

$$\min \left\{ \langle C, X \rangle : \langle A_i, X \rangle = b_i \,(i = 1..m),\; X \in \mathcal{K} \right\}, \qquad (3.5)$$

but for this general case an appropriate barrier function is by no means obvious. There is a fully developed theory of concordance which relates properties of the cone $\mathcal{K}$ to the existence of a tractable barrier which also would guarantee interior-point algorithms which deliver the solution to (3.5) up to any prescribed accuracy in polynomial time, see for instance [21] or [54].

In familiar cases this is true: in LPs where

$$\mathcal{K} = \mathcal{N} = \left\{ X = X^\top : X \geq O \right\},$$

with barrier $-\sum_{i,j} \log X_{ij}$, or in SDPs where

$$\mathcal{K} = \mathcal{P} = \left\{ X = X^\top : X \succeq O \right\}$$

with barrier $-\sum_i \log \lambda_i(X)$.

For all these methods, a duality theory for conic optimization is indispensable. Thus we need the dual cone of $\mathcal{K}$,

$$\mathcal{K}^* = \left\{ S = S^\top : \langle S, X \rangle \geq 0 \text{ for all } X \in \mathcal{K} \right\}$$

It is easy to see (left as an exercise below) that the cones $\mathcal{P}$ and $\mathcal{N}$ above are all *self-dual* which means $\mathcal{K}^* = \mathcal{K}$.

In general, a primal-dual pair of conic optimization problems is of the form

$$\begin{aligned}
p^* &= \inf \left\{ \langle C, X \rangle : \langle A_i, X \rangle = b_i \,(i = 1..m),\; X \in \mathcal{K} \right\}, \\
d^* &= \sup \left\{ b^\top y : S = C - \sum_i y_i A_i \in \mathcal{K}^* \right\}.
\end{aligned} \qquad (3.6)$$

*Weak duality* says that the *duality gap* $p^* - d^*$ is always nonnegative, i.e., $d^* \leq p^*$. In other words, for every primal-dual-feasible pair $(X, y)$ we have due to $\langle S, X \rangle \geq 0$ as $(X, S) \in \mathcal{K} \times \mathcal{K}^*$,

$$\begin{aligned}
b^\top y &= \sum_i \langle A_i, X \rangle y_i \\
&= \langle \sum_i y_i A_i, X \rangle \\
&= \langle C, X \rangle - \langle S, X \rangle \\
&\leq \langle C, X \rangle.
\end{aligned}$$

Strong duality results establish equality instead of inequality, for optimal solutions $(X^*, y^*)$ to (3.6), so that the duality gap is zero:

**Theorem 23 (Strong duality for conic programming)** [21]:

*Suppose there are strictly feasible $(X, y)$, i.e., $X \in \text{int } \mathcal{K}$ with $\langle A_i, X \rangle = b_i$, all $i$, and $S = C - \sum_i y_i A_i \in \text{int } \mathcal{K}^*$ (Slater's condition). Then there is a primal-dual-optimal pair $(X^*, y^*)$,*

$$d^* = b^\top y^* = \langle C, X^* \rangle = p^*,$$

*for $S^* = C - \sum_i y_i^* A_i$ which then satisfies the* complementary slackness *condition $\langle S^*, X^* \rangle = 0$.*

## 3.3 Copositive Programming

Here we consider a matrix cone different from $\mathcal{P}$ and $\mathcal{N}$, which is not self-dual:

$$\mathcal{K} = \text{conv } \left\{ xx^\top : x \in \mathbb{R}_+^n \right\}, \tag{3.7}$$

the cone of *completely positive matrices*, with its dual cone

$$\mathcal{K}^* = \left\{ S = S^\top \text{ is copositive } \right\} \neq \mathcal{K}. \tag{3.8}$$

**Exercise:** Show that with respect to the duality $\langle X, S \rangle = \text{trace } (SX)$, the cones defined in (3.7) and (3.8) are indeed dual to each other. Also show $\mathcal{P}^* = \mathcal{P}$ and $\mathcal{N}^* = \mathcal{N}$.

It is immediate from the definitions that the following inclusions hold:

$$\mathcal{K} \subseteq \mathcal{P} \cap \mathcal{N} \quad \text{and} \quad \mathcal{K}^* \supseteq \mathcal{P} + \mathcal{N}. \tag{3.9}$$

For $n \geq 5$, these inclusions are strict [37, they cite A. Horn] and [30]. However, (3.9) show the strict inclusion

$$\mathcal{K} \subseteq \mathcal{P} \cap \mathcal{N} \subset \mathcal{P} + \mathcal{N} \subseteq \mathcal{K}^*,$$

prohibiting self-duality even for small $n$.

Slightly abusing terminology, we speak of *copositive programming* or a *copositive optimization problem* (COP) whenever optimizing over $\mathcal{K}$ or $\mathcal{K}^*$.

**Theorem 24 (COP reformulation of StQPs)** [12]:
*Any StQP*

$$\alpha_Q = \min \left\{ x^\top Q x : e^\top x = 1, x \in \mathbb{R}_+^n \right\}$$

*can be expressed as COP (where $E = ee^\top$ the $n \times n$ all-ones matrix):*

$$\alpha_Q = \min \left\{ \langle Q, X \rangle : \langle E, X \rangle = 1, X \in \mathcal{K} \right\}$$

*or its dual*

$$\alpha_Q = \max\left\{y \in \mathbb{R} : Q - yE \in \mathcal{K}^*\right\}.$$

*Proof.* First let $x \in \Delta$ be such that $x^\top Q x = \alpha_Q$. Then $X = xx^\top \in \mathcal{K}$ satisfies $\langle E, X \rangle = 1$ and $\langle Q, X \rangle = x^\top Q x$, hence the optimal value of the COP $\min\{\langle Q, X \rangle : X \in \mathcal{K}, \ \langle E, X \rangle = 1\}$ cannot exceed $\alpha_Q$. To show the converse inequality, recall that $X \in \mathcal{K}$ means $X = \sum_k y_k y_k^\top$ with $y_k \in \mathbb{R}_+^n \setminus \{o\}$. Then $\lambda_k = (e^\top y_k)^2 > 0$ satisfy $\sum_k \lambda_k = \langle E, X \rangle = 1$. Put $x_k = \frac{1}{e^\top y_k} y_k \in \Delta$. Thus $\langle Q, X \rangle = \sum_k \lambda_k x_k^\top Q x_k \geq \alpha_Q$. Strong duality follows by Theorem 23, as we can always choose $y < \min_{i,j} q_{ij}$ so that $Q - yE \in \operatorname{int} \mathcal{K}^*$, whereas of course any $Z \in \operatorname{int} \mathcal{K}$ can be scaled such that $X = \frac{1}{\langle E, Z \rangle} Z \in \operatorname{int} \mathcal{K}$ still, and $\langle E, X \rangle = 1$.

By the above result, we see that one cannot expect to find a good barrier for $\mathcal{K}^*$, since this would reduce the NP-hard StQP to a mere line search ! This is also a nice example of a *convex minimization* problem which is NP-hard, due to complexity of the constraints defining $\mathcal{K}^*$.

Therefore we have to resort to alternative solution strategies. One of them employs *copositive relaxation bounds for StQPs*. Unfortunately, the inner relaxation of $\mathcal{K}$,

$$\mathcal{K}_+ = \left\{X \in \mathcal{P} : \sqrt{X} \in \mathcal{N}\right\},$$

where $\sqrt{X}$ denotes the symmetric square-root factorization of a symmetric psd matrix $X$, is not very helpful as $\mathcal{K}_+$ is not convex. The convex hull of $\mathcal{K}_+$ is exactly $\mathcal{K}$. See [33] for a very recent characterization of interior points of $\mathcal{K}$.

**Exercise:** Prove the preceding assertions about $\mathcal{K}_+$.

A more tractable approach is provided by an outer relaxation of $\mathcal{K}$ which we already know:

$$\begin{aligned}
\alpha_Q &= \min\left\{x^\top Q x : x \in \Delta\right\} \\
&= \min\left\{\langle Q, X \rangle : \langle E, X \rangle = 1, \ X \in \mathcal{K}\right\} \\
&= \max\left\{y \in \mathbb{R} : Q - yE \in \mathcal{K}^*\right\} \\
&\geq \max\left\{y \in \mathbb{R} : Q - yE \in \mathcal{P} + \mathcal{N}\right\} \\
&= \min\left\{\langle Q, X \rangle : \langle E, X \rangle = 1, \ X \in \mathcal{P} \cap \mathcal{N}\right\} \qquad (3.10)
\end{aligned}$$

The last expression was introduced by [1] as *COP relaxation bound* for StQPs and can be determined in a straightforward way by SDP methods.

We now discuss a recent and more general COP representation result by Burer for mixed-binary QPs in general form under very mild conditions (note that the boundedness assumption on $M$ can be dropped [22] but we retain it here to keep the arguments simpler).

**Theorem 25 (COP representation of mixed-binary QPs)** [22]:
*Let* $M = \{x \in \mathbb{R}_+^n : Ax = b\}$ *and* $B \subseteq \{1, \ldots, n\}$. *Consider the following mixed-binary QP*

$$\min\left\{\tfrac{1}{2}x^\top Q x + c^\top x : x \in M, \, x_j \in \{0,1\}, \quad \text{all} j \in B\right\} \tag{3.11}$$

*If $M$ is nonempty and bounded and if the continuous constraints $x \in M$ imply already $x_j \leq 1$ for $j \in B$ (if not, then one may add these explicitly), then (3.11) can be expressed as COP*

$$\min\left\{\tfrac{1}{2}\langle \hat{Q}, \hat{X}\rangle : \hat{\mathcal{A}}(\hat{X}) = \hat{b}, \, \hat{X} \in \mathcal{K}\right\}, \tag{3.12}$$

*where $\hat{X}$ and $\hat{Q}$ are $(n+1) \times (n+1)$ matrices, and the size of $(\hat{\mathcal{A}}, \hat{b})$ is polynomial in the size of $(A, b)$. Here, $\hat{\mathcal{A}}(\hat{X}) = \hat{b}$ stands for a generic system of linear equations in $\hat{X}$: $\langle A_i, \hat{X}\rangle = \hat{b}_i$, $1 \leq i \leq m$.*

*Proof.* First, we again pass from vector variables $x$ to matrix variables:

$$\hat{X} = \begin{bmatrix} 1 & x^\top \\ x & X \end{bmatrix} \quad \text{and} \quad \hat{Q} = \begin{bmatrix} 0 & c^\top \\ c & Q \end{bmatrix}$$

with $X = xx^\top$ so that $\langle \hat{Q}, \hat{X}\rangle = x^\top Q x + 2c^\top x$. The central idea is to replace the nonlinear constraint $X = xx^\top$ with an *equivalent* linear system. This is accomplished by defining the feasible set of (3.12) as

$$\hat{M} = \left\{\hat{X} \in \mathcal{K} : x \in M, \, \mathcal{A}(X) = b^2, \, x_j = X_{jj}, \, \text{all } j \in B\right\},$$

where $\mathcal{A}(X) = [a_1^\top X a_1, \ldots, a_m^\top X a_m]^\top$ and $b^2 = [b_1^2, \ldots, b_m^2]^\top$. Next we assess linear feasibility of points in $\hat{M}$. Now any $\hat{X} = \begin{bmatrix} 1 & x^\top \\ x & X \end{bmatrix} \in \hat{M} \subseteq \mathcal{K}$ can be written, by definition (3.7) of $\mathcal{K}$, in the form

$$\hat{X} = \sum_k \begin{bmatrix} \zeta_k \\ z_k \end{bmatrix} \begin{bmatrix} \zeta_k \\ z_k \end{bmatrix}^\top$$

with $[\zeta_k | z_k^\top]^\top \in \mathbb{R}_+^{n+1} \setminus \{o\}$. Fix $i$ and put $u = [\zeta_k]_k$ as well as $v = [a_i^\top z_k]_k$. Then $\|u\|^2 = u^\top u = \sum_k \zeta_k^2 = 1$ implies

$$(u^\top v)^2 = \left(\sum_k \zeta_k a_i^\top z_k\right)^2 = (a_i^\top x)^2 = b_i^2 = a_i^\top X a_i = a_i^\top \sum_k z_k z_k^\top a_i = \|u\|^2 \|v\|^2,$$

i.e., equality in the Cauchy-Schwarz-Bunyakovsky inequality obtains, which means that $u$ and $v$ are linearly dependent, i.e. (recall $\|u\|^2 = 1$)

$$\text{there is some } \beta_i \in \mathbb{R} \quad \text{such that} \quad a_i^\top z_k = \beta_i \zeta_k \quad \text{for all } k. \tag{3.13}$$

We now show that in fact all $\zeta_k > 0$. Indeed, suppose the contrary. Then (3.13) would imply $a_i^\top z_k = 0$ across all $i$, or $Az_k = o$, but also $z_k \neq o$ as $[\zeta_k | z_k^\top]^\top \neq o$, in contradiction to the assumption that $M$ is nonempty and bounded (recall that otherwise $x + t z_k \in M$ for all $t \in \mathbb{R}$ if $x \in M$). Thus $\zeta_k > 0$ for all $k$. Then define $x_k = \frac{1}{\zeta_k} z_k \in \mathbb{R}_+^n$, so that $a_i^\top x_k = \beta_i$ for all $k$, by (3.13). Putting $\lambda_k = \zeta_k^2$, we obtain the convex combination representation

$$\hat{X} = \sum_k \lambda_k \begin{bmatrix} 1 \\ x_k \end{bmatrix} \begin{bmatrix} 1 \\ x_k \end{bmatrix}^\top .$$

Next we show that $\hat{X} \in \hat{M}$ yields $x_k \in M$ for all $k$: indeed, $a_i^\top x_k = \beta_i$ for all $k$ implies

$$\beta_i^2 = \sum_k \lambda_k (a_i^\top x_k)^2 = a_i^\top X a_i = b_i^2, \quad \text{for all } i,$$

and hence $Ax_k = b$, i.e., $x_k \in M$. By assumption, $(x_k)_j \in [0,1]$ for all $j \in B$. It remains to show integrality for these $(x_k)_j$. But the constraint $x_j = X_{jj}$ in $\hat{M}$ implies

$$0 \leq \sum_k \lambda_k [1 - (x_k)_j] (x_k)_j = 0,$$

so that $[1 - (x_k)_j] (x_k)_j = 0$ or $(x_k)_j \in \{0,1\}$ for all $j \in B$. Hence

$$\hat{M} \subseteq \text{conv} \left\{ \begin{bmatrix} 1 \\ x \end{bmatrix} \begin{bmatrix} 1 \\ x \end{bmatrix}^\top : x \in M, \ x_j \in \{0,1\}, \text{ all } j \in B \right\}. \tag{3.14}$$

The opposite inclusion is easy. Hence an optimal solution of (3.11) is given by an optimal solution of (3.12), as a linear objective always attains its minimum at an extremal point of the feasible set in (3.14).

Special cases are purely continuous QPs where $B = \emptyset$, for instance StQPs with $A = e^\top$ and $b = 1$, or binary QPs with $B = \{1, \ldots, n\}$. For example, every Maximum-Cut Problem [61] is a COP.

# 4 Approaches to Copositivity

## 4.1 Copositivity Detection

We start with explicit copositivity criteria in low dimensions:

**Theorem 26 (Copositivity for $n = 2, 3$) [36]:**

(a) For $n = 2$, $C = C^\top$ is copositive if and only if

(a1) $c_{ii} \geq 0$ for all $i$, and

(a2) $\det C \geq 0$ or $c_{12} \geq 0$.

(b) For $n = 3$, $C = C^\top$ is copositive if and only if

(b1) $c_{ii} \geq 0$ for all $i$,

(b2) $c_{ij} \geq -\sqrt{c_{ii}c_{jj}}$ for all $i, j$, and

(b3) $\det C \geq 0$ or $c_{12}\sqrt{c_{33}} + c_{23}\sqrt{c_{11}} + c_{13}\sqrt{c_{22}} + \sqrt{c_{11}c_{22}c_{33}} \geq 0$.

It is instructive to relate above conditions to Sylvester's criterion for positive definiteness.

To proceed towards recursive copositivity detection, let us fix the last variables vector $z = [x_2, \ldots, x_n]^\top$, and vary only the first coordinate $x_1 = t$. Decompose

$$C = \begin{bmatrix} \alpha & a^\top \\ a & B \end{bmatrix} \quad \text{where } a \in \mathbb{R}^{n-1} \text{ and } B \text{ is } (n-1) \times (n-1).$$

**Theorem 27 (Root of recursive copositivity detection) [2]:**

Let $\Gamma_a^\pm = \{z \in \mathbb{R}_+^{n-1} : \pm a^\top z \geq 0\}$.

Then $C = C^\top$ is $\mathbb{R}_+^n$-copositive if and only if either

$$\alpha = 0, \quad \cdot a \in \mathbb{R}_+^{n-1}, \quad \text{and} \quad B \text{ is copositive;} \tag{5.1}$$

or

$$\alpha > 0, \quad B \text{ is } \Gamma_a^+\text{-copositive, and} \quad \alpha B - aa^\top \text{ is } \Gamma_a^-\text{-copositive.} \tag{5.2}$$

*Proof.* Consider

$$\phi(t|z) = \alpha t^2 + 2(a^\top z)t + z^\top B z = x^\top C x. \tag{5.3}$$

Then $\phi(t|z) \geq 0$ for all $t \geq 0$ if and only if

either $\alpha = 0$, $a^\top z \geq 0$, and $z^\top B z \geq 0$;

or $\alpha > 0$ and $\phi(t_z|z) \geq 0$ with $t_z = \operatorname{argmin}_{t \geq 0} \phi(t|z) = \max\left\{0, -\frac{a^\top z}{\alpha}\right\}$. The first option gives (5.1), the second (5.2).

This result gives rise to a branch-and-bound procedure for copositivity detection: the root of the problem tree is labeled by the data $(C, \mathbb{R}_+^n)$; branching (and sometimes pruning) consists of checking signs of the first row of $Q$: if $\alpha < 0$, we can prune the rest of the tree with the negative information. The same for $\alpha = 0$ and $a \notin \mathbb{R}_+^{n-1}$. If $\alpha \geq 0$ and $a \in \mathbb{R}_+^{n-1}$, we generate one successor node labeled by $(B, \mathbb{R}_+^{n-1})$, i.e., we just drop nonnegative rows and columns. If $\alpha > 0$ but $a \notin \mathbb{R}_+^{n-1}$, the next generation consists of the two nodes labeled by $(B, \Gamma_a^+)$ and $(\alpha B - aa^\top, \Gamma_a^-)$: we have reduced problem dimension by one, but added linear constraints. This is the reason why we have to keep track not only of the resulting matrices, but also of the polyhedral cones w.r.t. with we have to check copositivity. We prune a node $(G, \Gamma)$ if $\Gamma = \{o\}$ is trivial (in addition, one may check whether $\Gamma$ is contained in the linear subspace spanned by eigenvectors to nonnegative eigenvalues of $G$, if $\Gamma \neq \{o\}$). The leaves in the problem tree are characterized by $1 \times 1$ matrices on $\pm \mathbb{R}_+$, so a simple sign check does the job here.

**Exercise:** Suppose $\alpha > 0$ and $a \in \mathbb{R}_+^{n-1}$. Explain why we need only the one successor cone detailed above. Hint: $\Gamma_a^- \subseteq a^\perp$ in this case.

For continuing this branching, we thus need a generalization of Theorem 27 for general $\Gamma$-copositivity if $\Gamma$ is a polyhedral cone given by a finite set of linear inequalities. The approach is completely the same and as we deal with a block generalization below, we refer to [2] rather than detail it here. In general, there can be much more successor nodes than just two in the root, so, e.g., we may have much more granddaughters of the root than in a binary problem tree. To reduce the number of successors, one may employ also a one-step look ahead strategy as in strong branching [2].

**Exercise:** Determine the extremal rays $r_1, \ldots, r_q$ of $\Gamma_a^+$ and put $R = [r_1, \ldots, r_q]$. Then $B$ is $\Gamma_a^+$-copositive iff $RBR^\top$ is $\mathbb{R}_+^q$-copositive. Explain why this approach is not helpful for dimensional reduction in general.

We now present a block decomposition variant of the above recursive method from [4], to obtain a tree with quite smaller height. The idea is simple but exact formulation requires some notation. Again, we decompose $x = [y|z]$, but now with $y \in \mathbb{R}_+^k$; similarly as above, we have to discuss the condition $\phi(y_z|z) \geq 0$ where $y_z = \operatorname{argmin}_{z \in \mathbb{R}_+^k} \phi(z|y)$, which results in copositivity conditions for $\mathcal{O}(2^k)$ matrices with $n - k$ rows. But, as explained above, we will need the more general case of checking copositivity of $Q$ with respect to the polyhedral cone $\Gamma = \{Dx \in \mathbb{R}_+^m\}$, so we need the block structure

$$Q = \begin{bmatrix} A & B \\ B^\top & C \end{bmatrix} \quad \text{and} \quad D = \begin{bmatrix} E & F \end{bmatrix}$$

where $A$ is a nonsingular $k \times k$ matrix and $E$ an $m \times k$ matrix. Thus, $y_z$ will be the solution of the subproblem

$$\min \left\{ \phi(y|z) = y^\top A y + 2y^\top B z + z^\top C z : E y \geq -Fz \right\}, \qquad (5.4)$$

and as in (5.3) we have to check $v^\top Q v = \phi(y_z|z) \geq 0$ if $v = [y_z^\top|z^\top]^\top \in \Gamma$.

To ensure the existence of $y_z$, we have to control directions of unbounded recession. Thus introduce the following polyhedral cones:

$$\Gamma_0 = \{w \in \mathbb{R}^k : Ew \geq 0\}, \tag{5.5}$$

and the cone given in dual form

$$\Lambda = \{w \in \mathbb{R}^k : w^\top(Ay + Bz) \geq 0 \text{ if } Ey + Fz \geq 0, \; y \in \mathbb{R}^k, \; z \in \mathbb{R}^{n-k}\}. \tag{5.6}$$

The block variant of [2, Theorem 5] is part (a) of Theorem 28 below. However, here we also incorporate in part (b) a backtracking step in the case of a negative answer. This step must be performed recursively, to enrich a $z \in \Gamma_I \subseteq \mathbb{R}^{n-k}$ by the corresponding vector $y_z \in \mathbb{R}^k$ to obtain a violating direction $v \in \Gamma$, i.e., satisfying $v^\top Q v < 0$. Remember that these directions are needed to escape from inefficient local solutions to (1.1), see Theorem 2. In order to express $y_z$ in terms of $z$, we introduce the $m \times (n-k)$ matrix

$$H = EA^{-1}B - F, \tag{5.7}$$

For a (possibly empty) index set $I \subseteq \{1, \ldots, m\}$ and its complement $J = \{1, \ldots, m\} \setminus I$, partition $E$, $F$, and $H$ accordingly:

$$E = \begin{bmatrix} E_I \\ E_J \end{bmatrix}; \quad F = \begin{bmatrix} F_I \\ F_J \end{bmatrix}; \quad \text{and } H = \begin{bmatrix} H_I \\ H_J \end{bmatrix} = \begin{bmatrix} E_I A^{-1}B - F_I \\ E_J A^{-1}B - F_J \end{bmatrix}.$$

Furthermore let $A_I$ denote the nonsingular matrix

$$A_I = E_I A^{-1} E_I^\top. \tag{5.8}$$

By investigating KKT conditions for (5.4), we see [4, Lemma 2] that if $y_z$ exists, then there is a possibly empty index set $I$ with the following properties:

(a) the matrix $E_I$ has full row rank and $E_I y_z + F_I z = o$;

(b1) if $I \neq \emptyset$, there is a multiplier vector $\lambda_I \geq o$ with $Ay_z + Bz = \frac{1}{2}E_I^\top \lambda_I$;

(b2) if $I = \emptyset$, then $Ay_z + Bz = o$.

For such an $I$, we get

$$y_z(I) = A^{-1}(E_I^\top A_I^{-1} H_I - B)z. \tag{5.9}$$

Hence put

$$\mathcal{I} = \{I \subseteq \{1, \ldots, m\} : E_I \text{ has full row rank}\}. \tag{5.10}$$

It remains to establish a simple connection between $z$ and the set $I$ with properties (a), (b1), (b2) above: to this end, we have to introduce the following

polyhedral cones which take into account both primal and dual feasibility
in (5.4):

$$\Gamma_I = \left\{ z \in \mathbb{R}^{n-k} : \begin{bmatrix} A_I^{-1} H_I \\ E_J A^{-1} E_I^\top A_I^{-1} H_I - H_J \end{bmatrix} z \geq o \right\}. \qquad (5.11)'$$

Finally, define the symmetric $(n-k) \times (n-k)$ matrices $Q_I$ as follows:

$$Q_I = C - B^\top A^{-1} B + H_I^\top A_I^{-1} H_I. \qquad (5.12)$$

Note that using the usual conventions for empty matrices, we obtain for $I = \emptyset$

$$\begin{aligned} Q_\emptyset &= C - B^\top A^{-1} B, \\ \Gamma_\emptyset &= \{z \in \mathbb{R}^{n-k} : (F - EA^{-1}B)z \geq 0\}, \text{ and} \\ y_z(\emptyset) &= -A^{-1} Bz. \end{aligned} \qquad (5.13)$$

**Theorem 28 (Block recursive copositivity criterion)** [4]:
*Define $\Gamma_0$, $\Lambda$, $\mathcal{I}$, $\Gamma_I$, $Q_I$ and $y_z(I)$ as in  (5.5), (5.6), (5.10), (5.11), (5.12)
and (5.9), respectively. (a) Then $Q$ is $\Gamma$-copositive if and only if
(a1) $A$ is $\Gamma_0$-copositive, and $w^\top Aw = 0$ with $w \in \Gamma_0$ implies $w \in \Lambda$;*

*(a2) the $(n-k) \times (n-k)$ matrices $Q_I$ are $\Gamma_I$-copositive for all $I \in \mathcal{I}$.*

*(b) If one of the conditions (a1), (a2) does not hold, a violating direction
$v \in \Gamma$ can be obtained as follows:*

*(b1) If $w \in \Gamma_0 \setminus \{o\}$ satisfies $w^\top Aw < 0$, then*

$$v = \begin{pmatrix} w \\ 0 \end{pmatrix} \in \Gamma \setminus \{o\} \quad yields \quad v^\top Qv < 0.$$

*If $w \in \Gamma_0$ satisfies $w^\top Aw = 0$ but $w^\top(Ay + Bz) < 0$ for some $y, z$ with
$Ey + Fz \geq 0$, then*

$$v = \begin{pmatrix} y + tw \\ z \end{pmatrix} \in \Gamma \quad yields \quad v^\top Qv < 0,$$

*provided one chooses*

$$t = \begin{cases} 1, & \text{if } \phi(y|z) \leq 0, \\ -\phi(y|z)/w^\top(Ay + Bz), & \text{if } \phi(y|z) > 0. \end{cases}$$

*(b2) If $z \in \Gamma_I \setminus \{o\}$ satisfies $z^\top Q_I z < 0$ for some $I \in \mathcal{I}$, then*

$$v = \begin{bmatrix} y_z(I) \\ z \end{bmatrix} \in \Gamma \setminus \{o\} \quad yields \quad v^\top Qv < 0.$$

Note that a positive-definite block $A$ renders automatically (a1), leaving (b1) void. However, finding large positive-definite principal blocks in $Q$ can be hard or impossible. Anyhow, in this case there is a nice corollary which improves upon [44, Theorem 4].

**Theorem 29 (Block copositivity criterion)** [10]:
*Let $Q$ be a symmetric $n \times n$ matrix with block structure*

$$Q = \begin{bmatrix} A & B \\ B^\mathsf{T} & C \end{bmatrix} \quad and \quad I_n = \begin{bmatrix} E & F \end{bmatrix}$$

*where $A$ is a symmetric positive-definite $k \times k$ matrix, and $E$ is a $n \times k$ matrix. Define $Q_\emptyset$ and $\Gamma_\emptyset$ as in (5.13). Then*

*(a) if $Q$ is copositive, then $Q_\emptyset$ is $\Gamma_\emptyset$-copositive;*

*(b) if $Q_\emptyset$ is copositive, then $Q$ is copositive.*

*Further, if $-A^{-1}B \geq O$, then $\mathbb{R}_+^{n-k} \subseteq \Gamma_\emptyset$, so that (a) and (b) together imply the following criterion:*

$$Q \text{ is copositive if and only if } Q_\emptyset \text{ is copositive.}$$

*Proof.* (a) follows immediately from Theorem 28 and (5.13). To derive (b), observe that the cones $\Gamma_I$ defined in (5.11) are contained in $\mathbb{R}_+^{n-k}$, see [4, Theorem 8]. Further, the matrices $Q_I$ defined in (5.12) are such that $Q_I - Q_\emptyset$ are positive-semidefinite. Hence copositivity of $Q_\emptyset$ implies $\Gamma_I$-copositivity of $Q_I$ for all $I \subset \{1, \ldots, k\}$. A similar argument holds for the case $I = \{1, \ldots, k\}$, see [4, p.175]. Assertion (b) follows more directly from the proof of [44, Theorem 4], valid without assuming $-A^{-1}B \geq O$. On the other hand, this assumption implies $\mathbb{R}_+^{n-k} \subseteq \Gamma_\emptyset$, which establishes the last assertion.

We close this section with a very simple linear-time copositivity detection procedure for tridiagonal matrices.

### Algorithm

Input: Tridiagonal $n \times n$ matrix $Q$.
For $i = 1$ to $n$ do
    if $q_{ii} < 0$, stop ("$Q$ is not copositive");
        else if $i = n$, stop ("$Q$ is copositive");
        else if $q_{i,i+1} < 0$ update $q_{jk} := q_{ii}q_{jk} - q_{ij}q_{ik}$ if $|j - k| \leq 1$;
        endif
    endif
endfor.

**Theorem 30 (Linear-time copositivity check in tridiagonal case)** [7]:

*The above algorithm delivers the exact answer after at most n for-loops. Also a violating vector can be extracted in linear time.*

*Proof.* In the decomposition of Theorem 27, suppose that $a = [\rho, 0, \ldots, 0]^\top \in \mathbb{R}^{n-1}$. By the branch-and-bound argumentation after Theorem 27, we need consider only the case $\rho < 0$. Then

$$\Gamma_a^- = \mathbb{R}_+^{n-1} \supseteq \left\{ y \in \mathbb{R}_+^{n-1} : y_1 = 0 \right\} = \Gamma_a^+,$$

and

$$y^\top (\alpha B - aa^\top) y = \alpha y^\top B y - (a^\top y)^2 \leq \alpha y^\top B y,$$

so that $\Gamma_a^-$-copositivity of $\alpha B - aa^\top$ is sufficient for using Theorem 27 if $\alpha > 0$. Now the update formula in the algorithm exactly delivers $\alpha B - aa^\top$ in the tridiagonal case. To construct a violating vector one has to specialize (b2) from Theorem 28, see [7]. $\square$

Variants and extensions of the above procedure can be found for block-tridiagonal matrices in [7], and for acyclic matrices in [43].

## 4.2 Approximation Hierarchies

Recall (3.9) that the copositive cone $\mathcal{K}^* = \left\{ S = S^\top : y^\top S y \geq 0 \text{ for all } y \in \mathbb{R}_+^n \right\}$ is larger than the completely positive cone $\mathcal{K} = \text{conv} \left\{ xx^\top : x \in \mathbb{R}_+^n \right\}$. So it is quite natural to search for outer approximation of $\mathcal{K}$ and an inner approximation of $\mathcal{K}^*$. We already know an approximation of order zero; the inclusions below are strict for $n \geq 5$:

$$\mathcal{K} \subset \mathcal{P} \cap \mathcal{N} \quad \text{and} \quad \mathcal{K}^* \supset \mathcal{P} + \mathcal{N}.$$

A higher-order approximation was provided by Parrilo [56, 57]. He got rid of the constraint $y \geq o$ by squaring coordinates, and observed that $S \in \mathcal{K}^*$ if and only if $y^\top S y \geq 0$ for all $y$ such that $y_i = x_i^2$ for some $x \in \mathbb{R}^n$, which is guaranteed if the following $n$-variable polynomial of degree $2(r+2)$

$$p_S^{(r)}(x) = \left( \sum x_i^2 \right)^r z^\top S z = \left( \sum x_i^2 \right)^r \sum_{j,k} S_{jk} x_j^2 x_k^2$$

is nonnegative for all $x \in \mathbb{R}^n$. The integer power $2r$ of the Euclidean norm factor will determine the order $r$ of the approximation as follows.

Positivity of the polynomial $p_S^{(r)}$ is in turn guaranteed if

(a) $p_S^{(r)}$ has no negative coefficients; or if

(b) $p_S^{(r)}$ is a sum-of-squares (s.o.s.):

$$p_S^{(r)}(x) = \sum_i [f_i(x)]^2, \quad f_i \text{ some polynomials.}$$

This yields the following convex approximation cones for $\mathcal{K}^*$:

$$\mathcal{C}^{(r)} = \{S \text{ symmetric } n \times n : p_S^{(r)} \text{ satisfies (a)}\},$$

$$\mathcal{K}^{(r)} = \{S \text{ symmetric} n \times n : p_S^{(r)} \text{ satisfies (b)}\}.$$

These cones increase with $r$ and $\mathcal{C}^{(r)} \subset \mathcal{K}^{(r)} \subset \mathcal{K}^*$. Moreover, the following *exhaustivity result* holds:

**Theorem 31 (Exhaustive approximation for strict copositivity)** [29]: *For any $S \in int\, \mathcal{K}^*$ there are $r_\mathcal{K} \leq r_\mathcal{C}$ with*

$$S \in \mathcal{C}^{(r_\mathcal{C})} \cap \mathcal{K}^{(r_\mathcal{K})}.$$

It is not obvious to find a tractable description of these approximation cones. Here we will only illustrate the first order $r = 1$ case; for proofs and higher order $r$ see [13].

**Theorem 32 (First-order polyhedral approximation of $\mathcal{K}^*$)** [13]: $S \in \mathcal{C}^{(1)}$ *if and only if*

$$S_{ii} \geq 0, \quad i = 1, \ldots, n,$$
$$S_{jj} + 2S_{ij} \geq 0, \quad i \neq j$$
$$S_{jk} + S_{ik} + S_{ij} \geq 0, \quad i < j < k.$$

Similarly, we also get an explicit first-order approximation which involves positive-semidefiniteness conditions – sometimes conditions of these type are called *linear matrix inequalities* (LMIs); these representations can be treated by SDP methods.

**Theorem 33 (First-order LMI approximation of $\mathcal{K}^*$)** [13]: $S \in \mathcal{K}^{(1)}$ *if and only if there are symmetric $S^{(i)}$ such that*

$$S - S^{(i)} \in \mathcal{P} + \mathcal{N}, \quad i = 1, \ldots, n,$$
$$S_{ii}^{(i)} \geq 0, \quad i = 1, \ldots, n,$$
$$S_{jj}^{(i)} + 2S_{ij}^{(j)} \geq 0, \quad i \neq j,$$
$$S_{jk}^{(i)} + S_{ik}^{(j)} + S_{ij}^{(k)} \geq 0, \quad i < j < k.$$

In a similar vein, other approximation hierarchies were introduced; among them the *moment-matrix relaxation hierarchy* [45, 46], and a recursively defined hierarchy [60]. All have in common that they need, essentially, matrices of size $n^{r+1} \times n^{r+1}$ for treating order $r$. Hence they are of more theoretical interest. The next subsection will deal with complexity implications from these hierarchies.

## 4.3 Complexity Issues

For simplicity we will employ the LP hierarchy $\mathcal{C}^{(r)}$. Remember

$$
\begin{aligned}
\alpha_Q &= \min \left\{ x^\top Q x : x \in \Delta \right\} \\
&= \min \left\{ \langle Q, X \rangle : \langle E, X \rangle = 1, \, X \in \mathcal{K} \right\} \\
&= \max \left\{ y \in \mathbb{R} : Q - yE \in \mathcal{K}^* \right\} \\
&\geq \max \left\{ y \in \mathbb{R} : Q - yE \in \mathcal{C}^{(r)} \right\} =: \alpha_Q^{\mathcal{C}^{(r)}}.
\end{aligned}
$$

From these COP approximations we can go back to approximation of StQP as follows:

**Theorem 34 (Explicit polyhedral approximation result for StQPs)**
[13]: *For any order $r$, denote by $q_r = \frac{1}{r+2} \, diag\,(Q)$ and by*

$$
\Delta(r) = \{ y \in \Delta : (r+2)y \in \mathcal{N}_0^n \}
$$

*the rational grid approximation of $\Delta$. Then*

$$
\alpha_Q^{\mathcal{C}^{(r)}} = \tfrac{r+2}{r+1} \min \left\{ y^\top Q y - q_r^\top y : y \in \Delta(r) \right\}.
$$

The naïve counterpart to the above result simply optimizes over the finite rational grid approximation of $\Delta$:

$$
\alpha_Q^{\Delta(r)} = \min \left\{ y^\top Q y : y \in \Delta(r) \right\} \geq \alpha_Q \geq \alpha_Q^{\mathcal{C}^{(r)}}.
$$

This way, we enclose the desired value $\alpha_Q$ from below and above. The following result gives an approximation error bound:

**Theorem 35 (Polyhedral StQP approximation error)** [13]:
*Put $\beta_Q = \max \left\{ x^\top Q x : x \in \Delta \right\}$. Then $\beta_Q - \alpha_Q$ is the span of $Q$ over $\Delta$, and we have*

$$
0 \leq \alpha_Q - \alpha_Q^{\mathcal{C}^{(r)}} \leq \tfrac{1}{r+1} (\beta_Q - \alpha_Q).
$$

*Similarly,*

$$
0 \leq \alpha_Q^{\Delta(r)} - \alpha_Q \leq \tfrac{1}{r+2} (\beta_Q - \alpha_Q).
$$

As a consequence, we now deduce that StQPs belong to the PTAS class: this result improves [53] who first showed that an StQP allows for polynomial-time implementable $\frac{2}{3}$-approximation.

**Theorem 36 (StQPs belong to the class PTAS)** [13]:

*For arbitrarily small $\mu(= \frac{1}{r+2}) > 0$, StQP allows for polynomial-time implementable $\mu$-approximation.*

*Proof.* $\Delta(r)$ has $\binom{n+r+1}{r+1} = \mathcal{O}\left(n^{r+1}\right)$ elements, hence both $\alpha_Q^{\mathcal{C}^{(r)}}$ and $\alpha_Q^{\Delta(r)}$ are obtained in polynomial time.

This complexity result singles out another special feature of StQPs among all QPs. For further complexity results in this direction see [28].

## 4.4 SDP Relaxation Bounds for StQPs, Revisited

We recall the construction of COP relaxation bounds in (3.10):

$$
\begin{aligned}
\alpha_Q &= \min\left\{x^\top Q x : x \in \Delta\right\} && \text{by definition} \\
&= \min\left\{\langle Q, X\rangle : \langle E, X\rangle = 1,\ X \in \mathcal{K}\right\} && \text{by StQP} \Leftrightarrow \text{COP} \\
&= \max\left\{y \in \mathbb{R} : Q - yE \in \mathcal{K}^*\right\} && \text{by strong duality} \\
&\geq \max\left\{y \in \mathbb{R} : Q - yE \in \mathcal{P} + \mathcal{N}\right\} && \text{by } \mathcal{K}^* \supseteq \mathcal{P} + \mathcal{N} \\
&= \min\left\{\langle Q, X\rangle : \langle E, X\rangle = 1,\ X \in \mathcal{P} \cap \mathcal{N}\right\} && \text{by strong duality} \\
&= \alpha_Q^{\text{cvd}}.
\end{aligned}
$$

For obvious reasons, the bound $\alpha_Q^{\text{cvd}}$ is called COP relaxation bound in [1]. Less obvious is the fact that $\alpha_Q^{\text{cvd}}$ coincides with the tightest convex/vertex-optimal decomposition bound (this is the reason for the notation), which arises if we search for the largest bound arising from the decomposition

$$
\alpha_Q \geq \alpha_S + \alpha_{Q-S},
$$

where $S \succeq O$ but now, in contrast to $\alpha_Q^{\text{dcd}}$, the second term $T = Q - S$ need not be negative-semidefinite, only still must retain the property that $\min_{x \in \Delta} x^\top T x$ is attained at a vertex, so that $\alpha_{Q-S} = \min_i t_{ii}$ is again cheap. Obviously, we get an improvement $\alpha_Q^{\text{cvd}} \geq \alpha_Q^{\text{dcd}}$ by construction. That this improvement is strict in some instances, is already shown in [1], but also in [16] which contains a hierarchy of cheap bounds.

**Theorem 37 (The cvd bound is the COP-relaxation bound)** [16]:

$$
\alpha_Q^{\text{cvd}} = \max\left\{\alpha_S + \alpha_{Q-S} : S \succeq O,\ \alpha_{Q-S} = \min_i[q_{ii} - s_{ii}]\right\}.
$$

We can tighten this bound if we require instead of vertex-optimality only vertex/edge optimality, which means stipulating that $T = Q - S$ is such that $\alpha_T$ is attained at a vertex or an edge of the simplex $\Delta$. Surprisingly enough, the tightest bound of this type, which is called $\alpha_Q^{\text{cved}}$ for obvious reasons, emerges if we add one valid cut to the copositive program defining $\alpha_Q^{\text{cvd}}$. This cut is performed by help of the adjacency matrix $A_c$ of the $n$−cycle, a graph with clique number two. Because of Theorem 11, we get

$$\tfrac{1}{2} = \alpha_{E-A_c} \quad \text{or, equivalently,} \quad x^\top A_c x \leq \tfrac{1}{2} \quad \text{for all } x \in \Delta.$$

**Theorem 38 (Convex/vertex-edge decomposition bound)** [15]:
*If $x^\top A x \leq \tfrac{1}{2}$ for all $x \in \Delta$, we have*

$$\alpha_Q \geq \alpha_Q(A) := \min\left\{ \langle Q, X \rangle : \langle E, X \rangle = 1,\ \langle A, X \rangle \leq \tfrac{1}{2},\ X \in \mathcal{P} \cap \mathcal{N} \right\} \geq \alpha_Q^{\text{cvd}}.$$

*If, in particular, $A = A_c$ the adjacency matrix of the $n$−cycle, then we get*

$$\alpha_Q(A_c) = \alpha_Q^{\text{cved}} = \max\left\{ \alpha_S + \alpha_{Q-S} : S \in \mathcal{T}^Q \right\}$$

*with $\mathcal{T}^Q = \left\{ S \in \mathcal{P} : \min_{x \in \Delta} x^\top (Q - S)x \text{ is attained at } x \text{ on an edge of } \Delta \right\}$.*

# References

1. Anstreicher, K., and S. Burer, *D.C. Versus Copositive Bounds for Standard QP*, J. Global Optimiz., 33, 2005, 299–312.
2. Bomze, I.M., *Remarks on the recursive structure of copositivity*, J. Inf. & Optimiz. Sciences, 8, 1987, 243–260.
3. Bomze, I.M., *Copositivity conditions for global optimality in indefinite quadratic programming problems*, Czechoslovak J. Operations Research 1, 1992, 7–19.
4. Bomze, I.M., *Block pivoting and shortcut strategies for detecting copositivity*, Linear Alg. Appl. 248, 1996, 161–184.
5. Bomze, I.M., *Evolution towards the maximum clique*, J. Global Optimiz., 10, 1997, 143–164.
6. Bomze, I.M., *On standard quadratic optimization problems*, J. Global Optimiz., 13, 1998, 369–387.
7. Bomze, I.M., *Linear-time detection of copositivity for tridiagonal matrices and extension to block-tridiagonality*, SIAM J. Matrix Anal. Appl. 21, 2000, 840–848.
8. Bomze, I.M., *Branch-and-bound approaches to standard quadratic optimization problems*, J. Global Optimiz., 22, 2002, 17–37.
9. Bomze, I.M., *Portfolio selection via replicator dynamics and projection of indefinite estimated covariances*, Dynamics of Continuous, Discrete and Impulsive Systems B 12, 2005, 527–564.
10. Bomze, I.M., *Perron-Frobenius property of copositive matrices, and a block copositivity criterion*, Linear Algebra and its applications, 429, 2008, 68–71.
11. Bomze, I.M., and G. Danninger, *A global optimization algorithm for concave quadratic problems*, SIAM J. Optimiz., 3, 1993, 836–842.

12. Bomze, I.M., M. Dür, E. de Klerk, A. Quist, C. Roos, and T. Terlaky, *On copositive programming and standard quadratic optimization problems,* J. Global Optimiz., 18, 2000, 301–320.

13. Bomze, I.M., and E. de Klerk, *Solving standard quadratic optimization problems via linear, semidefinite and copositive programming,* J. Global Optimiz., 24, 2002, 163–185.

14. Bomze, I.M., F. Frommlet, and M. Rubey, *Improved SDP bounds for minimizing quadratic functions over the $\ell^1$-ball,* Optimiz. Letters, 1, 2007, 49–59.

15. Bomze, I.M., M. Locatelli, and F. Tardella, *Efficient and cheap bounds for (standard) quadratic optimization,* Technical Report dis tr 2005/10, 2005, Dipartimento di Informatica e Sistemistica "Antonio Ruberti", Universitá degli Studi di Roma "La Sapienza", available at www.optimization-online.org/DB_HTML/2005/07/1176.html, last accessed 12 May 2006.

16. Bomze, I.M., M. Locatelli, and F. Tardella, *New and old bounds for standard quadratic optimization: dominance, equivalence and incomparability,* Math. Programming. 115, 2008, 31–64.

17. Bomze, I.M., and L. Palagi, *Quartic formulation of standard quadratic optimization problems,* J. Global Optimiz., 32, 2005, 181–205.

18. Bomze, I.M., and V. Stix, *Genetical engineering via negative fitness: evolutionary dynamics for global optimization,* Annals of O.R. 89, 1999, 279–318.

19. Bonnans, J. F. and C. Pola, *A trust region interior point algorithm for linearly constrained optimization,* SIAM J. Optim., 7, 1997, 717–731.

20. Borwein, J.M., *Necessary and sufficient conditions for quadratic minimality,* Numer. Funct. Anal. and Optimiz., 5, 1982, 127–140.

21. Boyd, S., and L. Vandenberghe, *Convex Optimization,* Cambridge Univ. Press, Cambridge, UK., 2004.

22. Burer, S., *On the copositive representation of binary and continuous nonconvex quadratic programs,* preprint, 2006, Univ. of Iowa, available at http://www.optimization-online.org/DB_FILE/2006/10/1501.pdf, last accessed Jan. 08, to appear in: Math Programming (2009).

23. Contesse B., L., *Une caractérisation complète des minima locaux en programmation quadratique,* Numer. Math. 34, 1980, 315–332.

24. Cottle, R.W., J.-S. Pang, and R.E. Stone, *The Linear Complementarity Problem.* Academic Press, New York., 1992.

25. Danninger, G., *A recursive algorithm for determining (strict) copositivity of a symmetric matrix,* in: U. Rieder et al. (eds.), Methods of Operations Research 62, 1990,45–52. Hain, Meisenheim.

26. Danninger, G., *Role of copositivity in optimality criteria for nonconvex optimization problems,* J. Optimiz. Theo. Appl. 75, 1992,535–558.

27. Danninger, G., and I.M. Bomze, *Using copositivity for global optimality criteria in concave quadratic programming problems,* Math. Programming 62, 1993, 575–580.

28. de Klerk, E., *The complexity of optimizing over a simplex, hypercube or sphere: a short survey,,* Central European J. OR 16, 2008,111–128.

29. de Klerk, E., and D.V. Pasechnik, *Approximation of the stability number of a graph via copositive programming,* SIAM J. Optimiz. 12, 2002, 875–892.

30. Diananda, P.H., *On non-negative forms in real variables some or all of which are non-negative,* Proc. Cambridge Philos. Soc. 58, 1962, 17–25.

31. Dür, M., *Duality in Global Optimization – Optimality conditions and algorithmical aspects.* Shaker, Aachen, 1999.

32. Dür, M., *A parametric characterization of local optimality,* Math. Methods Oper. Res. 57, 2003, 101–109.

33. Dür, M., and G. Still, *Interior points of the completely positive cone,* Electronic J. Linear Algebra 17, 2008,48–53.

34. Fisher, R.A., *The Genetical Theory of Natural Selection.* Clarendon Press, Oxford, 1930.

35. Gonzaga, C.C. and L.A. Carlos, *A primal affine scaling algorithm for linearly constrained convex programs*, Tech. Report ES-238/90, COPPE Federal Univ.Rio de Janeiro, 1990.

36. Hadeler, K.P. *On copositive matrices*, Linear Alg. Appl. 49, 1983, 79–89.

37. Hall, M. Jr., and M. Newman *Copositive and Completely Positive Quadratic Forms*, Proc. Cambridge Philos. Soc. 59, 1963, 329–339.

38. Hiriart-Urruty, J.-B., *From Convex Optimization to Nonconvex Optimization, Part I: Necessary and sufficient conditions for Global Optimality,* in: F.H. Clarke et al. (eds.), Nonsmooth Optimization and Related Topics, 1989, 219–239, Plenum Press, New York.

39. Hiriart-Urruty, J.-B., and C. Lemaréchal, *Testing necessary and sufficient conditions for global optimality in the problem of maximizing a convex quadratic function over a convex polyhedron,* Preliminary report, University Paul Sabatier, Toulouse, 1990.

40. Hiriart-Urruty, J.-B., and C. Lemaréchal, *Convex Analysis and Minimization Algorithms II.* Grundlehren der Mathematik 306, Springer, Berlin, 1996.

41. Hofbauer, J. and Sigmund, K., *Evolutionary Games and Population Dynamics*, Cambridge University Press, Cambridge, UK, 1998.

42. Horst, R., and H. Tuy, *Global optimization – deterministic approaches*. Springer, Berlin, 1991.

43. Ikramov, Kh.D., *Linear-time algorithm for verifying the copositivity of an acyclic matrix,* Comput. Math. Math. Phys. 42, 2002, 1701–1703.

44. Johnson, C.R., and R. Reams, *Spectral theory of copositive matrices*, Linear Algebra and its Applications 395, 2005, 275–281.

45. Lasserre, J.B., *Global optimization with polynomials and the problem of moments*, SIAM Journal on Optimization 11, 2001, 796–817.

46. Lasserre, J.B., *An explicit exact SDP relaxation for nonlinear 0-1 programming*, In: K. Aardal and A.H.M. Gerards, eds., Lecture Notes in Computer Science 2081, 2001, 293–303.

47. Lyubich, Yu., G. D. Maistrowskii, and Yu. G. Ol'khovskii, *Selection-induced convergence to equilibrium in a single-locus autosomal population,* Problems of Information Transmission, 16, 1980, 66–75.

48. Majthay, A., *Optimality conditions for quadratic programming,* Math. Programming, 1, 1971, 359–365.

49. Monteiro, R. D. C. and Y. Wang, Y., *Trust region affine scaling algorithms for linearly constrained convex and concave programs,* Math. Programming. 80, 1998, 283–313.

50. Moran, P. A. P., The Statistical Processey, Oxford, Clarendon Press, 1962.

51. Motzkin, T.S., and E.G. Straus, *Maxima for graphs and a new proof of a theorem of Turán,* Canadian J. Math. 17, 1965, 533–540.

52. Murty, K.G., and S.N. Kabadi, *Some NP-complete problems in quadratic and linear programming,* Math. Programming 39, 1987, 117–129.

53. Nesterov, Y.E., *Global Quadratic Optimization on the Sets with Simplex Structure,* Discussion paper 9915, CORE, Catholic University of Louvain, Belgium, 1999.

54. Nesterov, Y.E., and A. Nemirovski, *Interior Point Polynomial Algorithms in Convex Programming.* SIAM Publications. SIAM, Philadelphia, USA, 1994.

55. Ozdaglar, A., and P. Tseng, *Existence of Global Minima for Constrained Optimization,* J. Optimiz. Theo. Appl. 128, 2006, 523–546.

56. Parrilo, P.A., *Structured Semidefinite Programs and Semi-algebraic Geometry Methods in Robustness and Optimization,* PhD thesis, California Institute of Technology, Pasadena, USA, 2000. Available at: http://www.cds.caltech.edu/~pablo/.

57. Parrilo, P.A., *Semidefinite programming relaxations for semi-algebraic problems,* Math. Programming 696B, 2003, 293–320.

58. Pavan, M., and M. Pelillo, *Dominant sets and pairwise clustering*, IEEE Trans. on Pattern Analysis and Machine Intelligence, 29, 2007, 167–172.

59. Pelillo, M., *Matching free trees, maximal cliques, and monotone game dynamics*, IEEE Trans. Pattern Anal. Machine Intell., 24, 2002, 1535–1541.

60. Peña J., J. Vera, and L. Zuluaga, *Computing the stability number of a graph via linear and semidefinite programming,* SIAM J. Optimiz., 18, 2007, 87–105.

61. Poljak, S., and Z. Tuza, *Maximum cuts and largest bipartite subgraphs,* in: W. Cook, L. Lovasz, and P. Seymour (eds.), *Combinatorial Optimization,* 1995, pp. 181–244, American Mathematical Society, Baltimore.

62. Scozzari, A., and F. Tardella, *A clique algorithm for standard quadratic programming,* 2008, to appear in Discrete Applied Mathematics.

63. Sun, J., *A convergence proof for an affine-scaling algorithm for convex quadratic programming without nondegeneracy assumptions,* Math. Programming 60, 1993, 69–79.

64. Tseng, P., *Convergence properties of Dikin's affine scaling algorithm for nonconvex quadratic minimization,* J. Global Optim., 30, 2004, 285–300.

65. Tseng, P., *A scaled projected reduced-gradient method for linearly constrained smooth optimization,* preprint, 2007, Univ. of Washington.

66. Tseng, P., I.M. Bomze, W. Schachinger, *A first-order interior-point method for Linearly Constrained Smooth Optimization,* preprint, 2007, to appear in: Math Programming (2009).

# Nonsmooth Optimization

Vladimir F. Demyanov

*Dedicated to Alexander M. Rubinov*

## 1 Introduction

In the classical *Mathematical Analysis*, the functions under study are, mostly, differentiable. *Nonsmooth Analysis* came into being in the 60's of the XX-th century. Its appearance was requested by practical problems of industry, airspace engineering, economics, other sciences where nonsmooth mathematical models were employed to describe more adequately the processes to be investigated. One of the most important problems in Mathematical Analysis is that of finding extremal values of a functional. The same is true in Nonsmooth Analysis.

In the present notes, the problem of finding extremal values of a functional defined on some space is discussed. If there are no constraints on the variables, the problem is called the *unconstrained optimization problem*. If constraints are present, the problem becomes the *constrained optimization one*.

In Section 1 the finite dimensional case is considered. The unchallenged main notion in Mathematical Analysis is that of gradient. In the nonsmooth case there are several candidates to replace the gradient. Some of them are described in Section 1. Several classes of nonsmooth functions and related tools are presented. In particular, the class of quasidifferentiable functions is discussed in detail.

V.F. Demyanov (✉)
Applied Mathematics Dept. St.-Petersburg State University, Staryi Peterhof, St.-Petersburg, 198504 Russia
e-mail: vfd@ad9503.spb.edu

G. Di Pillo et al. (eds.), *Nonlinear Optimization*, Lecture Notes in Mathematics 1989, DOI 10.1007/978-3-642-11339-0_2,
© Springer-Verlag Berlin Heidelberg 2010

Unconstrained optimization problems in metric and normed spaces are treated in Section 2. Section 3 deals with constrained optimization problems in metric and normed spaces.

The existing approaches to solving constrained optimization problems can be demonstrated by the mathematical programming problem in the finite dimensional case.

*The mathematical programming problem* (m.p.p.) is the problem of minimizing a function $f : \mathbb{R}^n \to \mathbb{R}$ on the set

$$\Omega = \{x \in \mathbb{R}^n \mid h_i(x) \leq 0 \quad \forall i \in I\}, \tag{1.1}$$

where $I = 1 : N$, the functions $f$ and $h_i : \mathbb{R}^n \to \mathbb{R}$, $i \in I$, are continuously differentiable on $\mathbb{R}^n$. Assume that the set $\Omega$ is nonempty.

If $f$ and $h_i$ are linear functions, the m.p.p. is called the *linear programming problem* (l.p.p.). For solving l.p.p.'s there exist efficient and well-developed methods (simplex-method and its modifications and extensions) (see, for example, [109]).

For solving m.p.p.'s there exists a rich armory of tools, using the idea of linearizing the function $f$ and the set $\Omega$ by means of cones (methods of feasible directions and their modifications) (see, e.g., [37, 87, 112, 113]).

Another approach is based on the idea of reducing the constrained optimization problem to an unconstrained one. Such a reduction can be performed in many ways. One of the most popular exploits the Lagrange method. Under very natural assumptions, it is possible to show that there exist coefficients $\lambda_i \geq 0 (i \in I)$ such that if $x^* \in \Omega$ is a minimizer of $f$ on the set $\Omega$, then

$$f'(x^*) + \sum_{i=1}^{N} \lambda_i h_i'(x^*) = 0_{\bar{n}}. \tag{1.2}$$

Here $f'$ and $h_i'$ are the gradients of, respectively, the functions $f$ and $h_i$, $0_n = (0, ..., 0)$ is the zero element of the space $\mathbb{R}^n$. Put $\lambda = (\lambda_1, ..., \lambda_N) \in \mathbb{R}^N$.

The relation (1.2) implies that $x^*$ is a stationary point of the function $L_\lambda(x) = f(x) + \sum_{i=1}^{N} \lambda_i h_i(x)$ (remind that a point $x^*$ is called a *stationary point* of a continuously differentiable function $F(x)$ if $F'(x^*) = 0_n$). The function $L_\lambda(x)$ will be referred to as the *Lagrange function*, and the coefficients $\lambda_1, ..., \lambda_N$ are the *Lagrange coefficients* (for the function $f$ on the set $\Omega$). Now, one can draw the following conclusion: to find a minimizer of $f$ on the set $\Omega$ it is possible to find all stationary points of the function $L_\lambda(x)$. Let $S(f)$ be the set of stationary points of $L_\lambda$. Then $x^* \in S(f)$. However, for practical implementation of this idea it is required to know the coefficients $\lambda_i$. In some cases this problem can be solved directly (getting simultaneously the point $x^*$ and the coefficients $\lambda_i$). Sometimes it is possible to construct a method of successive approximations where at each step approximations of both the minimizer and the Lagrange coefficients are calculated.

Another group of methods uses the so-called penalty functions. The set $\Omega$ can be written in the form

$$\Omega = \{x \in \mathbb{R}^n \mid \varphi(x) = 0\}, \tag{1.3}$$

where $\varphi(x) = \max_{i \in 0:N} h_i(x)$;   $h_0(x) = 0 \, \forall x \in \mathbb{R}^n$. Clearly, $\varphi(x) \geq 0 \, \forall x \in \mathbb{R}^n$. The function $F_\lambda(x) = f(x) + \lambda\varphi(x)$ is called a *penalty function* (for the given problem). For the sake of simplicity, let us assume that for every $\lambda \geq 0$ there exists a point $x_\lambda \in \mathbb{R}^n$ such that

$$F_\lambda(x_\lambda) = \inf_{x \in \mathbb{R}^n} F_\lambda(x).$$

Under natural assumptions, it is easy to demonstrate that every limit point of the set $\{x_\lambda\}$ is a minimizer of the function $f$ on the set $\Omega$. Recall that a point $x_0$ is a limit point of the family of points $\{x_\lambda\}$, if there exists a sequence $\{\lambda_k\}$, such that $\lambda_k \to \infty, x_{\lambda_k} \to x_0$ as $k \to \infty$. Hence, the constrained minimization problem is reduced to an unconstrained one. Note that the set $\Omega$ can be represented in the form (1.3) by means of other functions $\varphi$ (including smooth ones).

Of special interest is the case where the function $\varphi$ has the following property: there exists a $\lambda^*$ such that, for $\lambda > \lambda^*$, any minimizer of the function $F_\lambda$ on $\mathbb{R}^n$ is a minimizer of the function $f$ on $\Omega$. If it happens the function $F_\lambda$ is called an *exact penalty function*. Its advantage is the fact that the original constrained minimization problem is reduced to solving only one unconstrained minimization problem. However, the new problem is essentially *nonsmooth* (since usually the function $\varphi$ is nondifferentiable). At present there exist very efficient methods for solving nonsmooth extremal problems. The idea of exact penalties was first stated by I.I.Eremin in 1966. For practical application of exact penalty functions it is required to represent the set $\Omega$ in the form (1.3) where the function $\varphi$ is such that $F_\lambda$ is an exact penalty function for sufficiently large $\lambda$.

The above described Lagrange function is also an exact penalty function (and even smooth). The problem is that the coefficients $\lambda_i$ are not known in advance.

There is a wide-spread opinion that the optimization problem is solved if necessary (and/or sufficient) conditions are stated. In some cases it is true: the necessary conditions allow to find a solution. However, for the majority of extremal problems the deduction of necessary conditions represents only the first (and often not the most difficult and important) step. Besides, there exist "useless" necessary conditions: their verification only indicates whether the point is suspicious for being an extremum point. If the answer is negative, no additional information, allowing to improve the current approximation, is provided.

We are interested in "constructive necessary conditions" to be able to get a "better" point.

In Section 3 a general problem of finding extremal values of an arbitrary functional on an arbitrary set of an arbitrary metric space (the so-called *constrained optimization problem*) is considered. The following general approach to solving constrained optimization problems is developed in Section 3: the original constrained optimization problem is reduced to optimizing some (generally speaking, different) functional on the entire space (*an unconstrained optimization problem*). Such a reduction is performed by means of exact penalty functions. When an exact penalty function is constructed, one is able to use the available unconstrained solvers for finding an extremum.

For deeper and more active study of the material the reader is advised to solve problems formulated in exercises (though the basic material is provided with necessary proofs).

The author is grateful to many colleagues for their encouragement and collaboration. First of all, thanks are due to the late Alexander M. Rubinov with whom the author performed a 42-year long journey to the Land of Nondifferentiable optimization, to Professors F. Giannessi, D. Pallaschke, G. Di Pillo and to Prof. F. Facchinei who directed the author to the area of Exact penalization.

These Notes are based on lectures presented at the International Summer School on Nonlinear Optimization organized by Fondazione CIME – Centro Internazionale Matematico Estivo – in Cetraro (Cosenza, Italy) (July 1 – 7, 2007). The author is thankful to Professors Fabio Schoen, Elvira Mascolo and Pietro Zecca – CIME Director – for perfect conditions and friendly atmosphere during the School.

### Glossary of Main Notations and Abbreviations

$\emptyset$ – empty set

$\mathcal{B}$ – unit closed ball centered at zero (also denoted by $B$)

$\mathcal{B}_\varepsilon$ – closed ball centered at zero with radius $\varepsilon \geq 0$ (also denoted by $B_\varepsilon$)

$\mathcal{B}_\varepsilon(x)$ – closed ball centered at $x$ with radius $\varepsilon \geq 0$ (also denoted by $B_\varepsilon(x)$)

$S$ – unit sphere centered at zero

$S_\delta$ – sphere centered at zero with radius $\delta$

$S_\delta(x)$ – sphere centered at $x$ with radius $\delta$

cl $\Omega$ – closure of a set $\Omega$

int $\Omega$ – set of interior points of set $\Omega$

co $A$ – convex hull of a set $A$

$\mathbb{R}^n$ – $n$-dimensional space

$\mathbb{R}^n_+$ – cone of vectors from $\mathbb{R}^n$ with nonnegative coordinates

$\mathbb{R} = \mathbb{R}^1 = (-\infty, +\infty)$

$\overline{\mathbb{R}} = \mathbb{R} \cup \{-\infty, +\infty\}$

$x^{(i)}$ – $i$-th coordinate of a vector $x$ (sometimes denoted by $x_i$ if it does not cause any confusion)

$(x, y)$ – scalar product of vectors $x$ and $y$

$\alpha \downarrow 0$ means that $\alpha \to 0$, $\alpha > 0$

$x \not\to b$ means that $x$ doesn't tend to $b$

$\|x\|$ – norm of vector $x$ (and, if the otherwise is not stated, $\|x\| = (\sum (x^i)^2)^{1/2} = \sqrt{(x, x)}$ – Euclidean norm)

$x \downarrow a$ $(x \uparrow b)$ means that $x \to a$, $x > 0$ $(x \to b, x < b)$

$f'(x, g)$ – derivative of a function $f$ at a point $x$ in a direction $g$

$\nabla f(x) = f'(x)$ – gradient of a function $f$ at a point $x$

$f'_D(x, g)$ – Dini derivative of a function $f$ at a point $x$ in a direction $g$

$f_D^\uparrow(x, g) = \limsup\limits_{\alpha \downarrow 0} \dfrac{1}{\alpha}[f(x + \alpha g) - f(x)]$ – Dini upper derivative of a function $f$ at a point $x$ in a direction $g$

$f_D^\downarrow(x, g) = \liminf\limits_{\alpha \downarrow 0} \dfrac{1}{\alpha}[f(x + \alpha g) - f(x)]$ – Dini lower derivative of a function $f$ at a point $x$ in a direction $g$

$f_H^\uparrow(x, g) = \limsup\limits_{\alpha \downarrow 0, q \to g} \dfrac{1}{\alpha}[f(x + \alpha q) - f(x)]$ – Hadamard upper derivative of a function $f$ at a point $x$ in a direction $g$

$f_H^\downarrow(x, g) = \liminf\limits_{\alpha \downarrow 0, q \to g} \dfrac{1}{\alpha}[f(x + \alpha q) - f(x)]$ – Hadamard lower derivative of a function $f$ at a point $x$ in a direction $g$

$\{x_i\}_1^\infty$ – sequence $x_1, x_2, x_3, \ldots$, sometimes denoted by $\{x_i\}$ or $x_i$ (if no confusion is possible)

$\{x\}$ – set consisting of one point $x$ (singleton)

$A^+$ – cone, conjugate to a set $A$

$1 : N$ – set of natural numbers from 1 to $N$

$0_X$ – zero of a space $X$

$0_n = (0, \ldots, 0)$ – zero of the space $\mathbb{R}^n$ (sometimes denoted by 0, if it causes no confusion)

$\arg\min\limits_{x \in X} f$ – set of minimizers of a function $f$ on a set $X$

$\arg\max\limits_{x \in X} f$ – set of maximizers of a function $f$ on a set $X$

$dom\, f$ – effective set of a function $f$

$\lim\sup$ – upper limit (also denoted by $\overline{\lim}$)

$\lim\inf$ – lower limit (also denoted by $\underline{\lim}$)

$[a, b]$ – element of the space $A \times B$ (where $a \in A$, $b \in B$)

$(a, b)$ – scalar product of vectors $a$ and $b$ (sometimes it is an element of $\mathbb{R}^2$ with coordinates $a$ and $b$, the same as $[a, b]$)

$a := b$ – $a$ is equal to $b$ by definition

$a =: b$ – $b$ is equal to $a$ by definition

$2^A$ – set of all subsets of a set $A$

$\square$ – end of proof

d.d. – directionally differentiable

$D$-d.d. – Dini directionally differentiable

$H$- Hadamard directionally differentiable

s.d.d. – steepest descent direction

s.a.d.– steepest ascent direction

$A = \{x \in B \mid \exists y \in G : G(x,y) = 0\}$ means: $A$ is a set of elements $x \in B$ such that there exists $y \in G$ satisfying the equality $G(x,y) = 0$.

## 1.1   The Smooth Case: The Kingdom of Gradient

Let a function $f : \Omega \to \mathbb{R}$ be continuously differentiable on $\Omega$ where $\Omega \subset \mathbb{R}^n$ is an open set. Then the gradient mapping $f' : \Omega \to \mathbb{R}^n$ is defined and continuous on $\Omega$. Fix any $x \in \Omega$. By means of the gradient one is able:

- 1. To find the directional derivative of $f$ at $x$ in any direction $g \in \mathbb{R}^n$:

$$f'(x; g) := \lim_{\alpha \downarrow 0} \frac{f(x + \alpha g) - f(x)}{\alpha} = (f'(x), g). \tag{1.4}$$

Here the notation $(a, b)$ stands for the scalar product of vectors $a$ and $b$, $\alpha \downarrow 0$ means that $\alpha \to +0$.

- 2. To construct a first-order approximation of $f$ near $x$:

$$f(x + \Delta) := f(x) + (f'(x); \Delta) + o_x(\Delta), \tag{1.5}$$

wheres

$$\lim_{\alpha \downarrow 0} \frac{o_x(\alpha \Delta)}{\alpha} = 0 \quad \forall \Delta \in \mathbb{R}^n. \tag{1.6}$$

- 3. To formulate first-order necessary conditions for an extremum:

3a: For a point $x^* \in \Omega$ to be a minimizer of $f$ on $\mathbb{R}^n$ it is necessary that

$$f'(x^*) = 0_n, \tag{1.7}$$

3b: For a point $x^{**} \in \Omega$ to be a maximizer of $f$ on $\mathbb{R}^n$ it is necessary that

$$f'(x^{**}) = 0_n. \tag{1.8}$$

Here $0_n = (0, ..., 0)$ is the zero element of $\mathbb{R}^n$. Note that the conditions (1.7) and (1.8) coincide. A point $x_0$, satisfying (1.7) and (1.8), is called a *stationary point.*

- 4. To find the directions of steepest descent and ascent (if $x$ is not a stationary point):

4a: the direction $g_0(x) = -\frac{f'(x)}{||f'(x)||}$ is the steepest descent direction of the function $f$ at a point $x$,

4b: the direction $g_1(x) = \frac{f'(x)}{||f'(x)||}$ is the steepest ascent direction.

Here $\|a\|$ is the Euclidean norm of a vector $a \in \mathbb{R}^n$.

Note that in the smooth case there exist only one steepest descent direction and only one steepest ascent direction and that

$$g_0(x) = -g_1(x). \tag{1.9}$$

- 5. To construct numerical methods for finding extremal values of $f$ (numerous versions of the *gradient method*).

  The following properties also hold:
- 6. If $f'(x; g) < 0(>0)$, then $f'(x; -g) > 0(<0)$, i.e., if a function $f$ decreases (increases) in some direction it necessarily increases (decreases) in the opposite direction, and, due to (1.4),

$$f'(x; g) = -f'(x; -g). \tag{1.10}$$

- 7. If $f'(x; g) < 0$ then $f'(x'; g) < 0$ for all $x'$ near the point $x$ (that is, the direction $g$ is a robust direction of descent: if $f$ is decreasing in a direction $g$, it is also decreasing in the same direction at all points from some neighbourhood of $x$). The same is true with respect to ascent directions.
- 8. The function $F(x, \Delta) = (f'(x), \Delta) = f'(x; \Delta)$, which is an approximation of the increment $f(x + \Delta) - f(x)$, is continuous as a function of $x$.
- 9. The following mean-value theorem holds:

  *If $co\{x_1, x_2\} \subset \Omega$ then there exists an $\alpha \in (0, 1)$, such that*

$$f(x_2) - f(x_1) = (f'(x_1 + \alpha(x_2 - x_1)); x_2 - x_1). \tag{1.11}$$

- 10. A crucial property is the existence of Calculus: having once calculated the gradients (derivatives in the one-dimensional case) of some basic (elementary) functions and using the *composition theorem*, one is able to find the gradients of a wide family of smooth functions. Many other important problems can successfully be treated by means of the gradient, for example, implicit and inverse function theorems, a fixed point theorem.
- 11. It is worth noting that one can study the above properties (and many others) by means of only $n + 1$ numbers (the value of $f$ at $x$ and $n$ partial derivatives constituting the gradient at $x$). Therefore it is only necessary to compute and to store the mentioned $n + 1$ numbers.

Evaluating the aforesaid, one may conclude that *the gradient is the principal actor performing on the stage of classical (smooth) Mathematical Analysis.* Thus, the gradient is an unchallenged ruler in the kingdom of smooth functions. What can replace it in the absence of differentiability? A candidate for a successor has to preserve as many properties of the gradient as possible.

## 1.2  In the Search for a Successor

### 1.2.1  Directionally Differentiable Functions

First of all, consider the class of directionally differentiable (d.d.) functions. Recall that a function $f : \Omega \to \mathbb{R}$ is called *Dini directionally differentiable* $(D. - d.d.)$ at $x \in \Omega$ if the limit

$$f'_D(x; g) := \lim_{\alpha \downarrow 0} \frac{f(x + \alpha g) - f(x)}{\alpha} \tag{1.12}$$

exists and is finite for every $g \in \mathbb{R}^n$. The quantity $f'_D(x, g)$ is called the *Dini derivative* of $f$ at $x$ in the direction $g$.

A function $f : \Omega \to \mathbb{R}$ is called *Hadamard directionally differentiable* $(H. - d.d.)$ at $x$ if the limit

$$f'_H(x; g) := \lim_{[\alpha, g'] \to [+0, g]} \frac{f(x + \alpha g') - f(x)}{\alpha} \tag{1.13}$$

exists and is finite for every $g \in \mathbb{R}^n$. The quantity $f'_H(x; g)$ is called the *Hadamard derivative* of $f$ at $x$ in the direction $g$.

The Hadamard directional differentiability implies the Dini directional differentiability, while the opposite is not true. Since the functions $f'_D(x; g)$ and $f'_H(x; g)$ are positively homogeneous (p.h.) (as functions of $g$) of degree one, it is sufficient to consider only $g \in S_1 := \{g \in \mathbb{R}^n \mid ||g|| = 1\}$. (Recall that a function $h(g)$ is p.h. if $h(\lambda g) = \lambda h(g) \quad \forall \, \lambda > 0$). If $f$ is $H. - d.d.$ then the function $h(g) = f'_H(x; g)$ is continuous.

Examining problems 1-8 of Section 1.1, we observe that the directional derivatives allows us:

- 1. To find the directional derivatives (by the definition),
- 2. To construct a first-order approximation of $f$ near $x$:

$$f(x + \Delta) := f(x) + f'_D(x; \Delta) + o_{x1}(\Delta),$$

$$f(x + \Delta) := f(x) + f'_H(x; \Delta) + o_{x2}(\Delta), \tag{1.14}$$

where

$$\lim_{\alpha \downarrow 0} \frac{o_{x1}(\alpha \Delta)}{\alpha} = 0 \quad \forall \Delta \in \mathbb{R}^n, \tag{1.15}$$

$$\lim_{\alpha \downarrow 0} \frac{o_{x2}(||\Delta||)}{||\Delta||} = 0. \tag{1.16}$$

- 3. To formulate first-order necessary conditions for an extremum:
  3a: For a point $x^* \in \Omega$ to be a minimizer of $f$ on $\mathbb{R}^n$ it is necessary that

$$f'_D(x^*; g) \geq 0 \quad \forall g \in \mathbb{R}^n, \tag{1.17}$$

$$f'_H(x^*; g) \geq 0 \quad \forall g \in \mathbb{R}^n. \tag{1.18}$$

If

$$f'_H(x^*; g) > 0 \quad \forall g \in \mathbb{R}^n \setminus \{0_n\}, \tag{1.19}$$

then $x^*$ is a strict local minimizer.

3b: For a point $x^{**} \in \Omega$ to be a maximizer of $f$ on $\mathbb{R}^n$ it is necessary that

$$f'_D(x^{**}; g) \leq 0 \quad \forall g \in \mathbb{R}^n, \tag{1.20}$$

$$f'_H(x^{**}; g) \leq 0 \quad \forall g \in \mathbb{R}^n. \tag{1.21}$$

If

$$f'_H(x^{**}; g) < 0 \quad \forall g \in \mathbb{R}^n \setminus \{0_n\}, \tag{1.22}$$

then $x^{**}$ is a strict local maximizer. Note that the necessary conditions for a minimum (1.17),(1.18) and for a maximum (1.20),(1.21) *do not coincide* any more, and the sufficient conditions (1.19) and (1.22) have no equivalence in the smooth case, *they are just impossible.*

A point $x^*$, satisfying (1.17), is called a *D-inf-stationary point* of $f$. A point $x^*$, satisfying (1.18), is called an *H-inf-stationary point* of $f$. A point $x^{**}$, satisfying (1.20), is called a *D-sup-stationary point*. A point $x^{**}$, satisfying (1.21), is called an *H-sup-stationary point*.

- 4. To define directions of steepest descent (if $x$ is not a $D$- or $H$-inf-stationary point) and steepest ascent (if $x$ is not a $D$- or $H$-sup-stationary point):

4a: a direction $g_0(x) \in S_1$ is called a *D-steepest descent direction* of $f$ at $x$, if

$$f'_D(x; g_0(x)) = \inf_{g \in S_1} f'_D(x; g),$$

a direction $g_0(x) \in S_1$ is called an *H-steepest descent direction* of $f$ at $x$, if

$$f'_H(x; g_0(x)) = \inf_{g \in S_1} f'_H(x; g),$$

4b: a direction $g_1(x) \in S_1$ is called a *D-steepest ascent direction* of $f$ at $x$ if

$$f'_D(x; g_1(x)) = \sup_{g \in S_1} f'_D(x; g),$$

a direction $g_1(x) \in S_1$ is called an *H-steepest ascent direction* of $f$ at $x$ if

$$f'_H(x; g_1(x)) = \sup_{g \in S_1} f'_H(x; g).$$

Note that in the nonsmooth case *directions of steepest ascent and descent are not necessarily unique*, and the property, similar to (1.9), doesn't hold now. The properties 6, 7 and 8 (Section 1.1) are not valid any more.

- 9. To formulate the following mean-value theorem:
  *If $co\{x_1, x_2\} \subset \Omega$ then there exists an $\alpha \in (0,1)$, such that*

$$f(x_2) - f(x_1) = f'_H(x_1 + \alpha(x_2 - x_1); x_2 - x_1). \qquad (1.23)$$

It follows from the mean-value theorem that an $H.$-$d.d.$ function is uniquely (up to a constant) defined. The same is true for a $D.$-$d.d.$ function: if $S$ is a convex open set of $\mathbb{R}^n$ and

$$f'_1(x; g) = f'_2(x; g) \quad \forall g \in \mathbb{R}^n, \ \forall x \in S$$

then $f_1(x) = f_2(x) + C \quad \forall x \in S$.
- 10. There exists a Calculus of directional derivatives:
  For example, if functions $f_1$ and $f_2$ are $(D$ or $H)$-differentiable at a point $x \in \Omega$ in a direction $g \in \mathbb{R}^n$, then the functions

$$F_1 = \lambda_1 f_1 + \lambda_2 f_2 \ (where \ \lambda_j \in \mathbb{R}), \ F_2 = f_1 f_2, \ F_3 = \frac{f_1}{f_2} \ (if \ f_2(x) \neq 0)$$

are also $(D$ or $H)$-differentiable at the point $x$ in the direction $g$ and the following formulae hold:

$$F'_1(x; g) = \lambda_1 f'_1(x; g) + \lambda_2 f'_2(x; g),$$
$$F'_2(x; g) = f_2(x) f'_1(x; g) + f_1(x) f'_2(x; g),$$
$$F'_3(x; g) = f_2^{-2}(x)[f_2(x) f'_1(x; g) - f_1(x) f'_2(x; g)].$$

Here $F'_i(x; g)$ is either $F'_{iD}(x; g)$ or $F'_{iH}(x; g)$. The above formulae follow from the rules of smooth Calculus.

The most important (*and new!*) is the following property:
*If functions $f_1, ..., f_N$ are directionally differentiable at $x$ then the functions*

$$F_1 = \max_{i \in I} f_i, \quad F_2 = \min_{i \in I} f_i,$$

*where $I = 1 : N$, are also directionally differentiable at $x$ and*

$$F'_1(x; g) = \max_{i \in R(x)} f'_i(x; g), \quad F'_2(x; g) = \min_{i \in Q(x)} f'_i(x; g). \qquad (1.24)$$

Here

$$R(x) = \{i \in I \mid f_i(x) = F_1(x)\}, \quad Q(x) = \{i \in I \mid f_i(x) = F_2(x)\}.$$

The composition theorem holds as well.

Summing up, one may conclude that in the class of directionally differentiable functions the directional derivative is a good candidate to play the role of the gradient. The only problem remains: the directional derivative is a *function of direction*, and one should be able to compute it (the property 11 from Section 1.1 is not valid any more). In some cases is it not the problem (as in the smooth case: the directional derivative is calculated by formula (1.4)), other "simple" cases are described below.

### 1.2.2 Convex Functions

Let $\Omega \subset \mathbb{R}^n$ be a convex open set. A function $f : \Omega \to \mathbb{R}$ is called *convex* on $\Omega$ if

$$f(\alpha x_1 + (1-\alpha)x_2) \le \alpha f(x_1) + (1-\alpha)f(x_2) \quad \forall x_1, x_2 \in \Omega, \ \forall \alpha \in [0,1].$$

Every convex function is continuous on $\Omega$ and Hadamard-directionally differentiable at any point $x \in \Omega$. Moreover (see [91]), the following relation holds:

$$f'(x; g) := f'_H(x; g) = \max_{v \in \partial f(x)} (v, g), \tag{1.25}$$

where

$$\partial f(x) = \{v \in \mathbb{R}^n \mid f(z) - f(x) \ge (v, z - x) \quad \forall x \in \Omega\}. \tag{1.26}$$

The set $\partial f(x)$ is called the *subdifferential* of $f$ at $x$. This set is nonempty, convex, closed and bounded. Any element of $\partial f(x)$ is called a *subgradient* of $f$ at $x$.

Optimality conditions can be expressed in terms of the subdifferential. For example, the necessary condition for a minimum (1.18) is equivalent to the condition

$$0_n \in \partial f(x^*), \tag{1.27}$$

while the sufficient condition for a strict local minimum (1.19) is equivalent to the condition

$$0_n \in int \ \partial f(x^*). \tag{1.28}$$

However, in the convex case *condition (1.27) is not only necessary but also sufficient.* The necessary condition for a maximum (1.21) is equivalent to the condition

$$\partial f(x^{**}) = \{0_n\}, \tag{1.29}$$

while the sufficient condition for a maximum (1.22) is impossible. Since condition (1.27) is necessary and sufficient for a point $x^*$ to be a minimizer, it follows from (1.29) that (unless $f$ is a constant function) $f$ can not attain its maximum value on $\Omega$, only on the boundary (recall that $\Omega$ is an *open* set, hence, $x^{**}$ is an interior point of $\Omega$). To be more precise, there exists no point $x^{**} \in \Omega$ such that

$$\sup_{x \in \Omega} f(x) = f(x^{**}).$$

If $\Omega = \mathbb{R}^n$ and the function $f$ is not a constant one then there exists no maximizer of $f$ on $\mathbb{R}^n$.

Note that the necessary condition for a minimum (1.27) is a dual formulation of condition (1.18) while the necessary condition for a maximum (1.29) is a dual formulation of condition (1.21).

If $0_n \notin \partial f(x)$, then the direction $g_0(x) = -\frac{z(x)}{||z(x)||}$, where

$$||z(x)|| = \min_{v \in \partial f(x)} ||z||,$$

is the direction of steepest descent of $f$ at $x$. *The steepest descent direction is unique.*

If $\partial f(x) \neq \{0_n\}$, then the direction $g_1(x) = \frac{z_1(x)}{||z_1(x)||}$, where

$$||z_1(x)|| = \max_{v \in \partial f(x)} ||z||,$$

is a direction of steepest ascent of $f$ at $x$. Note that, *a steepest ascent direction is not necessarily unique.*

### 1.2.3  Max-Type Functions

Let $\Omega \subset \mathbb{R}^n$ be an open set, $\varphi(x, y) : \Omega \times G \to \mathbb{R}$ be continuously differentiable in $x$ for every fixed $y \in G$, $G$ be a bounded closed set in some normed space $Y$. It is assumed that the functions $\varphi(x, y)$ and $\varphi'(x, y)$ are continuous jointly in both variables on $\Omega \times G$. Put

$$f(x) = \max_{y \in G} \varphi(x, y). \tag{1.30}$$

The function $f$ is continuous and, under some additional conditions, Hadamard-directionally differentiable and

$$f'(x; g) := f'_H(x; g) = \max_{v \in \partial f(x)} (v, g) \tag{1.31}$$

where

$$\begin{aligned} \partial f(x) &= co\ \{f'_x(x, y) \mid y \in R(x)\}, \\ R(x) &= \{y \in G \mid \varphi(x, y) = f(x)\}. \end{aligned} \tag{1.32}$$

The formula (1.31) was independently discovered by J. M. Danskin [8], V. F. Demyanov [15] and I. V. Girsanov [48]. The set $\partial f(x)$ thus defined will also be referred to as the *subdifferential of f at x*. This set is nonempty, convex, closed and bounded. The problem of minimizing the function $f$ defined by (1.30) is called the *minimax problem.*

Optimality conditions can be expressed in terms of the subdifferential. Thus, the necessary condition for a minimum (1.18) is equivalent to the condition

$$0_n \in \partial f(x^*), \tag{1.33}$$

while the sufficient condition for a strict local minimum (1.19) is equivalent to the condition

$$0_n \in int\ \partial f(x^*). \tag{1.34}$$

The necessary condition for a maximum (1.21) is equivalent to the condition

$$\partial f(x^{**}) = \{0_n\} \tag{1.35}$$

while the sufficient condition for a maximum (1.22) is impossible.

Note again that the necessary condition for a minimum (1.33) is a dual formulation of condition (1.18), while the sufficient condition for a strict local minimum (1.34) is a dual formulation of condition (1.19). The necessary condition for a maximum (1.35) is a dual formulation of condition (1.21).

Like in the convex case, if $0_n \notin \partial f(x)$, then the direction $g_0(x) = -\frac{z(x)}{||z(x)||}$, where

$$||z(x)|| = \min_{v \in \partial f(x)} ||z||, \tag{1.36}$$

is the direction of steepest descent of $f$ at $x$. *The steepest descent direction is unique.*

If $\partial f(x) \neq \{0_n\}$, then the direction $g_1(x) = \frac{z_1(x)}{||z_1(x)||}$, where

$$||z_1(x)|| = \max_{v \in \partial f(x)} ||z||, \tag{1.37}$$

is a direction of steepest ascent of $f$ at $x$. Note that, *a steepest ascent direction is not necessarily unique.*

The subdifferential provides a property, analogous to property 11 from Section 1.1. A Calculus of subdifferentials also exists, though it is a limited one since, e.g., the multiplication of a convex (or max-type) function by -1 produces a function which is not any more a convex (or max-type) one.

The problems related to maximum-type functions are treated by *Minimax Theory* (see [9, 26]).

*Remark 1.2.1.* A very important observation can be made now: for each convex or max-type function *with every point* $x \in \Omega$ *a convex compact set* $\partial f(x) \subset \mathbb{R}^n$ *is associated* (and in the smooth case this set is reduced to a point – the gradient $f'(x)$). This observation opened a hunting season for set-valued candidates (not necessarily convex).

## 1.3  Set-Valued Tools

### 1.3.1  Subdifferentiable Functions

The above properties of convex and max-type functions inspired B. N. Pshenichnyi (see [86]) to introduce a new class of nonsmooth functions – the *class of subdifferentiable functions* (B. N. Pshenichnyi called them *quasidifferentiable*).

A function $f : \Omega \to \mathbb{R}$ is called *Dini subdifferentiable* (or $D$-subdifferentiable) at $x \in \Omega$ if it is Dini directionally differentiable at $x$ and if there exists a convex compact set $\partial f(x) \subset \mathbb{R}^n$ such that

$$f_D'(x; g) := \lim_{\alpha \downarrow 0} \frac{f(x + \alpha g) - f(x)}{\alpha} = \max_{v \in \partial f(x)} (v, g) \quad \forall g \in \mathbb{R}^n. \qquad (1.38)$$

A function $f : \Omega \to \mathbb{R}$ is called *Hadamard subdifferentiable* (or $H$- subdifferentiable) at $x \in \Omega$ if it is Hadamard directionally differentiable at $x$ and if there exists a convex compact set $\partial f(x) \subset \mathbb{R}^n$ such that

$$f_H'(x; g) := \lim_{[\alpha, g'] \to [0, g]} \frac{f(x + \alpha g') - f(x)}{\alpha} = \max_{v \in \partial f(x)} (v, g) \quad \forall g \in \mathbb{R}^n. \qquad (1.39)$$

The set $\partial f(x) \subset \mathbb{R}^n$ such that (1.38) (or (1.39)) holds is called *D-subdifferential* (respectively, *H-subdifferential*) of $f$ at $x$.

The class of subdifferentiable functions contains the families of smooth functions, convex functions, max-type functions. The sum of two subdifferentiable functions and the product of a positive constant and a subdifferentiable function are also subdifferentiable. The family of subdifferentiable functions enjoys all the properties of d.d. functions.

As above, optimality conditions can be expressed in terms of the subdifferential. Thus, for a point $x^* \in \Omega$ to be a minimizer of a $D$- or $H$-subdifferentiable function it is necessary that

$$0_n \in \partial f(x^*). \qquad (1.40)$$

If $f$ is $H$-subdifferentiable at $x^*$ then the condition

$$0_n \in int \ \partial f(x^*) \qquad (1.41)$$

is sufficient for $x^*$ to be a strict local minimizer of $f$.

For a point $x^{**} \in \Omega$ to be a maximizer of a $D$- or $H$-subdifferentiable function it is necessary that

$$\partial f(x^*) = \{0_n\}. \qquad (1.42)$$

If $0_n \notin \partial f(x)$, then the direction $g_0(x) = -\frac{z(x)}{||z(x)||}$, where

$$||z(x)|| = \min_{v \in \partial f(x)} ||z||,$$

is a direction of steepest descent of $f$ at $x$ ($D$- or $H$-steepest descent depending on the type of subdifferentiability of $f$). *The steepest descent direction is unique.*

If $\partial f(x) \neq \{0_n\}$, then the direction $g_1(x) = \frac{z_1(x)}{||z_1(x)||}$, where

$$||z_1(x)|| = \max_{v \in \partial f(x)} ||z||,$$

is a direction of $D$-$(H)$-steepest ascent of $f$ at $x$. Note that, *a $D$-$(H$-$)$steepest ascent direction is not necessarily unique.*

### 1.3.2 The Shor Subdifferential

In 1972 N. Z. Shor introduced the notion of almost-gradient (see [100]):

Let a function $f$ be locally Lipschitz on $\Omega$. Then $f$ is almost everywhere (a.e.) differentiable. Fix $x \in \Omega$ and construct the set

$$\partial_{sh} f(x) = \{v \in \mathbb{R}^n \mid \exists \{x_k\} : \ x_k \in T(f), \ x_k \to x, \ f'(x_k) \to v\}, \qquad (1.43)$$

where $T(f) \subset \Omega$ is the set of points where $f$ is differentiable. Since $f$ is a.e. differentiable, then *meas* $\Omega \setminus T(f) = 0$. The set $\partial_{sh} f(x)$ thus defined will be called the *Shor subdifferential,* any element of it being called an *almost gradient.*

N. Z. Shor used almost-gradients to construct numerical methods for minimizing Lipschitz functions (see [101]). The Shor subdifferential is a compact set but not convex. In the convex case $co \ \partial_{sh} f(x) = \partial f(x)$. The set $\partial_{sh} f(x)$ is not convex, therefore it provides more "individual" information than the subdifferential $\partial f(x)$.

### 1.3.3 The Clarke Subdifferential

In 1973 F. Clarke proposed the notion of generalized gradient (see [6,7]).

Let $f : \Omega \to \mathbb{R}$, $x \in \Omega$, $g \in \mathbb{R}^n$. Put

$$f_{cl}^\uparrow(x; g) = \limsup_{[\alpha, x'] \to [+0, x]} \frac{1}{\alpha} [f(x' + \alpha g) - f(x')], \qquad (1.44)$$

$$f_{cl}^\downarrow(x; g) = \liminf_{[\alpha, x'] \to [+0, x]} \frac{1}{\alpha} [f(x' + \alpha g) - f(x')]. \qquad (1.45)$$

The value $f_{cl}^{\uparrow}(x; g)$ will be referred to as the *Clarke upper derivative* of $f$ in the direction $g$, the value $f_{cl}^{\downarrow}(x; g)$ will be referred to as the *Clarke lower derivative* of $f$ in the direction $g$. Observe that the limits in (1.44) and (1.45) always exist (though can be infinite). The functions $f_{cl}^{\uparrow}(x; g)$ and $f_{cl}^{\downarrow}(x; g)$ are positively homogeneous in $g$. If $f$ is locally Lipschitz, these values are finite. Consider this case in more detail.

Let a function $f$ be locally Lipschitz on $\Omega$. Put

$$\partial_{cl} f(x) = co\{v \in \mathbb{R}^n \mid \exists\{x_k\} : \ x_k \in T(f), \ x_k \to x, \ f'(x_k) \to v\}. \quad (1.46)$$

Clearly,

$$\partial_{cl} f(x) = co \ \partial_{sh} f(x). \quad (1.47)$$

The set $\partial_{cl} f(x)$ is called the *Clarke subdifferential*. This set is convex and compact (in the Lipschitz case). Any $v \in \partial_{cl} f(x)$ is called a *generalized gradient*. Every almost-gradient is a generalized gradient, but the opposite is not true. F. Clarke proved that

$$f_{cl}^{\uparrow}(x; g) = \max_{v \in \partial_{cl} f(x)} (v, g), \quad (1.48)$$

$$f_{cl}^{\downarrow}(x; g) = \min_{w \in \partial_{cl} f(x)} (w, g). \quad (1.49)$$

It is now easy to show that the condition

$$0_n \in \partial_{cl} f(x_0) \quad (1.50)$$

is a necessary condition for a point $x_0 \in \Omega$ to be a (local or global) minimizer, the same condition is necessary for a point $x_0$ to be a (local or global) maximizer as well. A point $x_0 \in \Omega$ satisfying (1.50) is called a *Clarke stationary point*. Condition (1.50) does not distinguish between maximizers and minimizers, the set of Clarke stationary points may contain points which are neither minimizers nor maximizers. The functions

$$h_1(x, \alpha g) = f(x) + \alpha f_{cl}^{\uparrow}(x, g), \ \ h_2(x, \alpha g) = f(x) + \alpha f_{cl}^{\downarrow}(x, g)$$

have nothing to do with approximations of the function $f(x + \alpha g)$. Most properties, indicated in Section 1.1, do not hold. Even if $f$ is directionally differentiable, its directional derivative does not necessarily coincide with (1.44) or (1.45). Nevertheless, the Clarke subdifferential describes some properties of the function $f$. For example, if $0 \notin \partial_{cl} f(x_0)$ then the direction $g_0 = -\frac{z_0}{||z_0||}$, where

$$||z_0|| = \min_{v \in \partial_{cl} f(x_0)} ||v||,$$

is a direction of descent of $f$ at $x_0$ while the direction $g_1 = -g_0$ is a direction of ascent of $f$ at $x_0$ (cf. property 6, Section 1.1).

Another very useful property: if $f_{cl}^{\uparrow}(x_0; g) < 0$ then there exists a neighbourhood $B_\varepsilon = \{x \in \mathbb{R}^n \| \; \|x - x_0\| < \varepsilon\}$ (with $\varepsilon > 0$) of $x_0$ such that

$$f_{cl}^{\uparrow}(x_0; g) < 0 \quad \forall x \in B_\varepsilon(x_0),$$

i.e. the direction $g$ is a robust descent direction.

R. T. Rockafellar (see [90]) introduced the following directional derivative:

$$f_R^{\uparrow}(x; g) = \limsup_{[\alpha, x', g'] \to [+0, x, g]} \frac{1}{\alpha}[f(x' + \alpha g') - f(x')]. \tag{1.51}$$

If $f$ is locally Lipschitz then

$$f_R^{\uparrow}(x; g) = f_{cl}^{\uparrow}(x; g).$$

The mentioned nice properties of the Clarke subdifferential triggered and intensified the search for other "subdifferentials".

### 1.3.4  The Michel–Penot Subdifferential

P. Michel and J. -P. Penot in [68] suggested the following generalized derivative:

$$f_{mp}^{\uparrow}(x; g) = \sup_{q \in \mathbb{R}^n} \left\{ \limsup_{\alpha \downarrow 0} \frac{1}{\alpha}[f(x + \alpha(g + q)) - f(x + \alpha q)] \right\}. \tag{1.52}$$

This quantity will be called the *Michel-Penot upper derivative* of $f$ at $x$ in the direction $g$. The quantity

$$f_{mp}^{\downarrow}(x; g) = \inf_{q \in \mathbb{R}^n} \left\{ \liminf_{\alpha \downarrow 0} \frac{1}{\alpha}[f(x + \alpha(g + q)) - f(x + \alpha q)] \right\}. \tag{1.53}$$

is called the *Michel-Penot lower derivative* of $f$ at $x$ in the direction $g$. If $f$ is locally Lipschitz then there exists a convex compact set $\partial_{mp} f(x) \subset \mathbb{R}^n$ such that

$$f_{mp}^{\uparrow}(x; g) = \max_{v \in \partial_{mp} f(x)} (v, g), \tag{1.54}$$

$$f_{mp}^{\downarrow}(x; g) = \min_{w \in \partial_{mp} f(x)} (w, g). \tag{1.55}$$

Recall that if $f$ is also d.d. then $\partial_{mp} f(x)$ is the Clarke subdifferential of the function $h(g) = f'(x, g)$ at the point $g = 0_n$. The set $\partial_{mp} f(x)$ is often called [55] the *small subdifferential*. In general,

$$\partial_{mp} f(x) \subset \partial_{cl} f(x) \tag{1.56}$$

and in some cases
$$\partial_{mp} f(x) \neq \partial_{cl} f(x). \tag{1.57}$$

At the same time the set $\partial_{mp} f(x)$ preserves some properties of $\partial_{cl} f(x)$. For example, the condition
$$0_n \in \partial_{mp} f(x_0) \tag{1.58}$$

is a necessary condition for an extremum (both for a maximum and a minimum). The condition (1.58) is stronger than the condition (1.50).

If $f$ is d.d. and the condition (1.58) is not yet satisfied then there exist directions $g$'s such that
$$f_{mp}^{\uparrow}(x; g) < 0, \tag{1.59}$$

and for such a direction

$$f'(x; g) \leq f_{mp}^{\uparrow}(x; g) < 0, \tag{1.60}$$

while
$$f'(x; -g) \geq f_{mp}^{\downarrow}(x; -g) = -f_{mp}^{\downarrow}(x; g) > 0, \tag{1.61}$$

i.e. all directions, satisfying (1.59), are descent directions while the opposite directions are ascent ones. However, the robustness of a direction satisfying (1.59) and (1.60) is not guaranteed. This is due to the fact that the mapping $\partial_{mp} f$ (unlike the mapping $\partial_{cl} f$) is not upper semicontinuous in $x$. Since $\partial_{mp} f$ is "smaller" than $\partial_{cl} f$, it has some preferences: (i) the necessary condition (1.58) is stronger than the condition (1.50); (ii) the set of directions satisfying simultaneously (1.59) and (1.60) is, in general, greater than the similar set obtained by means of the Clarke subdifferential.

Another advantage of the Michel-Penot subdifferential is the fact that if $f$ is quasidifferentiable (see Section 1.4) then one can construct a Calculus for computing "small" subdifferentials (see [19]).

*Remark 1.3.1.* Many other subdifferentials (convex as well as nonconvex) were proposed later (see, e.g., [54,57,63,69,76]). They are useful for the study of some properties of specific classes of nonsmooth functions. Their common property is that the function under consideration is investigated by means of one set (convex or nonconvex). The study of nonsmooth functions by means of several sets was started by V. F. Demyanov and A. M. Rubinov who in 1979 introduced the so-called quasidifferentiable (q.d.) functions. The class of q.d. functions will be discussed in Section 1.4. In 1980 B. N. Pschenichny (see [85]) proposed the notions of upper convex and lower concave approximations of the directional derivative. In 1982 A. M. Rubinov (see [18]) described exhaustive families of upper convex and lower concave approximations. This led to the notions of upper and lower exhausters [16,22], and finally the study of directional derivatives ("usual" and generalized) was reduced to the usage of a family of convex sets. Some generalized subdifferentials were investigated via exhausters (see [94,95]).

## 1.4 The Approximation of the Directional Derivative by a Family of Convex Functions: Quasidifferentiable Functions

Let $f$ be a real-valued function defined on an open set $X \subset \mathbb{R}^n$, $x \in X$. The function $f$ is called Dini (Hadamard) quasidifferentiable (q.d) at $x$ if is Dini (Hadamard) directionally differentiable at $x$ and if its directional derivative $f'_\mathcal{D}(x,g)$ ($f'_H(x,g)$) can be represented in the form

$$f'_\mathcal{D}(x,g) = \max_{v \in \underline{\partial} f_\mathcal{D}(x)} (v,g) + \min_{w \in \overline{\partial} f_\mathcal{D}(x)} (w,g),$$

$$(f'_H(x,g) = \max_{v \in \underline{\partial} f_H(x)} (v,g) + \min_{w \in \overline{\partial} f_H(x)} (w,g))$$

where the sets $\underline{\partial} f_\mathcal{D}(x), \overline{\partial} f_\mathcal{D}(x), \underline{\partial} f_H(x), \overline{\partial} f_H(x)$ are convex compact sets of $\mathbb{R}^n$. The pair

$$\mathcal{D} f_\mathcal{D}(x) = [\underline{\partial} f_\mathcal{D}(x), \overline{\partial} f_\mathcal{D}(x)] \quad (\mathcal{D} f_H(x) = [\underline{\partial} f_H(x), \overline{\partial} f_H(x)])$$

is called a Dini ( Hadamard ) quasidifferential of $f$ at $x$. Most of the results stated below are valid both for Dini and Hadamard q.d. functions, therefore we shall use the notation $\mathcal{D} f(x) = [\underline{\partial} f(x), \overline{\partial} f(x)]$ for both $\mathcal{D}_\mathcal{D} f(x)$ and $\mathcal{D}_H f(x)$ will just be called a quasidifferential of $f$ at $x$. Analogously, the notation $f'(x,g)$ is used for both $f'_\mathcal{D}(x,g)$ and $f'_H(x,g)$.

The directional derivative $f'(x,g)$ is positively homogeneous (in $g$):

$$f'(x, \lambda g) = \lambda f'(x,g) \quad \forall \lambda > 0. \tag{1.62}$$

Note that the Hadamard quasidifferentiability implies the Dini quasidifferentiability, the converse not necessarily being true.

Thus for a quasidifferentiable (q.d.) function

$$f'(x,g) = \max_{v \in \underline{\partial} f(x)} (v,g) + \min_{w \in \overline{\partial} f(x)} (w,g) \; \forall g \in \mathbb{R}^n. \tag{1.63}$$

The set $\underline{\partial} f(x)$ is called a subdifferential of $f$ at $x$, the set $\overline{\partial} f(x)$ is called a superdifferential of $f$ at $x$. Note that a quasidifferential is not uniquely defined: If a pair $\mathcal{D} = [A, B]$ is a quasidifferential of $f$ at $x$ then, e.g., for any convex compact set $C \subset \mathbb{R}^n$ the pair $\mathcal{D}_1 = [A + C, B - C]$ is a quasidifferential of $f$ at $x$ (since, by (1.63), both these pairs produce the same function $f'(x,g)$). The equivalence class of pairs of convex compact sets $\mathcal{D} f(x) = [\underline{\partial} f(x), \overline{\partial} f(x)]$ producing the function $f'(x,g)$ by formula (1.63) is called the quasidifferential of $f$ at $x$, we shall use the same notation $\mathcal{D} f(x)$ for the quasidifferential of $f$ at $x$.

If there exists a quasidifferential $\mathcal{D}f(x)$ of the form $\mathcal{D}f(x) = [\underline{\partial}f(x), \{0_n\}]$ then $f$ is called subdifferentiable at $x$. If there exists a quasidifferential $\mathcal{D}f(x)$ of the form $\mathcal{D}f(x) = [\{0_n\}, \overline{\partial}f(x)]$ then $f$ is called superdifferentiable at $x$. Here $0_n = (0, \ldots, 0) \in \mathbb{R}^n$.

### 1.4.1  Examples of q.d. Functions

*Example 1.4.1.* If $f$ is a smooth function on $X$ then

$$f'(x, g) = (f'(x), g) \tag{1.64}$$

where $f'(x)$ is the gradient of $f$ at $x$. It is clear that

$$f'(x, g) = \max_{v \in \underline{\partial}f(x)} (v, g) + \min_{w \in \overline{\partial}f(x)} (w, g) \tag{1.65}$$

with

$$\underline{\partial}f(x) = \{f'(x)\}, \overline{\partial}f(x) = \{0_n\}.$$

Hence, $f$ is Hadamard quasidifferentiable and even subdifferentiable. Since in (1.65) one can also take

$$\underline{\partial}f(x) = \{0_n\}, \overline{\partial}f(x) = \{f'(x)\},$$

then $f$ is superdifferentiable as well.

*Example 1.4.2.* If $f$ ia a convex function on a convex open set $X \subset \mathbb{R}^n$ then (as is known from Convex Analysis) $f$ is H-d.d. on $X$ and

$$f'(x, g) = \max_{v \in \partial f(x)} (v, g)$$

where $\partial f(x)$ is the subdifferential of $f$ (in the sense on Convex Analysis):

$$\partial f(x) = \{v \in \mathbb{R}^n \mid f(z) - f(x) \geq (v, z - x) \quad \forall z \in X\}.$$

Therefore $f$ is Hadamard quasidifferentiable and one can take the pair $\mathcal{D}f(x) = [\partial f(x), \{0_n\}]$ as its quasidifferential. Thus, $f$ is even subdifferentiable.

*Example 1.4.3.* If $f$ is concave on a convex set $X$ then $f$ is H-d.d. and

$$f'(x, g) = \min_{w \in \overline{\partial}f(x)} (w, g)$$

where

$$\overline{\partial}f(x) = \{w \in \mathbb{R}^n \mid f(z) - f(x) \leq (w, z - x) \quad \forall z \in X\}.$$

Hence, $f$ is Hadamard quasidifferentiable and one can take the pair

$$\mathcal{D}f(x) = [\{0_n\}, \overline{\partial}f(x)]$$

as its quasidifferential. Thus, $f$ is even superdifferentiable.


## 1.4.2  Calculus of Quasidifferentials

The family of q.d.functions enjoys a well developed Calculus: First let us define the operation of addition of two pairs of compact convex sets and the operation of multiplication of a pair by a real number.

If $\mathcal{D}_1 = [A_1, B_1], \mathcal{D}_2 = [A_2, B_2]$ are pairs of convex compact sets in $\mathbb{R}^n$ then

$$\mathcal{D}_1 + \mathcal{D}_2 = [A, B] \text{ with } A = A_1 + A_2, \ B = B_1 + B_2.$$

If $\mathcal{D} = [A, B]$ where $A, B$ are convex compact sets, $\lambda \in \mathbb{R}$ then

$$\lambda \mathcal{D} = \begin{cases} [\lambda A, \lambda B], \ \lambda \geq 0, \\ [\lambda B, \lambda A], \ \lambda < 0. \end{cases}$$

Let $X \subset \mathbb{R}^n$ be an open set.

**Theorem 1.4.1.** *(see [29], Ch III).*

*1) If functions $f_1, \ldots, f_N$ are quasidifferentiable at a point $x \in X$, and $\lambda_1, \ldots, \lambda_N$ are real numbers then the function*

$$f = \sum_{i=1}^{N} \lambda_i f_i$$

*is also quasidifferentiable at $x$ with a quasidifferential $\mathcal{D}f(x) = [\underline{\partial}f(x), \overline{\partial}f(x)]$ where*

$$\mathcal{D}f(x) = \sum_{i=1}^{N} \lambda_i \mathcal{D}f_i(x), \tag{1.66}$$

*$\mathcal{D}f_i(x)$ being a quasidifferential of $f_i$ at $x$.*

*2) If $f_1$ and $f_2$ are quasidifferentiable at a point $x \in X$ then the function $f = f_1 \cdot f_2$ is also q.d. at $x$ and*

$$\mathcal{D}f(x) = f_1(x)\mathcal{D}f_2(x) + f_2(x)\mathcal{D}f_1(x). \tag{1.67}$$

*3) If $f_1$ and $f_2$ are quasidifferentiable at a point $x \in X$ and $f_2(x) \neq 0$ then the function $f = \frac{f_1}{f_2}$ is also q.d. at $x$ and*

$$\mathcal{D}f(x) = \frac{1}{f_2^2(x)} [f_2(x)\mathcal{D}f_1(x) - f_1(x)\mathcal{D}f_2(x)]. \tag{1.68}$$

*4) Let functions $f_1, \ldots, f_N$ be quasidifferentiable at a point $x \in X$. Construct the functions*

$$\varphi_1(x) = \max_{i \in 1:N} f_i(x), \quad \varphi_2(x) = \min_{i \in 1:N} f_i(x).$$

*Then the functions $\varphi_1$ and $\varphi_2$ are q.d. at $x$ and*

$$\mathcal{D}\varphi_1(x) = [\underline{\partial}\varphi_1(x), \overline{\partial}\varphi_1(x)], \quad \mathcal{D}\varphi_2(x) = [\underline{\partial}\varphi_2(x), \overline{\partial}\varphi_2(x)] \qquad (1.69)$$

*where*

$$\underline{\partial}\varphi_1(x) = co\left\{\underline{\partial}f_k(x) - \sum_{i \in R(x), i \neq k} \overline{\partial}f_i(x) \mid k \in R(x)\right\},$$

$$\overline{\partial}\varphi_1(x) = \sum_{k \in R(x)} \overline{\partial}f_k, \quad \underline{\partial}\varphi_2(x) = \sum_{k \in Q(x)} \underline{\partial}f_k,$$

$$\overline{\partial}\varphi_2(x) = co\left\{\overline{\partial}f_k(x) - \sum_{i \in Q(x), i \neq k} \underline{\partial}f_i(x) \mid k \in Q(x)\right\}.$$

*Here $[\underline{\partial}f_k(x), \overline{\partial}f_k(x)]$ is a quasidifferential of the function $f_k$ at the point $x$,*

$$R(x) = \{i \in 1:N \mid f_i(x) = \varphi_1(x)\}, \quad Q(x) = \{i \in 1:N \mid f_i(x) = \varphi_2(x)\}.$$

*The following composition theorem holds.*

**Theorem 1.4.2.** *(see [29], Ch.III). Let $X$ be an open set in $\mathbb{R}^n$, $Y$ be an open set in $\mathbb{R}^m$ and let a mapping $H(x) = (h_1(x), \ldots, h_m(x))$ be defined on $X$, take its values in $Y$ and its coordinate functions $h_i$ be quasidifferentiable at a point $x_0 \in X$. Assume also that a function $f$ is defined on $Y$ and Hadamard quasidifferentiable at the point $y_0 = H(x_0)$. Then the function*

$$\varphi(x) = f(H(x))$$

*is quasidifferentiable at the point $x_0$.*

   *The corresponding formula for the quasidifferential of $\varphi$ at $x_0$ is presented in [29], Th.III.2.3.*

*Remark 1.4.1.* Thus, the family of quasidifferentiable functions is a linear space closed with respect to all "smooth" operations and, what is most important, the operations of taking the point-wise maximum and minimum. Formulas (1.66)–(1.68) are just generalizations of the rules of Classical Differential Calculus. Most problems and results of Classical Differential Calculus may be formulated for nonsmooth functions in terms of quasidifferentials (see, e.g., [25, 29]). For example, a Mean-value theorem is valid [108].

### 1.4.3   Necessary and Sufficient Conditions for an Unconstrained Optimum

The following results hold due to the properties of directionally differentiable functions.

Let $X \subset \mathbb{R}^n$ be an open set, $f$ be a real-valued function defined and directionally differentiable on $X$.

**Theorem 1.4.3.** *For a point $x^* \in X$ to be a local or global minimizer of $f$ on $X$ it is necessary that*

$$f'(x^*, g) \geq 0 \quad \forall g \in \mathbb{R}^n. \tag{1.70}$$

*If $f$ is Hadamard d.d. at $x^*$ and*

$$f'_H(x^*, g) > 0 \quad \forall g \in \mathbb{R}^n, \quad g \neq 0_n, \tag{1.71}$$

*then $x^*$ is a strict local minimizer of $f$.*

*For a point $x^{**} \in X$ to be a local or global maximizer of $f$ on $X$ it is necessary that*

$$f'(x^{**}, g) \leq 0 \quad \forall g \in \mathbb{R}^n. \tag{1.72}$$

*If $f$ is Hadamard d.d. at $x^{**}$ and*

$$f'_H(x^{**}, g) < 0 \quad \forall g \in \mathbb{R}^n, \quad g \neq 0_n, \tag{1.73}$$

*then $x^{**}$ is a strict local maximizer of $f$.*

These conditions may be restated in terms of quasidifferentials. Let $f$ be quasidifferentiable on an open set $X \subset \mathbb{R}^n$.

**Theorem 1.4.4.** *(see [15, 29, 82]). For a point $x^* \in X$ to be a local or global minimizer of $f$ on $X$ it is necessary that*

$$-\overline{\partial} f(x^*) \subset \underline{\partial} f(x^*). \tag{1.74}$$

*If $f$ is Hadamard quasidifferentiable at $x^*$ and*

$$-\overline{\partial} f(x^*) \subset int\ \underline{\partial} f(x^*), \tag{1.75}$$

*then $x^*$ is a strict local minimizer of $f$.*

*For a point $x^{**} \in X$ to be a local or global maximizer of $f$ on $X$ it is necessary that*

$$-\underline{\partial} f(x^{**}) \subset \overline{\partial} f(x^{**}). \tag{1.76}$$

*If $f$ is Hadamard quasidifferentiable at $x^{**}$ and*

$$-\underline{\partial} f(x^{**}) \subset int\ \overline{\partial} f(x^{**}), \tag{1.77}$$

*then $x^{**}$ is a strict local maximizer of $f$.*

*Remark 1.4.2.* The quasidifferential represents a generalization of the notion of gradient to the nonsmooth case and therefore conditions (1.74)–(1.77) are first-order optimality conditions.

In the smooth case one can take

$$\mathcal{D}f(x) = [\underline{\partial}f(x), \overline{\partial}f(x)]$$

where

$$\underline{\partial}f(x) = [\partial f(x) = \{f'(x)\}, \quad \overline{\partial}f(x) = \{0_n\},$$

therefore the condition (1.74) is equivalent to

$$f'(x^*) = 0_n, \tag{1.78}$$

the condition (1.76) is equivalent to

$$f'(x^{**}) = 0_n, \tag{1.79}$$

and, since both sets $\underline{\partial}f$ and $\overline{\partial}f$ are singletons, the conditions (1.75) and (1.77) are impossible. Thus, conditions (1.75) and (1.77) are essentially "nonsmooth".

A point $x^* \in X$ satisfying (1.74) is called an inf- stationary point, a point $x^{**} \in X$ satisfying (1.76) is called a sup- stationary point of $f$. In the smooth case the necessary condition for a minimum (1.76) is the same as the necessary condition for a maximum (1.79).

### 1.4.4 Directions of Steepest Descent and Ascent

Let $x \in X$ be not an inf-stationary point of $f$ (i.e. condition (16) is not satisfied). Take $w \in \overline{\partial}f(x)$ and find

$$\min_{v \in \underline{\partial}f(x)} ||v + w|| = ||v(w) + w|| = \rho_1(w).$$

Since $\underline{\partial}f(x)$ is a convex compact set, the point $v(w)$ is unique. Find now

$$\max_{w \in \overline{\partial}f(x)} \rho_1(w) = \rho_1(w(x)).$$

The point $w(x)$ is not necessarily unique. As $x$ is not an inf - stationary point, then $\rho_1(w(x)) > 0$. The direction

$$g_1(x) = -\frac{v(w(x)) + w(x)}{||v(w(x)) + w(x)||} = -\frac{v(w(x)) + w(x)}{\rho_1(w(x))} \tag{1.80}$$

is a steepest descent direction of the function $f$ at the point $x$, i.e.

$$f'(x, g_1(x))) = \min_{||g||=1} f'(x, g).$$

Here $|| \cdot ||$ is the Euclidean norm. The quantity $f'(x, g_1(x)) = -\rho_1(w(x))$ is the rate of steepest descent of $f$ at $x$. It may happen that the set of steepest descent directions is not a singleton (and it is also not convex). Recall that in the smooth case the steepest descent direction is always unique (if $x$ is not a stationary point).

Similarly, if $x \in X$ is not a sup-stationary point of $f$ (i.e. condition (1.76) does not hold). Take $v \in \underline{\partial} f(x)$ and find

$$\min_{w \in \overline{\partial} f(x)} ||v + w|| = ||v + w(v)|| = \rho_2(v)$$

and

$$\max_{v \in \underline{\partial} f(x)} \rho_2(v) = \rho_2(v(x)).$$

The direction

$$g_2(x) = \frac{v(x) + w(v(x))}{||v(x) + w(v(x))||} = \frac{v(x) + w(v(x))}{\rho_2(v(x))} \tag{1.81}$$

is a steepest ascent direction of the function $f$ at $x$, i.e.

$$f'(x, g_2(x))) = \max_{||g||=1} f'(x, g).$$

The quantity $f'(x, g_2(x)) = \rho_2(v(x))$ is the rate of steepest ascent of $f$ at $x$. As above it may happen that there exist many steepest ascent directions.

*Remark 1.4.3.* Thus, the necessary conditions (1.74) and (1.76) are "constructive" : in the case where one of these conditions is violated we are able to find steepest descent or ascent directions.

Condition for a minimum (1.74) can be rewritten in the equivalent form

$$0_n \in \bigcap_{w \in \overline{\partial} f(x^*)} [\underline{\partial} f(x^*) + w] := L_1(x^*) \tag{1.82}$$

and condition for a maximum (1.76) can also be represented in the equivalent form

$$0_n \in \bigcap_{v \in \underline{\partial} f(x^{**})} [\overline{\partial} f(x^{**}) + v] := L_2(x^{**}). \tag{1.83}$$

However, if, for example, (1.82) is violated at a point $x$, we are unable to recover steepest descent directions, it may even happen that the set $L_1(x)$ is empty (see [29], Sections V.2 and V.3).

Therefore, condition (1.82) is not "constructive" : if a point $x$ not inf-stationary then condition (1.82) supplies no information about the behaviour of the function in a neighbourhood of $x$ and we are unable to get a "better" point (e.g., to decrease the value of the function). The same is true for the condition for a maximum (1.83). Nevertheless conditions (1.82) and (1.83) may be useful for some other purposes.

*Example 1.4.4.* Let $x = (x^{(1)}, x^{(2)}) \in \mathbb{R}^2$, $x_0 = (0,0)$, $f(x) = |x^{(1)}| - |x^{(2)}|$. We have $f(x) = f_1(x) - f_2(x)$ where $f_1(x) = \max\{f_3(x), f_4(x)\}$, $f_2(x) = \max\{f_5(x), f_6(x)\}$, $f_3(x) = x^{(1)}$, $f_4(x) = -x^{(1)}$, $f_5(x) = x^{(2)}$, $f_6(x) = -x^{(2)}$. The functions $f_3 - f_6$ are smooth therefore (see (1.65))

$$\mathcal{D}f_3(x) = [\underline{\partial}f_3(x), \overline{\partial}f_3(x)], \text{with } \underline{\partial}f_3(x) = \{(1,0)\}, \overline{\partial}f_3(x) = \{(0,0)\},$$
$$\mathcal{D}f_4(x) = [\underline{\partial}f_4(x), \overline{\partial}f_4(x)], \text{with } \underline{\partial}f_4(x) = \{(-1,0)\}, \overline{\partial}f_4(x) = \{(0,0)\},$$
$$\mathcal{D}f_5(x) = [\underline{\partial}f_5(x), \overline{\partial}f_5(x)], \text{with } \underline{\partial}f_5(x) = \{(0,1)\}, \overline{\partial}f_5(x) = \{(0,0)\},$$
$$\mathcal{D}f_6(x) = [\underline{\partial}f_6(x), \overline{\partial}f_6(x)], \text{with } \underline{\partial}f_6(x) = \{(0,-1)\}, \overline{\partial}f_6(x) = \{(0,0)\},$$

Applying (1.69) one gets $\mathcal{D}f_1(x_0) = [\underline{\partial}f_1(x_0), \overline{\partial}f_1(x_0)]$, where

$$\underline{\partial}f_1(x_0) = \text{co } \{\underline{\partial}f_3(x_0) - \overline{\partial}f_4(x_0), \underline{\partial}f_4(x_0) - \overline{\partial}f_3(x_0)\} = \text{co } \{(1,0), (-1,0)\},$$

$$\overline{\partial}f_1(x_0) = \{(0,0)\}, \mathcal{D}f_2(x_0) = [\underline{\partial}f_2(x_0), \overline{\partial}f_2(x_0)],$$

where

$$\underline{\partial}f_2(x_0) = \text{co } \{\underline{\partial}f_5(x_0) - \overline{\partial}f_6(x_0), \underline{\partial}f_6(x_0) - \overline{\partial}f_5(x_0)\} = \text{co } \{(0,1), (0,-1)\},$$
$$\overline{\partial}f_2(x_0) = \{(0,0)\}.$$

Finally, formula (1.66) yields

$$\mathcal{D}f(x_0) = [\underline{\partial}f(x_0), \overline{\partial}f(x_0)],$$

where

$$\underline{\partial}f(x_0) = \text{co } \{(1,0), (-1,0)\}, \quad \overline{\partial}f(x_0) = \text{co } \{(0,1), (0,-1)\}.$$

Since (see Fig. 1) conditions (1.74) and (1.76) are not satisfied, the point $x_0$ is neither an inf-stationary point nor sup-stationary. Applying (1.80) and (1.81) we conclude that there exist two directions of steepest descent: $g_1 = (0,1)$, $g_1' = (0,-1)$ and two directions of steepest ascent: $g_2 = (1,0)$, $g_2' = (-1,0)$.

It is also clear that the sets (see(1.82), (1.83))

$$L_1(x_0) = \bigcap_{w \in \overline{\partial}f(x_0)} [\underline{\partial}f(x_0) + w]$$

**Fig. 1** Example 1.4.4

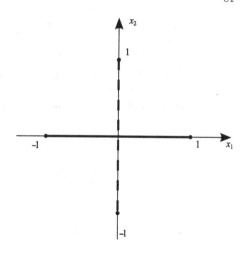

and

$$L_2(x_0) = \bigcap_{v \in \underline{\partial} f(x_0)} [\overline{\partial} f(x_0) + v]$$

are both empty.

*Remark 1.4.4.* If a function $f$ is directionally differentiable but not quasidifferentiable, and if its directional derivative $f'(x, g)$ is continuous as a function of direction (this is the case, e.g., if $f$ is directionally differentiable and Lipschitz) then (by the Stone-Weierstrass theorem) its directional derivative may be approximated (to within any given accuracy) by the difference of two positively homogeneous convex functions, i.e.

$$f'(x, g) \approx \max_{v \in A}(v, g) + \min_{w \in B}(w, g) \tag{1.84}$$

where $A$ and $B$ are convex compact sets in $\mathbb{R}^n$. Relation (1.84) shows that $f'$ can be studied by means of Quasidifferential Calculus (e.g., one is able to find an approximation of a steepest descent direction etc.). Corresponding results can be found in [29, 50].

*Remark 1.4.5.* In many cases of practical importance the quasidifferential of a function $f$ is a pair of sets each of them being the convex hull of a finite number of points or/and balls. If this happens it is easy to store and operate with the quasidifferential, to check necessary conditions, to find directions of descent or ascent, to construct numerical methods.

## 1.4.5 Codifferentiable Functions

In the above subsections necessary conditions for a minimum of $f$ were formulated in terms of quasidifferentials (q.d.) and a formula for computing

steepest descent directions was derived. However, it is difficult to apply steepest descent directions for constructing numerical methods for minimizing the
function $f$ since the quasidifferential mapping $\mathcal{D}f$ is, in general, discontinuous in the Hausdorff metric. This is why we need some other tool to overcome
the discontinuity of $\mathcal{D}f$.

A function $f : \mathbb{R}^n \to \mathbb{R}$ is called Dini codifferentiable at $x \in \mathbb{R}^n$ if there
exists a pair $Df(x) = \left[\underline{d}f(x), \overline{d}f(x)\right]$ of compact convex sets in $\mathbb{R}^{n+1}$ such
that

$$f(x + \Delta) = f(x) + \max_{[a,v] \in \underline{d}f(x)} [a + (v, \Delta)] + \min_{[b,w] \in \overline{d}f(x)} [b + (w, \Delta)] + o_x(\Delta).$$

$$(1.85)$$

$$\frac{o_x(\alpha\Delta)}{\alpha} \to 0 \quad \forall \Delta \in \mathbb{R}^n.$$

$$(1.86)$$

Here $a, b \in \mathbb{R}$; $v, w \in \mathbb{R}^n$. If in (1.85)

$$\frac{o_x(\Delta)}{\|\Delta\|} \underset{\|\Delta\| \to 0}{\longrightarrow} 0 \quad \forall \Delta \in \mathbb{R}^n.$$

$$(1.87)$$

then $f$ is called Hadamard codifferentiable at $x$.

Without loss of generality it may be assumed that

$$\max_{[a,v] \in \underline{d}f(x)} a = \min_{[b,w] \in \overline{d}f(x)} b = 0.$$

$$(1.88)$$

If it causes no misunderstanding, we shall use the term codifferentiable for
both Dini and Hadamard codifferentiable functions.

The pair $Df(x) = \left[\underline{d}f(x), \overline{d}f(x)\right]$ is called a codifferential of $f$ at $x$, $\underline{d}f(x)$
is a hypodifferential, $\overline{d}f(x)$ is a hyperdifferential. A codifferential (like quasidifferential) is not uniquely defined. If there exists a codifferential of the
form $Df(x) = \left[\underline{d}f(x), \{0_{n+1}\}\right]$, the function $f$ is called hypodifferentiable
at $x$. If there exists a codifferential of the form $Df(x) = \left[\{0_{n+1}\}, \overline{d}f(x)\right]$, the
function $f$ is called hyperdifferentiable at $x$.

It is easy to see that the class of Dini (Hadamard) codifferentiable functions
coincides with the class of Dini (Hadamard) quasidifferentiable functions.

For example, if $Df(x) = \left[\underline{d}f(x), \overline{d}f(x)\right]$ is a codifferential of $f$ at $x$ such
that (1.88) holds, then the pair $\mathcal{D}f(x) = \left[\underline{\partial}f(x), \overline{\partial}f(x)\right]$ where

$$\underline{\partial}f(x) = \{v \in \mathbb{R}^n \mid [0, v] \in \underline{d}f(x)\}, \quad \overline{\partial}f(x) = \{w \in \mathbb{R}^n \mid [0, w] \in \overline{d}f(x)\}$$

is a quasidifferential of $f$ at $x$.

A function $f$ is called continuously codifferentiable at $x$ if it is codifferentiable in some neighborhood of $x$ and there exists a codifferential mapping $Df$ which is Hausdorff continuous (in the sense of Hausdorff) at $x$.

*Remark 1.4.6.* Of course, it is possible to introduce the notion of continuously quasidifferentiable function, however, if $f$ is continuously q.d. at $x$ then it is just differentiable at $x$.

For a fixed $\Delta$ the functions (see (1.63) and (1.85))

$$\Phi_{1x}(\Delta) = f(x) + \max_{v \in \underline{\partial} f(x)} (v, \Delta) + \min_{w \in \overline{\partial} f(x)} (w, \Delta)$$

and

$$\Phi_{2x}(\Delta) = f(x) + \max_{[a,v] \in \underline{d} f(x)} [a + (v, \Delta)] + \min_{[b,w] \in \overline{d} f(x)} [b + (w, \Delta)]$$

are both first-order approximations of $f$ in a neighborhood of $x$. The function $F_1(\Delta) = \Phi_{1x}(x) - f(x)$ is positively homogeneous in $\Delta$ while the function $F_2(\Delta) = \Phi_{2x}(x) - f(x)$ is, in general, not positively homogeneous. The loss of homogeneity is the price to be paid for the continuity (if any) of the codifferential mapping.

Note again that the "value" of the mapping $Df$ at any $x$ is a pair of convex compact sets in $\mathbb{R}^{n+1}$.

If turns out that most known functions are continuously codifferentiable (see [16,28,29]). For example, all smooth, convex, concave and concavo-convex functions are continuously codifferentiable. The class of codifferentiable functions enjoys a very rich Calculus similar to that for q.d. functions (see Subsection 1.4.2) which is a generalization of the Classical Differential Calculus. The class of codifferentiable functions was introduced in [28,29].

For a codifferentiable function the following necessary condition holds:

**Theorem 1.4.5.** *For a point* $x^* \in \mathbb{R}^n$ *to be a minimizer of* $f$ *on* $\mathbb{R}^n$ *it is necessary that*

$$0_{n+1} \in \{\underline{d} f(x^*) + [0, w]\} \quad \forall [0, w] \in \overline{d} f(x^*) \tag{1.89}$$

*(it is assumed that condition (1.88) holds).*

A point $x^*$ satisfying (1.89) is called inf-stationary. Let $x$ be not inf-stationary. Then there exists $\bar{w} = [0, w] \in \overline{d} f(x)$ such that

$$0_{n+1} \notin \{\underline{d} f(x^*) + \bar{w}\} = L_{\bar{w}}(x). \tag{1.90}$$

Find

$$\min_{\bar{z} \in L_{\bar{w}}(x)} ||\bar{z}|| = ||\bar{z}_{\bar{w}}(x)||.$$

(1.90) implies that

$$\bar{z}_{\bar{w}}(x) = [\eta_{\bar{w}}(x), z_{\bar{w}}(x)] \neq 0_{n+1}, \quad (\eta_{\bar{w}}(x) \in \mathbb{R}, z_{\bar{w}}(x) \in \mathbb{R}^n).$$

It is also possible to show that $z_{\bar{w}}(x) \neq 0_n$ and that for the direction

$$g_{\bar{w}}(x) = -z_{\bar{w}}(x)/||z_{\bar{w}}(x)||$$

the inequality $f'(x, g_{\bar{w}}(x)) \leq -||z_{\bar{w}}(x)||$ holds.

This direction can be used to construct convergent numerical methods (if the function is continuously codifferentiable) (see [29]).


# 2  Unconstrained Optimization Problems

In Section 1 optimization problems in finite-dimensional spaces were treated. In the sequel we shall consider optimization problems in metric and normed spaces. An unconstrained optimization problem is the problem of finding extremal values of some functional in the absence of constraints on variables. This problem is discussed in Section 2. Constrained optimization problems are discussed in Section 3.


## 2.1  Optimization in a Metric Space

### 2.1.1  First-Order Conditions

Let $X$ be a metric space endowed with a metric $\rho$ and let a functional $f$ be defined on $X$ taking values from $\overline{\mathbb{R}} = \mathbb{R} \cup \{+\infty, -\infty\} = [-\infty, +\infty]$. It is denoted as follows: $f : X \to \overline{\mathbb{R}}$.

The set $\overline{\mathbb{R}}$ is the *extended real line*. Put

$$dom f = \{x \in X \mid f(x) \in \mathbb{R}\}$$

and assume that
$$dom f \neq \emptyset. \tag{2.1}$$

The set $dom f$ is called the *effective set* or *domain* of the functional $f$. Sometimes a functional $f$ will be referred to just as a *function* (though the notion of function is more general than that of functional).

Let $x \in dom f$. Put

$$f^{\downarrow}(x) = \liminf_{y \to x} \frac{f(y) - f(x)}{\rho(x, y)}. \tag{2.2}$$

Sometimes the notation $\underline{\lim}$ is used instead of $\lim\inf$.

If no sequence $\{y_k\}$ exists such that

$$y_k \in X, \ y_k \neq x \quad \forall k, \qquad y_k \to x,$$

then by definition $f^{\downarrow}(x) = +\infty$. Since $x \in dom f$, then the limit in (2.2) always exists (though it may be equal to $+\infty$ or $-\infty$).

The value $f^{\downarrow}(x)$ is called the *rate of steepest descent* of the function $f$ at the point $x$.

*Remark 2.1.1.* To be more precise, it is better to say " ... *the rate of steepest descent of the function f from the point x*" , however, we shall follow the existing tradition.

If the value $f^{\downarrow}(x)$ is finite then (2.2) implies the following expansion:

$$f(y) = f(x) + \rho(x, y)f^{\downarrow}(x) + \underline{o}(\rho(x, y)),$$

where

$$\liminf_{y \to x} \frac{\underline{o}(\rho(x, y))}{\rho(x, y)} = 0. \tag{2.3}$$

Analogously, for $x \in dom f$ the value

$$f^{\uparrow}(x) = \limsup_{y \to x} \frac{f(y) - f(x)}{\rho(x, y)} \tag{2.4}$$

is defined. Sometimes the notation $\overline{\lim}$ is used instead of $\lim\sup$.

If no sequence $\{y_k\}$ exists such that

$$y_k \in X, \ y_k \neq x \quad \forall k, \quad y_k \to x,$$

then by definition $f^{\uparrow}(x) = -\infty$. Since $x \in dom f$, then the limit in (2.4) always exists (though it may be equal to $+\infty$ or $-\infty$).

If the value $f^{\uparrow}(x)$ is finite then (2.4) implies the expansion

$$f(y) = f(x) + \rho(x, y)f^{\uparrow}(x) + \overline{o}(\rho(x, y)),$$

where

$$\limsup_{y \to x} \frac{\overline{o}(\rho(x, y))}{\rho(x, y)} = 0. \tag{2.5}$$

The value $f^{\uparrow}(x)$ is called the *rate of steepest ascent* of the function $f$ at the point $x$.

Put

$$f_* = \inf_{x \in X} f(x), \quad f^* = \sup_{x \in X} f(x).$$

Due to (2.1) one has $f_* < +\infty$, $f^* > -\infty$. The value $f_*$ is the *exact lower bound* of the function $f$ on the set $X$, and the value $f^*$ is the *exact upper bound* of the function $f$ on the set $X$.

If for some point $x_* \in X$ it turns out that $f(x_*) = f_*$, then the point $x_*$ is a *minimizer* (or *global minimizer*) of the function $f$ on $X$. Of course, it may happen that there exists no such a point $x_*$.

Put $A_* = \arg\min_{x \in X} f = \{x \in X \mid f(x) = f_*\}$. If $x_0 \notin dom f$, $f(x_0) = -\infty$, then $f_* = f(x_0) = -\infty$, $x_0 \in A_*$. If it happens that $f(x_1) = -\infty$, $f(x_2) = -\infty$ for points $x_1$, $x_2 \in X$ then we say that $f(x_1) = f(x_2) = f_*$.

Analogously, if for a point $x^* \in X$ one gets $f(x^*) = f^*$, then the point $x^*$ is called a *maximizer* (or *global maximizer*) of the function $f$ on $X$. Naturally, it may happen that there exists no such a point $x^*$.

Put $A^* = \arg\max_{x \in X} f = \{x \in X \mid f(x) = f^*\}$. If $x_0 \notin dom f$, $f(x_0) = +\infty$, then $f^* = f(x_0) = +\infty$, $x_0 \in A^*$. If it happens that $f(x_1) = +\infty$, $f(x_2) = +\infty$ for points $x_1$, $x_2 \in X$ then we say that $f(x_1) = f(x_2) = f^*$.

In other words, a point $x_* \in X$ is a (global) minimizer of $f$ on $X$, if

$$f(x_*) \le f(x) \qquad \forall x \in X, \tag{2.6}$$

and a point $x^* \in X$ is a (global) maximizer of $f$ on $X$, if

$$f(x^*) \ge f(x) \qquad \forall x \in X. \tag{2.7}$$

A point $x_* \in X$ is a *strict global minimizer* of a function $f$ on $X$, if

$$f(x_*) < f(x) \qquad \forall x \in X : x \ne x_*. \tag{2.8}$$

A point $x^* \in X$ is called a *strict global maximizer* of a function $f$ on $X$, if

$$f(x^*) > f(x) \qquad \forall x \in X : x \ne x^*. \tag{2.9}$$

A point $x_* \in X$ is a *local minimizer* of $f$ on $X$ if there exists a $\delta > 0$ such that

$$f(x_*) \le f(x) \qquad \forall x \in X : \quad \rho(x, x_*) < \delta. \tag{2.10}$$

If $\delta = +\infty$, then the point $x_*$ is a *global minimizer*. A point $x_* \in X$ is called a *strict local minimizer* if there exists a $\delta > 0$ such that

$$f(x_*) < f(x) \qquad \forall x \in X : \; x \ne x_*, \; \rho(x, x_*) < \delta. \tag{2.11}$$

A point $x^* \in X$ is called a *local maximizer* of a function $f$ on $X$ if there exists a $\delta > 0$ such that

$$f(x^*) \ge f(x) \qquad \forall x \in X : \; \rho(x, x^*) < \delta. \tag{2.12}$$

If $\delta = +\infty$ then the point $x^*$ is a *global maximizer*. A point $x^* \in X$ is called a *strict local maximizer* of a function $f$ on $X$ if there exists a $\delta > 0$ such that

$$f(x^*) > f(x) \qquad \forall x \in X : x \neq x^*, \quad \rho(x, x^*) < \delta. \qquad (2.13)$$

*Remark 2.1.2.* A global minimizer (maximizer) is a point of global minimum (maximum) independently on a metric. If two metrics are equivalent then a local minimizer in one metric is a local minimizer in the equivalent metric as well. If the metrics are not equivalent then a local minimizer in one metric is not necessarily a local minimizer in another metric.

However, if a metric $\rho_1$ majorizes a metric $\rho_2$ then a local minimizer (maximizer) in the metric $\rho_2$ is a local minimizer (maximizer) in the metric $\rho_1$ as well but not the other way around. Indeed, let a metric $\rho_1$ majorize a metric $\rho_2$, i.e.

$$\rho_1(x, y) \geq c\rho_2(x, y) \quad \forall x, y \in X,$$

where $c > 0$. If a point $x_*$ is a local minimizer in the metric $\rho_2$ then there exists a $\delta > 0$ such that

$$f(x_*) \leq f(x) \qquad \forall x \in X : \quad \rho_2(x, x_*) < \delta. \qquad (2.14)$$

For points $x \in X$ such that $\rho_1(x, x_*) < c\delta$ one has

$$\rho_2(x, x_*) \leq \frac{1}{c}\rho_1(x, x_*) < \delta.$$

It follows now from (2.14) that

$$f(x_*) \leq f(x) \qquad \forall x \in X : \quad \rho_1(x, x_*) < c\delta,$$

which implies that $x_*$ is a local minimizer of $f$ on $X$ in the metric $\rho_1$.

**Exercise 2.1.1.** Give an example of a function $f$ and metrics $\rho_1$ and $\rho_2$ such that a local minimizer in one metric is not a local minimizer in another one.

If for some point $\bar{x} \in X$ one has $f(\bar{x}) = +\infty$, then by definition $\bar{x}$ is a global maximizer of the function $f$ on $X$, and if $f(\bar{x}) = -\infty$, then by definition $\bar{x}$ is a global minimizer of the function $f$ on $X$.

**Theorem 2.1.1.** *For a point $x_* \in \mathrm{dom} f$ to be a global or local minimizer of a function $f$ on $X$, it is necessary that*

$$f^{\downarrow}(x_*) \geq 0. \qquad (2.15)$$

*If*

$$f^{\downarrow}(x_*) > 0, \qquad (2.16)$$

*then the point $x_*$ is a strict local minimizer of $f$ on $X$.*

*Proof.* **Necessity** follows from the definition. Indeed, let $x_*$ be a global or local minimizer. Then the relation (2.10) holds, hence,

$$f^{\downarrow}(x_*) = \liminf_{x \to x_*} \frac{f(x) - f(x_*)}{\rho(x, x_*)} \geq 0.$$

**Sufficiency.** Let the condition (2.16) hold at the point $x_*$. We have to show that a $\delta > 0$ exists such that (2.11) is valid. Assume the opposite. Choose a sequence $\{\delta_k\}$ such that $\delta_k \downarrow 0$. By the assumption, the point $x_*$ is not a strict local minimizer, therefore there exists a $x_k \in X$ such that $f(x_k) \leq f(x_*)$, $\rho(x_k, x_*) < \delta_k$. This yields

$$f^{\downarrow}(x_*) = \liminf_{x \to x_*} \frac{f(x) - f(x_*)}{\rho(x, x_*)} \leq \liminf_{k \to \infty} \frac{f(x_k) - f(x_*)}{\rho(x_k, x_*)} \leq 0,$$

which contradicts (2.16). Sufficiency is proved.                                         □

**Theorem 2.1.2.** *For a point $x^* \in \mathrm{dom} f$ to be a global or local maximizer of a function $f$ on $X$ it is necessary that*

$$f^{\uparrow}(x^*) \leq 0. \tag{2.17}$$

*If*

$$f^{\uparrow}(x^*) < 0, \tag{2.18}$$

*then the point $x^*$ is a strict local maximizer of the function $f$ on $X$.*

*Proof.* is analogous to that of Theorem 2.1.1.                                          □

**Definition.** A point $x_* \in X$ satisfying the condition (2.15) is called an inf-*stationary* point of the function $f$ on $X$. A point $x^* \in X$ satisfying the condition (2.17) is called a sup-*stationary* point of the function $f$ on $X$.

**Definition.** A sequence $\{x_k\}$ such that

$$x_k \in X, \ f(x_k) \to f_* = \inf_{x \in X} f(x),$$

is called a *minimizing sequence* (for the function $f$ on $X$).
    A sequence $\{x_k\}$ such that

$$x_k \in X, \ f(x_k) \to f^* = \sup_{x \in X} f(x),$$

is called a *maximizing sequence* (for the function $f$ on $X$).

## 2.1.2  k-th Order Conditions

Let $x \in dom f$, $k \in 0 : \infty$ (the set $0 : \infty$ contains zero and all positive numbers). Put

$$f_k^{\downarrow}(x) = \liminf_{y \to x} \frac{f(y) - f(x)}{\rho^k(x, y)}. \tag{2.19}$$

Analogously for $x \in dom f$ the value

$$f_k^{\uparrow}(x) = \limsup_{y \to x} \frac{f(y) - f(x)}{\rho^k(x, y)} \tag{2.20}$$

is defined.

Clearly,

$$f^{\downarrow}(x) = f_1^{\downarrow}(x), \ f^{\uparrow}(x) = f_1^{\uparrow}(x).$$

If $f_k^{\downarrow}(x) \in \mathbb{R}$ (i.e., it is finite), then (2.19) implies the expansion

$$f(y) = f(x) + \rho^k(x, y) f_k^{\downarrow}(x) + \underline{a}(\rho(x, y)), \tag{2.21}$$

where

$$\liminf_{y \to x} \frac{\underline{a}(\rho(x, y))}{\rho^k(x, y)} = 0. \tag{2.22}$$

Analogously, if $f_k^{\uparrow}(x) \in \mathbb{R}$, then (2.20) yields the expansion

$$f(y) = f(x) + \rho^k(x, y) f_k^{\uparrow}(x) + \overline{a}(\rho(x, y)), \tag{2.23}$$

where

$$\limsup_{y \to x} \frac{\overline{a}(\rho(x, y))}{\rho^k(x, y)} = 0. \tag{2.24}$$

The following conditions for an extremum hold.

**Theorem 2.1.3.** *For a point $x_* \in dom f$ to be a global or local minimizer of a function $f$ on $X$ it is necessary that*

$$f_k^{\downarrow}(x_*) \geq 0 \quad \forall k \in 0 : \infty. \tag{2.25}$$

*If for some $k \in 0 : \infty$*

$$f_k^{\downarrow}(x_*) > 0, \tag{2.26}$$

*then the point $x_*$ is a strict local minimizer of the function $f$ on $X$.*

*Proof.* is analogous to that of Theorem 2.1.1.                                    □

**Theorem 2.1.4.** *For a point* $x^* \in \text{dom} f$ *to be a global or local maximizer of* $f$ *on* $X$, *it is necessary that*

$$f_k^\uparrow(x^*) \leq 0 \quad \forall k \in 0 : \infty. \tag{2.27}$$

*If for some* $k \in 0 : \infty$

$$f_k^\uparrow(x^*) < 0, \tag{2.28}$$

*then* $x^*$ *is a strict local maximizer of* $f$ *on* $X$.

*Proof.* is analogous to that of Theorem 2.1.2. $\qquad\qquad\qquad\square$

**Definition.** We say that a point $x_0$ is a *k-th order* inf-*stationary point* of a function $f$, if

$$f_i^\downarrow(x_0) = 0 \quad \forall i \in 0 : k.$$

A function $f$ is called *lower semicontinuous* at a point $x_0$, if

$$\liminf_{x \to x_0} f(x) \geq f(x_0).$$

Clearly, if a function $f$ lower semicontinuous at a point $x_0$, then the point $x_0$ is a zero-th order inf-stationary point.

**Definition.** We say that a point $x_0$ is a *k-th order* sup-*stationary point* of a function $f$, if

$$f_i^\uparrow(x_0) = 0 \quad \forall i \in 0 : k.$$

A function $f$ is called *upper semicontinuous* at a point $x_0$, if

$$\limsup_{x \to x_0} f(x) \leq f(x_0).$$

Clearly, if a function $f$ upper semicontinuous at a point $x_0$, then the point $x_0$ is a zero-th order sup-stationary point.

*Remark 2.1.3.* It is easy to note that the following proposition is valid:

For a function $f$ to be continuous at a point $x_0$, it is necessary and sufficient that $f_0^\downarrow(x_0) = f_0^\uparrow(x_0) = 0$ (that is, the point $x_0$ is both zero-th order inf-stationary and zero-th order sup-stationary).

In other words, *a function $f$ is continuous at a point $x_0$ iff it is both upper and lower semicontinuous at this point.*

*Remark 2.1.4.* Theorems 2.1.3 and 2.1.4 yield the following property:

At any point $x \in \text{dom} f$ either

$$f_k^\downarrow(x) \geq 0 \quad \forall k \in 0 : \infty$$

*or*

$$f_k^\downarrow(x) \leq 0 \quad \forall k \in 0 : \infty.$$

If $f_{k_0}^\downarrow(x) > 0$ for some $k_0 \in 0 : \infty$, then the point $x$ is a strict local minimizer
and

$$f_k^\downarrow(x) = +\infty \quad \forall k > k_0.$$

If $f_{k_0}^\downarrow(x) < 0$ for some $k_0 \in 0 : \infty$ then the point $x$ is not a local minimizer
and

$$f_k^\downarrow(x) = -\infty \quad \forall k > k_0.$$

Analogously, at any point $x \in dom f$ either

$$f_k^\uparrow(x) \leq 0 \quad \forall k \in 0 : \infty$$

or

$$f_k^\uparrow(x) \geq 0 \quad \forall k \in 0 : \infty.$$

If $f_{k_0}^\uparrow(x) < 0$ for some $k_0 \in 0 : \infty$ then the point $x$ is a strict local maximizer
and

$$f_k^\uparrow(x) = -\infty \quad \forall k > k_0.$$

If $f_{k_0}^\uparrow(x) > 0$ for some $k_0 \in 0 : \infty$ then the point $x$ is not a local maximizer
and

$$f_k^\uparrow(x) = +\infty \quad \forall k > k_0.$$

And only in the case where

$$f_k^\downarrow(x) = 0 \quad \forall k \in 0 : \infty$$

or

$$f_k^\uparrow(x) = 0 \quad \forall k \in 0 : \infty,$$

we are unable to discover whether the point $x$ is an extremum point, or is
not. It follows from the examples below that in such cases any situation is
possible.

In Examples 2.1.1–2.1.5 below $X = \mathbb{R}$, $\rho(x, y) = |x - y|$, $x_0 = 0$.

*Example 2.1.1.* Let

$$f(x) = \begin{cases} x^2, & x \neq 0, \\ 1, & x = 0. \end{cases}$$

Clearly,

$$f_0^\downarrow(x_0) = -1, \ f_0^\uparrow(x_0) = -1, \qquad f_k^\downarrow(x_0) = f_k^\uparrow(x_0) = -\infty \quad \forall k \in 1 : \infty.$$

Thus, the sufficient condition for a local maximum holds at the point
$x_0 = 0$.

*Example 2.1.2.* Let

$$f(x) = \begin{cases} e^{-\frac{1}{|x|}}, & x \neq 0, \\ 0, & x = 0. \end{cases}$$

It is easy to find that $\quad f_k^{\downarrow}(x_0) = f_k^{\uparrow}(x_0) = 0 \quad \forall k \in 0 : \infty.$

This is just the case where we are unable to say whether the function $f$ attains its extremal value at the point $x_0$ (though $x_0$ is a global minimizer).

*Example 2.1.3.* Let

$$f(x) = \begin{cases} -e^{-\frac{1}{|x|}}, & x \neq 0, \\ 0, & x = 0. \end{cases}$$

One has

$$f_k^{\downarrow}(x_0) = f_k^{\uparrow}(x_0) = 0 \quad \forall k \in 0 : \infty.$$

And again Theorems 2.1.3, 2.1.4 are of no use to find out whether the function $f$ attains its extremal value at $x_0$ (though $x_0$ is a global maximizer).

*Example 2.1.4.* Let

$$f(x) = \begin{cases} e^{-\frac{1}{x}}, & x > 0, \\ -e^{\frac{1}{x}}, & x < 0, \\ 0, & x = 0. \end{cases}$$

One has $f_k^{\downarrow}(x_0) = f_k^{\uparrow}(x_0) = 0 \quad \forall k \in 0 : \infty.$

As in the previous two cases Theorems 2.1.3, 2.1.4 provide no information on the extremality of the point $x_0$ (though the point $x_0$ is neither a minimizer, nor a maximizer).

*Example 2.1.5.* Let $f : \mathbb{R} \to \mathbb{R}$, $f \in C^m(\mathbb{R})$ (where $C^m(\mathbb{R})$ is the space of $m$ times continuously differentiable on $\mathbb{R}$ functions). It is not difficult to see that

$$f_1^{\downarrow}(x) = -|f'(x)|, \quad f_1^{\uparrow}(x) = |f'(x)| \quad \forall x \in \mathbb{R},$$

$$f_2^{\downarrow}(x) = \begin{cases} -\infty, & f'(x) \neq 0, \\ \frac{1}{2}f''(x), & f'(x) = 0, \end{cases} \tag{2.29}$$

$$f_2^{\uparrow}(x) = \begin{cases} +\infty, & f'(x) \neq 0, \\ \frac{1}{2}f''(x), & f'(x) = 0. \end{cases} \tag{2.30}$$

**Exercise 2.1.2.** Prove the relations (2.29) and (2.30), find $f_k^{\downarrow}(x)$, $f_k^{\uparrow}(x)$ for $k \in 3 : m$.

*Remark 2.1.5.* It was assumed above that $k \in 0 : \infty$ (i.e. $k$ is zero or an integer). All the above statements hold for $k \in [0, \infty)$.

**Exercise 2.1.3.** Prove the following proposition.

**Theorem 2.1.5.** *If a function $f$ is lower (upper) semicontinuous on a compact set then $f$ is bounded from below (above) and attains there its exact lower (upper) bound.*

## 2.2 *Optimization in a Normed Space*

Let $X$ be a (linear) normed space with the norm $||x||$. Introduce the metric $\rho(x,y) = ||x - y||$. Let $f : X \to \mathbb{R}$ be given. Fix $x \in X$. Then

$$f^{\downarrow}(x) = \liminf_{y \to x} \frac{f(y) - f(x)}{\rho(x,y)} = \liminf_{||\Delta|| \to 0} \frac{f(x + \Delta) - f(x)}{||\Delta||}$$

$$= \liminf_{\alpha \downarrow 0, g \in S} \frac{f(x + \alpha g) - f(x)}{\alpha}, \qquad (2.31)$$

where $S = \{g \in X \mid ||g|| = 1.\}$ It follows from (2.31) that for all $g \in S$

$$f^{\downarrow}(x) \leq \liminf_{\alpha \downarrow 0} \frac{f(x + \alpha g) - f(x)}{\alpha}.$$

Hence,

$$f^{\downarrow}(x) \leq \inf_{g \in S} \liminf_{\alpha \downarrow 0} \frac{f(x + \alpha g) - f(x)}{\alpha}. \qquad (2.32)$$

Let $g \in X$. The value

$$f_D^{\downarrow}(x, g) = \liminf_{\alpha \downarrow 0} \frac{f(x + \alpha g) - f(x)}{\alpha} \qquad (2.33)$$

will be referred to as the *Dini lower derivative* of the function $f$ at the point $x$ in the direction $g$. Clearly,

$$f^{\downarrow}(x) \leq \inf_{||g|| = 1} f_D^{\downarrow}(x, g). \qquad (2.34)$$

The relation (2.31) also implies

$$f^{\downarrow}(x) \leq \inf_{g \in S} \liminf_{\substack{[\alpha, g'] \to [+0, g] \\ g' \in S}} \frac{f(x + \alpha g') - f(x)}{\alpha}. \qquad (2.35)$$

Let $g \in X$. The value

$$f_H^{\downarrow}(x, g) = \liminf_{[\alpha, g'] \to [+0, g]} \frac{f(x + \alpha g') - f(x)}{\alpha} \qquad (2.36)$$

is called the *Hadamard lower derivative* of the function $f$ at the point $x$ in the direction $g$.

Since

$$f_H^{\downarrow}(x, g) \leq f_D^{\downarrow}(x, g) \qquad \forall g \in X,$$

then (2.35) and (2.36) yield

$$f^{\downarrow}(x) \le \inf_{||g||=1} f_H^{\downarrow}(x,g) \le \inf_{||g||=1} f_D^{\downarrow}(x,g). \tag{2.37}$$

The limits in (2.33) and (2.36) always exist (though they may be equal to $+\infty$ or $-\infty$).

**Lemma 2.2.1.** *If the set $S = \{g \in X \mid ||g|| = 1\}$ is compact (that is, from any infinite sequence of points belonging to $S$, one can choose a converging subsequence whose limit point also belongs to $S$), then*

$$f^{\downarrow}(x) = \inf_{g \in S} f_H^{\downarrow}(x,g).$$

*Proof.* Denote $\inf_{g \in S} f_H^{\downarrow}(x,g) = A$. It follows from (2.37) that

$$f^{\downarrow}(x) \le A. \tag{2.38}$$

There exist sequences $\{\alpha_k\}$ and $\{g_k\}$ such that (see(2.31))

$$\alpha_k \underset{k \to \infty}{\downarrow} 0, \quad g_k \in S, \tag{2.39}$$

$$h_k = \frac{f(x + \alpha_k g_k) - f(x)}{\alpha_k} \underset{k \to \infty}{\longrightarrow} f^{\downarrow}(x). \tag{2.40}$$

Since the set $S$ is compact then without loss of generality one may assume that $g_k \to g_0 \in S$. The relations (2.36), (2.39) and (2.40) yield

$$f^{\downarrow}(x) = \lim_{k \to \infty} h_k \ge \liminf_{\substack{[\alpha, g'] \to [+0, g_0] \\ g' \in S}} \frac{f(x + \alpha g') - f(x)}{\alpha} = f_H^{\downarrow}(x, g_0). \tag{2.41}$$

Moreover

$$f^{\downarrow}(x) \ge f_H^{\downarrow}(x, g_0) \ge \inf_{g \in S} f_H^{\downarrow}(x, g) = A. \tag{2.42}$$

This inequality and (2.38) concludes the proof.                                         $\square$

**Definition.** A function $h(g) : X \to \mathbb{R}$ is called *positively homogeneous* (p.h.), if

$$h(\lambda g) = \lambda h(g) \quad \forall g \in \mathbb{R}, \quad \forall \lambda > 0.$$

Theorem 2.1.1 and relations (2.34) and (2.37) yield

**Theorem 2.2.1.** *For a point $x_* \in X$ to be a global or local minimizer of a function $f$ on $X$, it is necessary that*

$$f_D^{\downarrow}(x_*, g) \geq 0 \qquad \forall g \in X, \tag{2.43}$$

$$f_H^{\downarrow}(x_*, g) \geq 0 \qquad \forall g \in X. \tag{2.44}$$

*If the set $S$ is compact then the condition*

$$f_H^{\downarrow}(x_*, g) > 0 \qquad \forall g \in X, \quad g \neq 0 \tag{2.45}$$

*is sufficient for $x_*$ to be a strict local minimizer of the function $f$ on $X$.*

*Remark 2.2.1.* Due to the positive homogeneity of the functions $h_1(g) = f_D^{\downarrow}(x, g)$ and $h_2(g) = f_H^{\downarrow}(x, g)$, the conditions (2.43) and (2.44) hold for all $g \in X$ (not only for $g \in S$).

**Exercise 2.2.1.** Show that if the function $f$ is locally Lipschitz around the point $x \in X$, i.e. there exist $\delta > 0$ and $L < \infty$ such that

$$|f(x') - f(x'')| \leq L||x' - x''|| \quad \forall x', x'' \in B_\delta(x) = \{y \in X \mid ||x - y|| \leq \delta\},$$

then

$$f_D^{\downarrow}(x, g) = f_H^{\downarrow}(x, g) \qquad \forall g \in \mathbb{R}^n. \tag{2.46}$$

*Remark 2.2.2.* In the one-dimensional case $(X = \mathbb{R})$ the relation (2.46) always holds:

$$f_D^{\downarrow}(x, g) = f_H^{\downarrow}(x, g) \qquad \forall g \in \mathbb{R}.$$

The relation

$$f_D^{\uparrow}(x, g) = f_H^{\uparrow}(x, g) \quad \forall g \in \mathbb{R}$$

is valid too.

Analogously,

$$f^{\uparrow}(x) = \limsup_{y \to x} \frac{f(y) - f(x)}{\rho(x, y)} = \limsup_{\alpha \downarrow 0, g \in S} \frac{f(x + \alpha g) - f(x)}{\alpha}. \tag{2.47}$$

Let $g \in X$. The value

$$f_D^{\uparrow}(x, g) = \limsup_{\alpha \downarrow 0} \frac{f(x + \alpha g) - f(x)}{\alpha} \tag{2.48}$$

is called the *Dini upper derivative* of the function $f$ at the point $x$ in the direction $g$.

The value

$$f_H^{\uparrow}(x, g) = \limsup_{[\alpha, g'] \to [+0, g]} \frac{f(x + \alpha g') - f(x)}{\alpha} \tag{2.49}$$

will be referred to as the *Hadamard upper derivative* of the function $f$ at the point $x$ in the direction $g$. The functions $h_1(g) = f_D^\uparrow(x,g)$ and $h_2(g) = f_H^\uparrow(x,g)$ are positively homogeneous as functions of $g$. Clearly,

$$f_H^\uparrow(x,g) \geq f_D^\uparrow(x,g). \tag{2.50}$$

The limits (possibly, infinite) in (2.48) and (2.49) always exist.

*Remark 2.2.3.* If the set $S = \{g \in X \mid \|g\| = 1\}$ is compact, then, as in Lemma 2.2.1, it is possible to show that

$$f^\uparrow(x) = \sup_{g \in S} f_H^\uparrow(x,g). \tag{2.51}$$

**Exercise 2.2.2.** Prove the relations (2.51).

**Exercise 2.2.3.** Show that if a function $f$ is locally Lipschitz around a point $x$, then $f_D^\uparrow(x,g) = f_H^\uparrow(x,g)$      $\forall g \in \mathbb{R}^n$.
    From (2.50) one has

$$f^\uparrow(x) \geq \sup_{\|g\|=1} f_H^\uparrow(x,g) \geq \sup_{\|g\|=1} f_D^\uparrow(x,g). \tag{2.52}$$

Theorem 2.1.2 and the inequalities (2.52) yield

**Theorem 2.2.2.** *For a point $x^* \in X$ to be a global or local maximizer of a function $f$ on $X$, it is necessary that*

$$f_D^\uparrow(x^*,g) \leq 0 \qquad \forall g \in X, \tag{2.53}$$

$$f_H^\uparrow(x^*,g) \leq 0 \qquad \forall g \in X. \tag{2.54}$$

*If the set $S = \{g \in X \mid \|g\| = 1\}$ is compact and the condition*

$$f_H^\uparrow(x^*,g) < 0 \qquad \forall g \in X, \quad g \neq 0 \tag{2.55}$$

*holds then the point $x^*$ is a strict local maximizer of the function $f$.*

*Remark 2.2.4.* In each of Theorems 2.2.1 and 2.2.2 two necessary conditions are stated ((2.43)–(2.44) and (2.53)–(2.54)). It is often easier to compute the Dini derivatives that the Hadamard ones.
    To check the extremality of a point, it is possible first to check a weaker condition ((2.43) for a minimum and (2.53) for a maximum). If it holds then one checks a stronger condition ((2.44) for a minimum and (2.54) for a maximum). If the condition (2.43)((2.53)) doesn't hold, one may conclude that the point is not a minimizer (maximizer).

**Theorem 2.2.3.** *Assume that for every* $\Delta \in X$ *the values* $f_D^{\downarrow}(x,\Delta)$, $f_H^{\downarrow}(x,\Delta)$, $f_D^{\uparrow}(x,\Delta)$, $f_H^{\uparrow}(x,\Delta)$ *are finite. Then the following expansions hold:*

$$f(x+\Delta) = f(x) + f_D^{\downarrow}(x,\Delta) + \varrho_D(\Delta), \tag{2.56}$$

$$f(x+\Delta) = f(x) + f_H^{\downarrow}(x,\Delta) + \varrho_H(\Delta), \tag{2.57}$$

$$f(x+\Delta) = f(x) + f_D^{\uparrow}(x,\Delta) + \bar{o}_D(\Delta), \tag{2.58}$$

$$f(x+\Delta) = f(x) + f_H^{\uparrow}(x,\Delta) + \bar{o}_H(\Delta), \tag{2.59}$$

*where*

$$\liminf_{\alpha \downarrow 0} \frac{\varrho_D(\alpha\Delta)}{\alpha} = 0 \qquad \forall \Delta \in X, \tag{2.60}$$

$$\liminf_{[\alpha,\Delta'] \to [+0,\Delta]} \frac{\varrho_H(\alpha\Delta')}{\alpha} = \liminf_{[\alpha,\Delta'] \to [+0,\Delta]} \left[ \frac{f(x+\alpha\Delta') - f(x)}{\alpha} - f_H^{\downarrow}(x,\Delta') \right]$$
$$\leq 0, \tag{2.61}$$

$$\limsup_{\alpha \downarrow 0} \frac{\bar{o}_D(\alpha\Delta)}{\alpha} = 0 \qquad \forall \Delta \in X, \tag{2.62}$$

$$\limsup_{[\alpha,\Delta'] \to [+0,\Delta]} \frac{\bar{o}_H(\alpha\Delta')}{\alpha} = \limsup_{[\alpha,\Delta'] \to [+0,\Delta]} \left[ \frac{f(x+\alpha\Delta') - f(x)}{\alpha} - f_H^{\uparrow}(x,\Delta') \right]$$
$$\geq 0. \tag{2.63}$$

*Proof.* follows from (2.33),(2.36), (2.48) and (2.49). Indeed, the relations (2.56)– (2.60) and (2.58)–(2.62) are valid due to the definitions. Let us prove, for example, (2.57)–(2.61). The expansion (2.57) holds for

$$\varrho_H(\Delta) = f(x+\Delta) - f(x) - f_H^{\downarrow}(x,\Delta). \tag{2.64}$$

We have to show that $\varrho_H(\Delta)$ satisfies (2.61). Set

$$A := \liminf_{[\alpha,\Delta'] \to [+0,\Delta]} \frac{\varrho_H(\alpha\Delta')}{\alpha}$$
$$= \liminf_{[\alpha,\Delta'] \to [+0,\Delta]} \left[ \frac{f(x+\alpha\Delta') - f(x)}{\alpha} - f_H^{\downarrow}(x,\Delta') \right]. \tag{2.65}$$

Choose a sequence $\{[\alpha_k, \Delta_k]\}$ such that

$$[\alpha_k, \Delta_k] \xrightarrow[k \to \infty]{} [+0, \Delta],$$
$$h(\alpha_k, \Delta_k) := \frac{f(x+\alpha_k\Delta_k) - f(x)}{\alpha_k} \xrightarrow[k \to \infty]{} f_H^{\downarrow}(x,\Delta). \tag{2.66}$$

Without loss of generality one can assume that

$$f_H^\downarrow(x, \Delta_k) \xrightarrow[k \to \infty]{} B.$$

Then (2.65) and (2.66) yield

$$A \le \lim_{k \to \infty} [h(\alpha_k, \Delta_k) - f_H^\downarrow(x, \Delta_k)] = f_H^\downarrow(x, \Delta) - B. \qquad (2.67)$$

Take a sequence $\{\varepsilon_k\}$ such that $\varepsilon_k \downarrow 0$, and find $\beta_k, \Delta_k'$ such that

$$0 < \beta_k < \varepsilon_k, \quad \|\Delta_k - \Delta_k'\| < \varepsilon_k,$$

$$h(\beta_k, \Delta_k') := \frac{f(x + \beta_k \Delta_k') - f(x)}{\beta_k} = f_H^\downarrow(x, \Delta_k) + \gamma_k, \qquad (2.68)$$

with $|\gamma_k| < \varepsilon_k$. (Such $\beta_k$'s and $\Delta_k'$'s always exist.)
  Then for $k \to \infty$

$$h(\beta_k, \Delta_k') \longrightarrow B, \quad \beta_k \downarrow 0, \quad \Delta_k' \longrightarrow \Delta.$$

Therefore

$$f_H^\downarrow(x, \Delta) := \liminf_{[\alpha, \Delta'] \to [+0, \Delta]} \frac{f(x + \alpha \Delta') - f(x)}{\alpha}$$

$$\le \liminf_{k \to \infty} h(\beta_k, \Delta_k') = B. \qquad (2.69)$$

Now (2.69) and (2.67) imply the required $A \le 0$.
  The relations (2.57)–(2.61) are deduced in a similar way.                     □

**Exercise 2.2.4.** Prove the relations (2.57)–(2.61).

*Example 2.2.1.* Let $X = \mathbb{R}$,

$$f(x) = \begin{cases} x \sin \frac{1}{x}, & x \ne 0, \\ 0, & x = 0. \end{cases}$$

Since $X = \mathbb{R}$, then (see Remark 2.2.3)

$$f_D^\downarrow(x, g) = f_H^\downarrow(x, g), \quad f_D^\uparrow(x, g) = f_H^\uparrow(x, g) \qquad \forall g \in \mathbb{R}.$$

It is clear that at the point $x_0 = 0$

$$f_D^\downarrow(x_0, \Delta) = -|\Delta|, \quad f_D^\uparrow(x_0, \Delta) = |\Delta|.$$

*Remark 2.2.5.* Sometimes the definitions (2.36) and (2.49) are written in the form

$$f_H^{\downarrow}(x,g) = \liminf_{\alpha \downarrow 0, g' \to g} \frac{f(x + \alpha g') - f(x)}{\alpha} \qquad (2.70)$$

and

$$f_H^{\uparrow}(x,g) = \limsup_{\alpha \downarrow 0, g' \to g} \frac{f(x + \alpha g') - f(x)}{\alpha}. \qquad (2.71)$$

However, generally speaking, these relations do not hold (see [46]). In the definitions (2.36) and (2.49) $g'$ may take the value $g$, while in the definitions (2.70) and (2.71) $g' \neq g$. Put

$$f^{\downarrow}(x,g) = \liminf_{\alpha \downarrow 0, g' \to g} \frac{f(x + \alpha g') - f(x)}{\alpha}, \qquad (2.72)$$

$$f^{\uparrow}(x,g) = \limsup_{\alpha \downarrow 0, g' \to g} \frac{f(x + \alpha g') - f(x)}{\alpha}. \qquad (2.73)$$

**Lemma 2.2.2.** *(see [46]) If a function $f$ is continuous around a point $x$, then*

$$f^{\downarrow}(x,g) = f_H^{\downarrow}(x,g) \quad \forall g \in X, \qquad (2.74)$$

$$f^{\uparrow}(x,g) = f_H^{\uparrow}(x,g) \quad \forall g \in X. \qquad (2.75)$$

*Proof.* Let us show, e.g., the relation (2.74). Clearly,

$$f_H^{\downarrow}(x,g) \le f^{\downarrow}(x,g). \qquad (2.76)$$

Assume that (2.74) is not valid. Then (2.76) implies

$$f_H^{\downarrow}(x,g) < f^{\downarrow}(x,g) \qquad (2.77)$$

and, hence, there exists a sequence $\{\alpha_k\}$ such that $\alpha_k \downarrow 0$,

$$\frac{f(x + \alpha_k g) - f(x)}{\alpha_k} := h_k \longrightarrow f_H^{\downarrow}(x,g). \qquad (2.78)$$

Choose an arbitrary sequence $\{\beta_k\}$ such that $\beta_k \downarrow 0$. Since the function $f$ is continuous at the point $x + \alpha_k g$, then for any $k$ one can find a $g_k \neq g$ such that

$$\|g_k - g\| \le \beta_k, \quad |f(x + \alpha_k g_k) - f(x + \alpha_k g)| \le \alpha_k \beta_k.$$

Then

$$\begin{aligned}
\overline{h}_k &:= \frac{f(x + \alpha_k g_k) - f(x)}{\alpha_k} \\
&= \frac{f(x + \alpha_k g) - f(x)}{\alpha_k} + \frac{f(x + \alpha_k g_k) - f(x + \alpha_k g)}{\alpha_k} \\
&= h_k + \gamma_k, \qquad (2.79)
\end{aligned}$$

where $|\gamma_k| \le \beta_k$. Since $g_k \to g$, then (2.78) and (2.79) yield

$$f^{\downarrow}(x,g) \le \liminf_{k \to \infty} \frac{f(x + \alpha_k g_k) - f(x)}{\alpha_k}$$

$$= \liminf_{k \to \infty} \overline{h}_k = \lim_{k \to \infty} (h_k + \gamma_k) = f_H^{\downarrow}(x,g), \qquad (2.80)$$

which contradicts (2.77).                                                                □

*Remark 2.2.6.* Observe that the continuity of the function $f$ at the point $x$ is not required.

*Remark 2.2.7.* It is of interest to study the properties of the functions $f^{\downarrow}(x,g)$ and $f^{\uparrow}(x,g)$ defined by the relations (2.72) and (2.73).


## 2.3 Directional Differentiability in a Normed Space

Let $X$ be a normed space with the norm $\|x\|$ and the metric $\rho(x,y) = \|x - y\|$ and let a function $f : X \to \mathbb{R}$ be given. Fix $x \in X$ and $g \in X$.

If there exists the finite limit

$$f_D'(x,g) = \lim_{\alpha \downarrow 0} \frac{f(x + \alpha g) - f(x)}{\alpha}, \qquad (2.81)$$

then the function $f$ is called *Dini directionally differentiable at the point $x$ in the direction $g$*, and the value $f_D'(x,g)$ is called the *Dini directional derivative of the function $f$ at the point $x$ in the direction $g$*. If the limit in (2.81) exists and is finite for any $g \in X$, then the function $f$ is called *D-directionally differentiable at the point $x$* (D-d.d.).

If there exists the finite limit

$$f_H'(x,g) = \lim_{[\alpha,g'] \to [+0,g]} \frac{f(x + \alpha g') - f(x)}{\alpha}, \qquad (2.82)$$

then $f$ is called *Hadamard directionally differentiable at the point $x$ in the direction $g$*, and the value $f_H'(x,g)$ is the *Hadamard directional derivative of the function $f$ at the point $x$ in the direction $g$*. If the limit in (2.82) exists and is finite for any $g \in X$, then the function $f$ is called *H-directionally differentiable at the point $x$* (H-d.d.).

If a function $f$ is H-d.d., then it is also D-d.d., and

$$f_H'(x,g) = f_D'(x,g) \qquad \forall g \in X.$$

The functions $h(g) = f_H'(x,g)$ and $h(g) = f_D'(x,g)$ are positively homogeneous (as functions of $g$).

**Exercise 2.3.1.** Prove the positive homogeneity of the functions $h(g) = f'_H(x,g)$ and $h(g) = f'_D(x,g)$.

**Exercise 2.3.2.** Prove that if a function $f$ is $H$-d.d. at a point $x$, then $f$ is continuous at this point. Show that the analogous property for $D$-d.d. functions does not hold. Give an example.

**Lemma 2.3.1.** *If a function $f : X \to \mathbb{R}$ is $H$-d.d. at a point $x \in X$, then the function $h(g) = f'_H(x,g)$ is continuous (as a function of $g$) on $X$.*

*Proof.* Assume the opposite. Let, for example, the function $h(g)$ be discontinuous at a point $g_0 \in X$. Then there exist an $\varepsilon > 0$ and a sequence of points $\{g_k\}$ such that

$$g_k \to g_0, \qquad |h(g_k) - h(g_0)| > \varepsilon. \tag{2.83}$$

Choose a sequence $\{\delta_k\}$ such that $\delta_k \downarrow 0$ and for every $k$ find $\alpha_k$ and $g'_k$ such that

$$\|g'_k - g_k\| < \delta_k, \qquad \alpha_k < \delta_k,$$

$$\left| \frac{f(x + \alpha_k g'_k) - f(x)}{\alpha_k} - f'_H(x, g_k) \right| < \delta_k. \tag{2.84}$$

The relations (2.83) and (2.84) yield

$$\left| \frac{f(x + \alpha_k g'_k) - f(x)}{\alpha_k} - f'_H(x, g_0) \right| = \left| \left( \frac{f(x + \alpha_k g'_k) - f(x)}{\alpha_k} - f'_H(x, g_k) \right) \right.$$
$$\left. + \left( f'_H(x, g_k) - f'_H(x, g_0) \right) \right|$$
$$\geq \left| f'_H(x, g_k) - f'_H(x, g_0) \right|$$
$$- \left| \frac{f(x + \alpha_k g'_k) - f(x)}{\alpha_k} - f'_H(x, g_k) \right|$$
$$> \varepsilon - \delta_k. \tag{2.85}$$

Since $[\alpha, g_k] \to [+0, g_0]$, then (see (2.82))

$$\frac{f(x + \alpha_k g'_k) - f(x)}{\alpha_k} \longrightarrow f'_H(x, g_0),$$

which contradicts (2.85). □

*Remark 2.3.1.* Thus, the Hadamard derivative is continuous in $g$, while the Dini one is not necessarily continuous.

*Remark 2.3.2.* If a function $f$ is $D$-d.d., then

$$f_D^{\downarrow}(x,g) = f_D^{\uparrow}(x,g) = f'_D(x,g),$$

and if $f$ is $H$-d.d., then

$$f_H^{\downarrow}(x,g) = f_H^{\uparrow}(x,g) = f_H'(x,g).$$

Therefore the necessary and sufficient conditions (Theorems 2.2.1 and 2.2.2) can be reformulated in terms of $f_H'$, $f_D'$.

**Theorem 2.3.1.** *For a point $x_* \in X$ to be a global or local minimizer of the function $f$ on $X$, it is necessary that*

$$f_D'(x_*,g) \geq 0 \qquad \forall g \in X, \tag{2.86}$$

$$f_H'(x_*,g) \geq 0 \qquad \forall g \in X. \tag{2.87}$$

*If the set $S$ is compact then the condition*

$$f_H'(x_*,g) > 0 \qquad \forall g \in X, \quad g \neq 0, \tag{2.88}$$

*is a sufficient condition for a strict local minimum of the function $f$ on $X$.*

**Theorem 2.3.2.** *For a point $x^* \in X$ to be a global or local maximizer of the function $f$ on $X$, it is necessary that*

$$f_D'(x^*,g) \leq 0 \qquad \forall g \in X, \tag{2.89}$$
$$f_H'(x^*,g) \leq 0 \qquad \forall g \in X. \tag{2.90}$$

*If the set $S$ is compact then the condition*

$$f_H'(x^*,g) < 0 \qquad \forall g \in X, \quad g \neq 0, \tag{2.91}$$

*is a sufficient condition for a strict local maximum of the function $f$ on $X$.*

*Remark 2.3.3.* Sometimes in the literature the definition (2.82) is written in the form

$$f_H'(x,g) = \lim_{\alpha \downarrow 0, g' \to g} \frac{f(x + \alpha g') - f(x)}{\alpha}. \tag{2.92}$$

However, generally speaking, the expression in the right-hand part of (2.92) does not coincide with that in the right-hand part of (2.82). To be more precise, if the limit in (2.82) exists, then the limit in (2.92) also exists, and these limits coincide. On the other hand, it may happen that the both limits (2.81) and (2.92) exist, but the limit in (2.82) does not. Give a very simple example: Let $X = \mathbb{R}^2$, $x = (x_1, x_2)$, $x_0 = (0,0)$, $g_0 = (1,0)$,

$$f(x) = f(x_1,x_2) = \begin{cases} x_1, & x_2 \neq 0, \\ 0, & x_2 = 0. \end{cases}$$

It is clear, that

$$f'_D(x_0, g_0) = \lim_{\alpha \downarrow 0} \frac{f(x_0 + \alpha g_0) - f(x_0)}{\alpha} = 0,$$

$$f'(x_0, g_0) = \lim_{\alpha \downarrow 0, g' \to g_0} \frac{f(x_0 + \alpha g') - f(x)}{\alpha} = 1.$$

However, the limit

$$f'_H(x_0, g_0) = \lim_{[\alpha, g'] \to [+0, g_0]} \frac{f(x_0 + \alpha g') - f(x)}{\alpha}$$

does not exist.

Nevertheless, as in Lemma 2.2.2, it is possible to show that if the function $f$ is continuous then the limits the right-hand sides of (2.92) and (2.82) coincide (precisely, if there exists one of these limits, the other also exists, and both of them coincide).

## 2.4 The Gâteaux and Fréchet Differentiability

### 2.4.1 Differentiable Functions

Let $X$ be a normed space, $f : X \to \mathbb{R}$. If the function $f$ is $D$-d.d. at a point $x \in X$, then (2.81) implies the expansion

$$f(x + \Delta) = f(x) + f'_D(x, \Delta) + o_D(\Delta), \tag{2.93}$$

where

$$\frac{o_D(\alpha \Delta)}{\alpha} \xrightarrow[\alpha \downarrow 0]{} 0 \qquad \forall \Delta \in X. \tag{2.94}$$

If $f$ is $H$-d.d. at $x \in X$, then (2.82) yields the expansion

$$f(x + \Delta) = f(x) + f'_H(x, \Delta) + o_H(\Delta), \tag{2.95}$$

where

$$\frac{o_H(\Delta)}{||\Delta||} \xrightarrow[||\Delta|| \to 0]{} 0. \tag{2.96}$$

**Definition.** A function $f : X \to \mathbb{R}$ is called *Gâteaux differentiable at a point* $x \in X$, if there exists a linear functional $H_x : X \to \mathbb{R}$ such that

$$f(x + \Delta) = f(x) + H_x(\Delta) + o(\Delta), \tag{2.97}$$

where

$$\frac{o(\alpha\Delta)}{\alpha} \xrightarrow[\alpha\to 0]{} 0 \qquad \forall \Delta \in X. \tag{2.98}$$

In such a case the functional $H_x(\Delta)$ will be denoted by $f'_G(x)(\Delta)$.

A function $f : X \to \mathbb{R}$ is called *Fréchet differentiable at a point* $x \in X$ if there exists a linear functional $H_x : X \to \mathbb{R}$ such that

$$f(x + \Delta) = f(x) + H_x(\Delta) + o(\Delta), \tag{2.99}$$

where

$$\frac{o(\Delta)}{||\Delta||} \xrightarrow[||\Delta||\to 0]{} 0. \tag{2.100}$$

In such a case the functional $H_x(\Delta)$ will be denoted by $f'_F(x)(\Delta)$.

The function $H_x(\Delta) = f'_G(x)(\Delta)$ in (2.97) is continuous in $\Delta$ due to the properties of a linear functional (unlike the Dini derivative above).

It is not difficult to see that the functional $H_x$, satisfying (2.97)–(2.98) or (2.99)–(2.100), is unique. The Fréchet differentiability implies the Gâteaux one, but the reverse is not true.

Clearly, if a function $f$ is Gâteaux differentiable at a point $x$, then it is $D$-d.d. at this point with $f'_D(x, \Delta) = H_x(\Delta)$. If $f$ is Fréchet differentiable at $x$, then it is $H$-d.d. at this point and $f'_H(x, \Delta) = H_x(\Delta)$.


## 2.4.2  Conditions for an Extremum for Differentiable Functions

For Gâteaux or Fréchet differentiable functions the necessary conditions for an extremum on $X$ (Theorems 2.2.1 and 2.2.2) take the following form.

**Theorem 2.4.1.** *Let a function $f$ be Gâteaux differentiable at a point $x_* \in X$. For the point $x_* \in X$ to be a global or local minimizer of the function $f$ on $X$, it is necessary that*

$$f'_G(x_*)(g) = 0 \qquad \forall g \in X. \tag{2.101}$$

**Theorem 2.4.2.** *Let a function $f$ be Fréchet differentiable at a point $x_* \in X$. For the point $x_* \in X$ to be a global or local minimizer of the function $f$ on $X$, it is necessary that*

$$f'_F(x_*)(g) = 0 \qquad \forall g \in X. \tag{2.102}$$

**Theorem 2.4.3.** *Let a function* $f$ *be Gâteaux differentiable at a point* $x^* \in X$. *For the point* $x^* \in X$ *to be a global or local maximizer of the function* $f$ *on* $X$, *it is necessary that*

$$f'_G(x^*)(g) = 0 \qquad \forall g \in X. \qquad (2.103)$$

**Theorem 2.4.4.** *Let a function* $f$ *be Fréchet differentiable at a point* $x^* \in X$. *For the point* $x^* \in X$ *to be a global or local maximizer of the function* $f$ *on* $X$, *it is necessary that*

$$f'_F(x^*)(g) = 0 \qquad \forall g \in X. \qquad (2.104)$$

*Thus, the necessary conditions for a minimum and a maximum coincide and the linear functional should be equal zero. The sufficient conditions (2.88) for a minimum or (2.91) for a maximum are never satisfied.*

### 2.4.3 The Function of Maximum of Differentiable Functions

Let $f_1, f_2, \cdots, f_N : X \to \mathbb{R}$ be functionals which are Gâteaux or Fréchet differentiable at a point $x_0 \in X$. Then (see (1.24)) the function $f(x) = \max_{i \in 1:N} f_i(x)$ is directionally differentiable at the point $x_0$, and if all $f_i$'s are Gâteaux differentiable, then $f$ is $D$-d.d., while the Fréchet differentiability of all $f_i$'s implies that $f$ is $H$-d.d. at $x_0$ and

$$f'(x_0, g) = \max_{i \in R(x_0)} f'_i(x_0, g) \quad \forall g \in X, \qquad (2.105)$$

where $R(x) = \{i \in 1 : N \mid f_i(x) = f(x)\}$. Here $f'$ is either $f'_D$, or $f'_H$; $f'_i$ is a linear functional.

The relation (2.105) provides the following expansion:

$$f(x_0 + \Delta) = f(x_0) + f'(x_0, \Delta) + o(\Delta) = f(x_0) + \max_{i \in R(x_0)} f'_i(x_0, \Delta) + o(\Delta). \qquad (2.106)$$

Here $\Delta \in X$, and $o(\Delta)$ depends on $x_0$.

It is also possible to get another expansion:

$$f(x_0 + \Delta) = \max_{i \in 1:N} f_i(x_0 + \Delta) = \max_{i \in 1:N} [f_i(x_0) + f'_i(x_0, \Delta) + o_i(\Delta)]$$
$$= f(x_0) + \max_{i \in 1:N} [f_i(x_0) - f(x_0) + f'_i(x_0, \Delta)] + o(\Delta). \qquad (2.107)$$

Observe that the second summand in the right-hand part of (2.106) is a positively homogeneous function while the second summand in the right-hand part of (2.107) is not a p.h. function.

## 2.5  The Finite-Dimensional Case

### 2.5.1  Conditions for an Extremum

Consider in more details the case of the $n$-dimensional arithmetic Euclidean space.

Let $X = \mathbb{R}^n$. If $x = (x_1, \cdots, x_n)$, $y = (y_1, \cdots, y_n)$, then

$$\rho(x, y) = \sqrt{\sum_{i=1}^{n}(x_i - y_i)^2} = \|x - y\|.$$

By $(x, y)$ let us denote the scalar product of vectors $x$ and $y$ : $(x, y) = \sum_{i=1}^{n} x_i y_i$.

Let $f : \mathbb{R}^n \to \overline{\mathbb{R}} := [-\infty, +\infty]$ and assume that $\operatorname{dom} f \neq \emptyset$. Take $x \in \operatorname{dom} f$ and find (see (2.2))

$$f^{\downarrow}(x) = \liminf_{y \to x} \frac{f(y) - f(x)}{\rho(x, y)}. \tag{2.108}$$

**Lemma 2.5.1.** *The following relation*

$$f^{\downarrow}(x) = \inf_{\|g\|=1} \liminf_{\substack{[\alpha, g'] \to [+0, g] \\ \|g'\|=1}} \frac{f(x + \alpha g') - f(x)}{\alpha} \tag{2.109}$$

*holds.*

*Proof.* Denote by $A$ the right-hand part of (2.109). Putting $y = x + \Delta$ and taking into account that $\rho(x, y) = \|\Delta\|$, one can rewrite the expression (2.108) in the form

$$f^{\downarrow}(x) = \liminf_{\|\Delta\| \downarrow 0} \frac{f(x + \Delta) - f(x)}{\|\Delta\|}. \tag{2.110}$$

Denoting $\|\Delta\| = \alpha$, $\frac{\Delta}{\|\Delta\|} = g$, from (2.110) we have

$$f^{\downarrow}(x) = \liminf_{\alpha \downarrow 0, g \in S} \frac{f(x + \alpha g) - f(x)}{\alpha}, \tag{2.111}$$

where $S = \{g \in \mathbb{R}^n \mid \|g\| = 1\}$. The relation (2.111) implies that there exist sequences $\{\alpha_k\}$ and $\{g_k\}$ such that $\alpha_k \downarrow 0$, $\|g_k\| = 1$,

$$\frac{f(x + \alpha_k g_k) - f(x)}{\alpha_k} \longrightarrow f^{\downarrow}(x).$$

Since the set $S$ is compact, without loss of generality one can assume that $g_k \to g_0$. Then

$$h(g_0) = \liminf_{\substack{[\alpha,g'] \to [+0, g_0] \\ \|g'\| = 1}} \frac{f(x + \alpha g') - f(x)}{\alpha}$$

$$\leq \liminf_{k \to \infty} \frac{f(x + \alpha_k g_k) - f(x)}{\alpha_k} = f^{\downarrow}(x).$$

Furthermore,

$$A \leq h(g_0) \leq f^{\downarrow}(x). \tag{2.112}$$

Let a sequence $\{g_k\}$ be such that $\|g_k\| = 1$,

$$h(g_k) = \liminf_{[\alpha,g'] \to [+0, g_k]} \frac{f(x + \alpha g') - f(x)}{\alpha} \longrightarrow A. \tag{2.113}$$

Choose a sequence $\{\varepsilon_k\}$ such that $\varepsilon_k \downarrow 0$.
For any $k$ there exist $\alpha_k$ and $g'_k$ such that $\alpha_k < \varepsilon_k$, $\|g'_k\| = 1$, $\|g'_k - g_k\| < \varepsilon_k$,

$$\frac{f(x + \alpha_k g'_k) - f(x)}{\alpha_k} = h(g_k) + \mu_k, \tag{2.114}$$

where $|\mu_k| < \varepsilon_k$. The relation (2.111) implies

$$f^{\downarrow}(x) = \liminf_{\alpha \downarrow 0, g \in S} \frac{f(x + \alpha g) - f(x)}{\alpha} \leq \liminf_{k \to \infty} \frac{f(x + \alpha_k g'_k) - f(x)}{\alpha_k}. \tag{2.115}$$

It follows from (2.113), (2.114) and (2.115) that

$$f^{\downarrow}(x) \leq A. \tag{2.116}$$

Comparing (2.116) and (2.112), we get (2.109).  □

The following proposition is proved in the same way.

**Lemma 2.5.2.** *The relation*

$$f^{\uparrow}(x) = \sup_{\|g\| = 1} \limsup_{\substack{[\alpha,g'] \to [+0, g] \\ \|g'\| = 1}} \frac{f(x + \alpha g') - f(x)}{\alpha} \tag{2.117}$$

*is valid*

**Exercise 2.5.1.** Show that in (2.109) and (2.117) the condition $\|g'\| = 1$ (but not $\|g\| = 1$!) may be omitted.

**Definition.** Let $x \in dom f$, $g \in \mathbb{R}^n$. The value

$$f^{\downarrow}_D(x, g) = \liminf_{\alpha \downarrow 0} \frac{f(x + \alpha g) - f(x)}{\alpha} \tag{2.118}$$

is called the *Dini lower derivative* of the function $f$ at the point $x$ in the direction $g$. Note that the function $h(g) = f_D^\downarrow(x, g)$ is positively homogeneous (of the first degree) in $g$, i.e. $h(\lambda g) = \lambda h(g) \quad \forall \lambda \geq 0, \forall g \in \mathbb{R}^n$.

The value

$$f_H^\downarrow(x, g) = \liminf_{[\alpha, g'] \to [+0, g]} \frac{f(x + \alpha g') - f(x)}{\alpha} \tag{2.119}$$

is called the *Hadamard lower derivative* of the function $f$ at the point $x$ in the direction $g$. The function $h(g) = f_H^\downarrow((x, g)$ is also p.h. in $g$.

The value

$$f_D^\uparrow(x, g) = \limsup_{\alpha \downarrow 0} \frac{f(x + \alpha g) - f(x)}{\alpha} \tag{2.120}$$

is called the *Dini upper derivative* of the function $f$ at the point $x$ in the direction $g$.

The value

$$f_H^\uparrow(x, g) = \limsup_{[\alpha, g'] \to [+0, g]} \frac{f(x + \alpha g') - f(x)}{\alpha} \tag{2.121}$$

is called the *Hadamard upper derivative* of the function $f$ at the point $x$ in the direction $g$.

As above, the functions $h(g) = f_D^\uparrow(x, g)$ and $h(g) = f_H^\uparrow(x, g)$ are p.h. in $g$. Note that since $x \in dom f$, then all the limits in (2.118)–(2.121) exist (though may attain the values $+\infty$ or $-\infty$).

*Remark 2.5.1.* In the case $\mathbb{R}^n = \mathbb{R} = (-\infty, +\infty)$, we have $g \in \mathbb{R}$, and for all $g \in \mathbb{R}$ the relations $f_D^\uparrow(x, g) = f_H^\uparrow(x, g)$, $f_D^\downarrow(x, g) = f_H^\downarrow(x, g)$ hold.

**Exercise 2.5.2.** Show that if a function $f$ is locally Lipschitz around a point $x \in \mathbb{R}^n$, then

$$f_H^\uparrow(x, g) = f_D^\uparrow(x, g), \quad f_H^\downarrow(x, g) = f_D^\downarrow(x, g) \quad \forall g \in R^n.$$

Remind that $f$ is called *locally Lipschitz* around a point $x$, if there exist $\delta > 0$ and a constant $L < \infty$ (depending on $x$) such that

$$|f(x') - f(x'')| \leq L \|x' - x''\| \quad \forall x', x'' \in B_\delta(x),$$

where

$$B_\delta(x) = \{y \in \mathbb{R}^n \mid \|x - y\| \leq \delta\}.$$

Lemma 2.5.1 and Theorem 2.1.1 imply

**Theorem 2.5.1.** *For a point $x_* \in \mathbb{R}^n$, $x_* \in dom f$, to be a global or local minimizer of a function $f$ on $\mathbb{R}^n$, it is necessary that*

$$f_H^\downarrow(x_*, g) \geq 0 \qquad \forall g \in \mathbb{R}^n. \tag{2.122}$$

*If*

$$f_H^\downarrow(x_*, g) > 0 \qquad \forall g \in \mathbb{R}^n, \ g \neq 0_n, \tag{2.123}$$

*then the point $x_*$ is a strict local minimizer of the function $f$ on $\mathbb{R}^n$.*

*Remark 2.5.2.* Since (see the definitions (2.118) and (2.119))

$$f_H^\downarrow(x, g) \leq f_D^\downarrow(x, g), \tag{2.124}$$

then the condition

$$f_D^\downarrow(x_*, g) \geq 0 \qquad \forall g \in \mathbb{R}^n \tag{2.125}$$

is also a necessary condition for a minimum, however, the condition

$$f_D^\downarrow(x_*, g) > 0 \qquad \forall g \in \mathbb{R}^n, \ g \neq 0_n$$

is not sufficient for a strict local minimum.

The condition (2.125) is weaker than the condition (2.122) (the point $x_*$ may satisfy (2.125), but not the condition (2.122)).

Lemma 2.5.2 and Theorem 2.1.2 imply

**Theorem 2.5.2.** *For a point $x^* \in \mathbb{R}^n$, $x^* \in \mathrm{dom} f$, to be a global or local maximizer of a function $f$ on $\mathbb{R}^n$, it is necessary that*

$$f_H^\uparrow(x^*, g) \leq 0 \qquad \forall g \in \mathbb{R}^n. \tag{2.126}$$

*If*

$$f_H^\uparrow(x^*, g) < 0 \qquad \forall g \in \mathbb{R}^n, \ g \neq 0_n, \tag{2.127}$$

*then the point $x^*$ is a strict local maximizer of the function $f$ on $\mathbb{R}^n$.*

*Remark 2.5.3.* Since (see the definitions (2.120) and (2.121))

$$f_H^\uparrow(x, g) \geq f_D^\uparrow(x, g), \tag{2.128}$$

then the condition

$$f_D^\uparrow(x^*, g) \leq 0 \qquad \forall g \in \mathbb{R}^n \tag{2.129}$$

is also a necessary condition for a maximum, however, the condition

$$f_D^\uparrow(x^*, g) < 0 \qquad \forall g \in \mathbb{R}^n, \ g \neq 0_n$$

is not sufficient for a strict local maximum. The condition (2.129) is weaker than the condition (2.126).

*Remark 2.5.4.* The definitions (2.119), (2.121) and Lemmas 2.5.1, 2.5.2 yield

$$f^\downarrow(x) = \inf_{\|g\|=1} f_H^\downarrow(x, g), \quad f^\uparrow(x) = \sup_{\|g\|=1} f_H^\uparrow(x, g). \tag{2.130}$$

**Definitions.** If $f^{\downarrow}(x) < 0$, then a direction $g_0 \in \mathbb{R}^n$, such that $\|g_0\| = 1$, $f^{\downarrow}_H(x, g_0) = f^{\downarrow}(x)$, is called a *steepest descent direction (in the sense of Hadamard)* (H-s.d.d.) of the function $f$ at the point $x$. The value $f^{\downarrow}(x) = f^{\downarrow}_H(x, g_0)$ is called the *rate of H-steepest descent* of the function $f$ at the point $x$.

If $\inf\limits_{\|g\|=1} f^{\downarrow}_D(x, g) = f^{\downarrow}_D(x, g_0) < 0$ (this implies $f^{\downarrow}(x) < 0$), then the direction $g_0$ is called a *steepest descent direction (in the sense of Dini)* (D-s.d.d.) of the function $f$ at the point $x$. The value $f^{\downarrow}(x) = f^{\downarrow}_D(x, g_0)$ is called the *rate of D-steepest descent* of the function $f$ at the point $x$.

Analogously, if $f^{\uparrow}(x) > 0$, then a direction $g^0 \in \mathbb{R}^n$, such that $\|g^0\| = 1$, $f^{\uparrow}_H(x, g_0) = f^{\uparrow}(x)$, is called a *steepest ascent direction (in the sense of Hadamard)* (H-s.a.d.) of the function $f$ at the point $x$. The value $f^{\uparrow}(x) = f^{\uparrow}_H(x, g^0)$ is called the *rate of H-steepest ascent* of the function $f$ at the point $x$.

If $\sup\limits_{\|g\|=1} f^{\uparrow}_D(x, g) = f^{\uparrow}_D(x, g^0) > 0$ (this implies $f^{\uparrow}(x) > 0$), then the direction $g^0$ is called a *steepest ascent direction (in the sense of Dini)* (D-s.a.d.) of the function $f$ at the point $x$. The value $f^{\uparrow}(x) = f^{\uparrow}_D(x, g^0)$ is called the *rate of D-steepest ascent* of the function $f$ at the point $x$.

Note that it may happen that steepest descent and ascent directions either do not exist, or are not unique.

## 2.5.2   The Differentiable Case

Let us consider the case where $f$ is a differentiable function. Remind that a function $f : \mathbb{R}^n \to \mathbb{R}$ is called differentiable at a point $x \in \mathbb{R}^n$ if there exists a $V \in \mathbb{R}^n$ such that

$$f(x + \Delta) = f(x) + (V, \Delta) + o(\Delta), \qquad (2.131)$$

where $o(\Delta)$ is described below. If in (2.131)

$$\frac{o(\alpha\Delta)}{\alpha} \xrightarrow[\alpha \to 0]{} 0 \quad \forall \Delta \in \mathbb{R}^n, \qquad (2.132)$$

then the function $f$ is called *Gâteaux differentiable at the point $x$*. If (2.132) holds uniformly with respect to $\Delta \in S = \{\Delta \in \mathbb{R}^n \mid \|\Delta\| = 1\}$, then $f$ is called *Fréchet differentiable*. In this case

$$\frac{o(\Delta)}{\|\Delta\|} \xrightarrow[\|\Delta\| \to 0]{} 0. \qquad (2.133)$$

Clearly, the Fréchet differentiability implies the Gâteaux one, but the reverse is not true.

If a function $f$ is Gâteaux or Fréchet differentiable at a point $x$ then the vector $V$ in (2.131) is unique and represents the gradient of the function $f$ at the point $x$. It is the collection of partial derivatives of the function $f$:

$$V = f'(x) = \left( \frac{\partial f(x_1, \cdots, x_n)}{\partial x_1}, \cdots, \frac{\partial f(x_1, \cdots, x_n)}{\partial x_n} \right).$$

*Example 2.5.1.* Let us demonstrate a function which is Gâteaux differentiable but not Fréchet differentiable.

Let $X = \mathbb{R}^2$, $x = (x_1, x_2)$, $x_0 = (0,0)$, $S = \{x = (x_1, x_2) \mid x_1^2 + (x_2 - 1)^2 = 1\}$,

$$f(x) = \begin{cases} 0, & x \notin S, \\ x_1, & x \in S. \end{cases}$$

The set $S$ is shown in Fig. 2. We have $f(x_0 + \Delta) = f(x_0) + (A, \Delta) + o(\Delta)$,

where $A = (0,0)$, $\dfrac{o(\alpha\Delta)}{\alpha} \xrightarrow[\alpha\downarrow 0]{} 0$ $\quad \forall \Delta \in \mathbb{R}^2$.

The function $f$ is Gâteaux differentiable at the point $x_0 = (0,0)$, however, it is not Fréchet differentiable. Clearly,

$$f'(x_0) = 0_2 = (0,0), \quad f_D^{\downarrow}(x_0, g) = f_D^{\uparrow}(x_0, g) = 0$$

and the both necessary conditions for a minimum (2.125), and for a maximum (2.129) are satisfied, that is, the point $x_0$ is both $D$-inf-stationary, and $D$-sup-stationary.

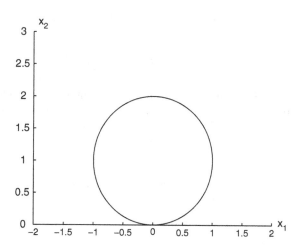

**Fig. 2**  Example 2.5.1

Find (see Lemma 2.5.1)

$$f^{\downarrow}(x_0) = \inf_{\substack{\|g\|=1}} \liminf_{\substack{[\alpha,g']\to[+0,g] \\ \|g'\|=1}} \frac{f(x+\alpha g') - f(x)}{\alpha} = \inf_{\|g\|=1} f_H^{\downarrow}(x_0, g).$$

Since

$$f_H^{\downarrow}(x_0, g) = \begin{cases} 0, & g \neq g_0, \ \|g\| = 1, \\ -1, & g = g_0, \end{cases}$$

where $g_0 = (-1, 0)$, then $f^{\downarrow}(x_0) = -1$.

Analogously, $f^{\uparrow}(x_0) = \sup_{\|g\|=1} f_H^{\uparrow}(x_0, g)$. Since

$$f_H^{\uparrow}(x_0, g) = \begin{cases} 0, & g \neq g_1, \ \|g\| = 1, \\ +1, & g = g_1, \end{cases}$$

where $g_1 = (1, 0)$, then $f^{\uparrow}(x_0) = +1$. Since $f^{\downarrow}(x_0) = -1 < 0$, $f^{\uparrow}(x_0) = +1 > 0$, then at the point $x_0 = 0$ neither the necessary condition for a minimum (2.15), nor the necessary condition for a maximum (2.17) is satisfied.

It is clear from (2.131) and the definitions (2.118) and (2.120) that if a function $f$ is Gâteaux differentiable at a point $x$ then

$$f_D^{\downarrow}(x, g) = f_D^{\uparrow}(x, g) = (f'(x), g), \tag{2.134}$$

and if $f$ is Fréchet differentiable at $x$, then the definitions (2.119) and (2.121) yield

$$f_H^{\downarrow}(x, g) = f_H^{\uparrow}(x, g) = (f'(x), g). \tag{2.135}$$

If $f$ is Fréchet differentiable then (2.130) and (2.135) imply

$$f^{\downarrow}(x) = \inf_{\|g\|=1} (f'(x), g) = -\|f'(x)\|, \tag{2.136}$$

$$f^{\uparrow}(x) = \sup_{\|g\|=1} (f'(x), g) = \|f'(x)\|. \tag{2.137}$$

If $f$ is Gâteaux differentiable at a point $x$ then it follows from (2.124), (2.128) and (2.130) that

$$f^{\downarrow}(x) \leq \inf_{\|g\|=1} f_D^{\downarrow}(x, g) = -\|f'(x)\|, \tag{2.138}$$

$$f^{\uparrow}(x) \geq \sup_{\|g\|=1} f_D^{\uparrow}(x, g) = \|f'(x)\|. \tag{2.139}$$

Theorem 2.1.1 and relations (2.136), (2.138) imply the following.

**Theorem 2.5.3.** *Let a function $f$ be Gâteaux or Fréchet differentiable at a point $x_* \in \mathbb{R}^n$. For the point $x_*$ to be a global or local minimizer of $f$ on $\mathbb{R}^n$, it is necessary that*

$$f'(x_*) = 0_n. \tag{2.140}$$

*Theorem 2.1.2 and the relations (2.137), (2.139) produce the following.*

**Theorem 2.5.4.** *Let a function $f$ be Gâteaux or Fréchet differentiable at a point $x^* \in \mathbb{R}^n$. For the point $x^*$ to be a global or local maximizer of $f$ on $\mathbb{R}^n$, it is necessary that*

$$f'(x_*) = 0_n. \tag{2.141}$$

*Remark 2.5.5.* It follows from (2.140) and (2.141) that the necessary condition for a minimum coincides with the necessary condition for a maximum. Thus, in this case the notion of inf-stationary point coincides with that of sup-stationary point. Therefore points satisfying (2.140) and (2.141) will be referred to as just *stationary* points.

The conditions (2.140) and (2.141) are well known from the classical ("smooth") analysis. Note that the sufficient conditions (2.16) and (2.17) never hold in the smooth case (they are essentially nonsmooth).

*Remark 2.5.6.* The relations (2.136) and (2.138) imply that if $f'(x) \neq 0_n$, then the direction $g_0 = -\frac{f'(x)}{\|f'(x)\|}$ is the unique steepest descent direction (in the sense of Dini and in the sense of Hadamard if $f$ is Fréchet differentiable), and the value $-\|f'(x)\|$ is the steepest descent direction of the function $f$ at the point $x$.

Analogously, we observe from (2.137) and (2.139) that if $f'(x) \neq 0_n$, then the direction $g^0 = \frac{f'(x}{\|f'(x)\|}$ is the unique steepest ascent direction (in the sense of Dini and in the sense of Hadamard if $f$ is Fréchet differentiable), and the value $\|f'(x)\|$ is the steepest ascent direction of the function $f$ at the point $x$.

Note that the mentioned uniqueness of the directions of steepest descent and ascent is due to the chosen metric $\rho(x, y) = \|x - y\|$. If one takes another metric, the uniqueness is not guaranteed.

**Exercise 2.5.3.** Show that if a function $f$ is locally Lipschitz around a point $x$, then the Gâteaux differentiability at the point $x$ is equivalent to the Fréchet differentiability at this point.

*Example 2.5.2.* Let $X = \mathbb{R} = (-\infty, \infty)$, $f(x) = |x|$. The graph of the function $f$ is depicted in Fig. 3. For $x \neq 0$ the function $f$ is differentiable and

$$f'(x) = \begin{cases} +1, & x > 0, \\ -1, & x < 0. \end{cases}$$

For $x > 0$, $\quad f'(x, g) = g$ and the direction $g = +1$ is the steepest ascent direction (s.a.d.), while the direction $g = -1$ is the steepest descent direction.

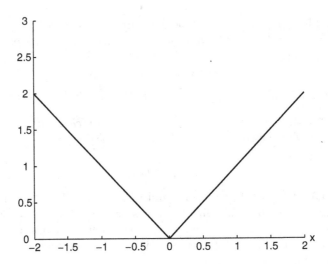

**Fig. 3** Example 2.5.2

Analogously, for $x < 0$, $f'(x, g) = -g$ and the direction $g = +1$ is the steepest descent direction, while $g = -1$ is the steepest ascent direction.

Consider the point $x_0 = 0$. We have

$$f_H^\downarrow(x_0; g) = f_D^\downarrow(x_0; g) = \liminf_{\alpha \downarrow 0} \frac{f(x_0 + \alpha g) - f(x_0)}{\alpha} = \liminf_{\alpha \downarrow 0} \frac{|\alpha g|}{\alpha} = +|g|,$$

$$f_H^\uparrow(x_0; g) = f_D^\uparrow(x_0; g) = \limsup_{\alpha \downarrow 0} \frac{|\alpha g|}{\alpha} = |g|.$$

This implies (see (2.130))

$$f^\downarrow(x_0) = \inf_{\|g\|=1} f_H^\downarrow(x_0; g) = +1, \quad f^\uparrow(x_0) = \sup_{\|g\|=1} f_H^\uparrow(x_0; g) = +1.$$

Since $f^\downarrow(x_0) = +1 > 0$, then by Theorem 2.5.1 one concludes that the point $x_0 = 0$ is a strict local minimizer of the function $f$ (in fact, the point $x_0$ is a global minimizer, however, our theory allows to guarantee only that $x_0$ is at least a strict local minimizer).

Since the condition (2.126) does not hold, then $x_0$ is not a maximizer, and the directions $g^0 = +1$ and $\overline{g^0} = -1$ are both steepest ascent directions, and the rate of steepest ascent is $f^\uparrow(x_0; g^0) = f_H^\uparrow(x_0; \overline{g^0}) = +1$.

*Example 2.5.3.* Let $x = \mathbb{R} = (-\infty, \infty)$,

$$f(x) = \begin{cases} x \sin \frac{1}{x}, & x \neq 0, \\ 0, & x = 0. \end{cases}$$

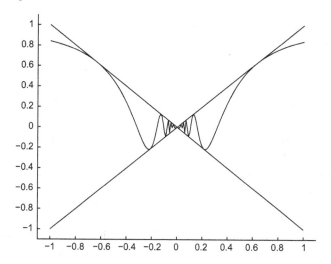

**Fig. 4** Example 2.5.3

The graph of the function $f$ is depicted in Fig. 4.

For $x \neq 0$ the function $f$ is differentiable and

$$f'(x) = \sin\frac{1}{x} - \frac{1}{x}\cos\frac{1}{x}.$$

Therefore (see (2.136), (2.137))

$$f^{\downarrow}(x) = -\left|\sin\frac{1}{x} - \frac{1}{x}\cos\frac{1}{x}\right|, \qquad f^{\uparrow}(x) = \left|\sin\frac{1}{x} - \frac{1}{x}\cos\frac{1}{x}\right|.$$

At the point $x_0 = 0$ one has

$$f^{\downarrow}(x_0) = \liminf_{x \to x_0}\frac{f(x) - f(x_0)}{\|x - x_0\|} = \liminf_{x \to 0}\frac{x\sin\frac{1}{x}}{|x|} = -1,$$

$$f^{\uparrow}(x_0) = \limsup_{x \to 0}\frac{x\sin\frac{1}{x}}{|x|} = +1.$$

Clearly, at the point $x_0 = 0$ neither the necessary condition for a minimum (2.15), nor the necessary condition for a maximum (2.17) holds.

It was shown in Remark 2.5.1 that in the case $\mathbb{R}^n = \mathbb{R}$ the Dini upper and lower derivatives coincide with the corresponding Hadamard derivatives. Therefore

$$f_H^{\downarrow}(x_0, g) = f_D^{\downarrow}(x_0, g) = \liminf_{\alpha \downarrow 0}\frac{f(x_0 + \alpha g) - f(x_0)}{\alpha} = \liminf_{\alpha \downarrow 0}\frac{\alpha g\sin\frac{1}{\alpha g}}{\alpha} = -|g|.$$

Clearly, the directions $g_0 = +1$ and $\overline{g}_0 = -1$ are steepest descent directions of the function $f$ at the point $x_0 = 0$.

Analogously, since

$$f_H^\uparrow(x_0, g) = f_D^\uparrow(x,g) = \limsup_{\alpha \downarrow 0} \frac{\alpha g \sin \frac{1}{\alpha g}}{\alpha} = |g|,$$

then the directions $g^0 = +1$ and $\overline{g}^0 = -1$ are steepest descent directions of the function $f$ at the point $x_0 = 0$.

Thus, the direction $g_0 = +1$ is both a steepest descent direction, and a steepest ascent direction.

The direction $\overline{g}_0 = -1$ is also simultaneously a steepest descent direction and a steepest ascent directions.

Note that the function $f(x)$ is not locally Lipschitz around the point $x_0 = 0$ (though it is locally Lipschitz in a neighborhood of any point $x \neq 0$).

**Exercise 2.5.4.** Study (as it was done in Example 2.5.3) the following functions:

$$f_1(x) = f_H^\uparrow(x_0, g) = \begin{cases} |x| \sin \frac{1}{x}, & x \neq 0, \\ 0, & x = 0, \end{cases}$$

$$f_2(x) = \begin{cases} |x \sin \frac{1}{x}|, & x \neq 0, \\ 0, & x = 0, \end{cases}$$

$$f_3(x) = \begin{cases} x \sin \frac{1}{x}, & x > 0, \\ 0, & x \leq 0, \end{cases}$$

$$f_4(x) = \begin{cases} 2x + x \sin \frac{1}{x}, & x \neq 0, \\ 0, & x = 0. \end{cases}$$

# 3 Constrained Optimization Problems via Exact Penalization

The constrained optimization problem is that of finding extremal values of a functional in the presence of constraints.

## 3.1  Optimization in a Metric Space in the Presence of Constraints

### 3.1.1  First-Order Conditions

Let $[X,\rho]$ be a metric space (i.e., on a set $X$ a metric $\rho$ is defined), $\Omega \subset X$ be a nonempty subset of this space. Assume that a functional $f : X \to \mathbb{R}$ is defined on $X$. It is required to find

$$\inf_{x \in \Omega} f(x) = f_\Omega^*. \tag{3.1}$$

This problem will be referred to as the *constrained optimization problem* (in the case $\Omega = X$ the problem (3.1) is called the *unconstrained optimization problem*). In the sequel we discuss, mainly, only the minimization problem since the problem of maximizing a function $f$ is equivalent to that of minimizing the function $f_1(x) = -f(x)$.

If one takes, instead of $X$, the set $\Omega$ with the same metric $\rho$, then the constrained minimization problem (3.1) becomes the problem of unconstrained minimization in the metric space $[\Omega, \rho]$. However, it is often more convenient and useful not to use such a reduction.

Points of global, strict global, local and strict local minima and maxima of a function $f$ on a set $\Omega$ are defined like in Section 2.1, only everywhere instead of $x \in X$ in the definitions there one should write $x \in \Omega$.

Using the same arguments as in the proofs of Theorems 2.1.1 and 2.1.2 we get the following extremality conditions.

**Theorem 3.1.1.** *For a point $x_* \in \Omega$ to be a global or local minimizer of a function $f$ on a set $\Omega$, it is necessary that*

$$f^\downarrow(x_*, \Omega) = \liminf_{\substack{y \in \Omega \\ y \to x_*}} \frac{f(y) - f(x_*)}{\rho(y, x_*)} \geq 0. \tag{3.2}$$

*If it turns out that $f^\downarrow(x_*, \Omega) > 0$, then the point $x_*$ is a strict local minimizer of the function $f$ on the set $\Omega$.*

**Theorem 3.1.2.** *For a point $x^* \in \Omega$ to be a global or local maximizer of a function $f$ on a set $\Omega$, it is necessary that*

$$f^\uparrow(x^*, \Omega) = \limsup_{\substack{y \in \Omega \\ y \to x^*}} \frac{f(y) - f(x^*)}{\rho(y, x^*)} \leq 0. \tag{3.3}$$

*It it turns out that $f^\uparrow(x^*, \Omega) < 0$, then the point $x^*$ is a strict local maximizer of the function $f$ on the set $\Omega$.*

A point $x_* \in \Omega$ satisfying (3.2) is called an *inf-stationary point* of the function $f$ on the set $\Omega$, while a point $x^* \in \Omega$ satisfying (3.3) is called a *sup-stationary point* of $f$ on $\Omega$.

*Assume that the set $\Omega$ is defined as*

$$\Omega = \{x \in X \mid \varphi(x) = 0\}, \tag{3.4}$$

*where*

$$\varphi : X \to \mathbb{R}, \quad \varphi(x) \geq 0 \quad \forall x \in X. \tag{3.5}$$

*Note that any set $\Omega \subset X$ can be written in the form (3.4), where $\varphi$ satisfies (3.5). For example, one can take*

$$\varphi(x) = \begin{cases} 0, & x \in \Omega, \\ 1, & x \neq \Omega. \end{cases} \tag{3.6}$$

*If the set $\Omega$ is closed then, as $\varphi$, we can take the function*

$$\varphi(x) = \rho(x, \Omega), \tag{3.7}$$

*with*

$$\rho(x, \Omega) = \inf_{y \in \Omega} \rho(x, y). \tag{3.8}$$

*The function $\rho(x, \Omega)$ is called the distance function from a point $x$ to the set $\Omega$.*

**Definition.** A sequence $\{x_k\}$ is called a *minimizing sequence* (m.s.) for a function $f$ on a set $\Omega$ if $x_k \in \Omega \quad \forall k, \quad f(x_k) \longrightarrow f_\Omega^* = \inf_{x \in \Omega} f(x)$.

A sequence $\{x_k\}$ is called a *generalized minimizing sequence* for a function $f$ on $\Omega$ (or g.m.s.) if

$$x_k \in X \ \forall k, \quad \rho(x_k, \Omega) = \inf_{y \in \Omega} \rho(x_k, y) \xrightarrow[k \to \infty]{} 0, \quad f(x_k) \xrightarrow[k \to \infty]{} f_\Omega^*. \tag{3.9}$$

If $\{x_k\}$ is a g.m.s. of a function $f$ on $\Omega$, and the function $f$ is uniformly continuous on $\Omega(\varepsilon) = \{x \in X \mid \rho(x, \Omega) < \varepsilon\}$ (for some $\varepsilon > 0$), then it is possible to construct a minimizing sequence. Indeed, choose a sequence $\{\varepsilon_k\}$ such that $\varepsilon_k \downarrow 0$. Without loss of generality one may assume that

$$\rho(x_k, \Omega) < \frac{\varepsilon}{2}, \quad \varepsilon_k < \frac{\varepsilon}{2} \quad \forall k,$$

i.e., $x_k \in \Omega(\varepsilon)$.

For any $x_k \in X$ let us find a point $y_k \in \Omega$ such that

$$\rho(x_k, y_k) < \rho(x_k, \Omega) + \varepsilon_k.$$

Then for the sequence $\{y_k\}$ one gets $y_k \in \Omega \ \forall k$,

$$f(y_k) = f(x_k) + [f(y_k) - f(x_k)] \xrightarrow[k \to \infty]{} f_\Omega^*,$$

since $f(y_k) - f(x_k) \longrightarrow 0$ due to the uniform continuity of $f$ on $\Omega(\varepsilon)$, $f(x_k) \to f^*_\Omega$ (see (3.9)).

In particular, if the function $f$ is Lipschitz on $\Omega(\varepsilon)$, then it is uniformly continuous there.

Let a set $\Omega$ be given in the form (3.4), where $\varphi$ satisfies (3.5). A sequence $\{x_k\}$ is called a $\varphi$-minimizing sequence for the function $f$ on the set $\Omega$ (or $\varphi$-m.s.), if

$$x_k \in X, \quad \varphi(x_k) \longrightarrow 0, \quad f(x_k) \longrightarrow f^*_\Omega.$$

## 3.2 The Constrained Optimization Problem in a Normed Space

Let $X$ be a normed space with the norm $\|x\|$, $\Omega \subset X$ be a nonempty set and let a functional $f : X \to \mathbb{R}$ be given.

The necessary conditions for a minimum and a maximum (Theorems 3.1.1 and 3.1.2) have the following form:

**Theorem 3.2.1.** For a point $x_* \in \Omega$ to be a global or local minimizer of the function $f$ on the set $\Omega$, it is necessary that

$$f^\downarrow(x_*, \Omega) = \liminf_{\substack{(x_* + \alpha g') \in \Omega \\ \alpha \downarrow 0, \ g' \in S}} \frac{f(x_* + \alpha g') - f(x_*)}{\alpha} \geq 0. \tag{3.10}$$

If

$$f^\downarrow(x_*, \Omega) > 0, \tag{3.11}$$

then the point $x_*$ is a strict local minimizer of the function $f$ on the set $\Omega$.

As usual, here $S = \{g \in X \mid \|g\| = 1\}$ is the unit sphere in $X$.

Let $x \in \Omega, g \in X$. The value

$$f^\downarrow_D(x, g; \Omega) = \liminf_{\substack{\alpha \downarrow 0 \\ x + \alpha g \in \Omega}} \frac{f(x + \alpha g) - f(x)}{\alpha} \tag{3.12}$$

is called the Dini conditional lower derivative of the function $f$ at the point $x \in \Omega$ in the direction $g$ with respect to the set $\Omega$.

If $x + \alpha g \notin \Omega$ for sufficiently small $\alpha > 0$ then put $f^\downarrow_D(x, g; \Omega) = +\infty$.

The value

$$f^\downarrow_H(x, g; \Omega) = \liminf_{\substack{[\alpha, g'] \to [+0, g] \\ x + \alpha g' \in \Omega}} \frac{f(x + \alpha g') - f(x)}{\alpha} \tag{3.13}$$

is called the Hadamard conditional lower derivative of the function $f$ at the point $x \in \Omega$ in the direction $g$ with respect to the set $\Omega$.

*If $x + \alpha g' \notin \Omega$ for sufficiently small $\alpha > 0$ and $g'$ sufficiently close to $g$, then put $f_H^\downarrow(x, g; \Omega) = +\infty$.*

*Note that*

$$f_H^\downarrow(x, g; \Omega) \leq f_D^\downarrow(x, g; \Omega). \tag{3.14}$$

*One has*

$$f^\downarrow(x, \Omega) \leq \inf_{g \in S} f_H^\downarrow(x, g; \Omega) \leq \inf_{g \in S} f_D^\downarrow(x, g; \Omega). \tag{3.15}$$

*If the set $S$ is compact (i.e., from any infinite sequence of points from $S$ it is possible to find a subsequence converging to a point of the set $S$) then*

$$f^\downarrow(x, \Omega) = \inf_{g \in S} f_H^\downarrow(x, g; \Omega). \tag{3.16}$$

*Theorem 3.2.1, the relations (3.15), (3.16) and positive homogeneity of the functions $h_1(g) = f_D^\downarrow(x, g; \Omega)$ and $h_2(g) = f_H^\downarrow(x, g; \Omega)$ imply*

**Theorem 3.2.2.** *For a point $x_* \in \Omega$ to be a global or local minimizer of the function $f$ on the set $\Omega$, it is necessary that*

$$f_D^\downarrow(x_*, g; \Omega) \geq 0 \qquad \forall g \in X, \tag{3.17}$$

$$f_H^\downarrow(x_*, g; \Omega) \geq 0 \qquad \forall g \in X. \tag{3.18}$$

*If the set $S$ is compact then the condition*

$$f_H^\downarrow(x_*, g; \Omega) > 0 \qquad \forall g \in X, \ g \neq 0 = 0_X \tag{3.19}$$

*is sufficient for a strict local minimum of the function $f$ on the set $\Omega$.*

*Here $0_X$ is the zero element of the space $X$.*

*Remark 3.2.1.* If the set $\Omega$ is convex, $x \in \Omega$, then

$$f^\downarrow(x, \Omega) = \liminf_{\substack{x + \alpha g \in \Omega \\ \|\alpha g\| \to 0}} \frac{f(x + \alpha g) - f(x)}{\alpha \|g\|}$$

$$= \liminf_{\substack{\alpha \|y - x\| \to 0 \\ y \in \Omega, y \neq x}} \frac{f(x + \alpha(y - x)) - f(x)}{\alpha \|y - x\|}$$

$$= \inf_{\substack{y \in \Omega \\ y \neq x}} \left( \frac{1}{\|y - x\|} f_H^\downarrow(x, y - x) \right) \leq \inf_{\substack{y \in \Omega \\ y \neq x}} \left( \frac{1}{\|y - x\|} f_D^\downarrow(x, y - x) \right).$$

(The definitions of $f_D^\downarrow(x, g)$ and $f_H^\downarrow(x, g)$ see in Section 2.2.) This implies that in the case of convexity of $\Omega$, the conditions (3.17) and (3.18) take the form

$$f_D^\downarrow(x_*, x - x_*) \geq 0 \qquad \forall x \in \Omega, \tag{3.20}$$

$$f_H^\downarrow(x_*, x - x_*) \geq 0 \quad \forall x \in \Omega. \tag{3.21}$$

Let $x \in \Omega, g \in X$. The value

$$f_D^\uparrow(x, g; \Omega) = \limsup_{\substack{\alpha \downarrow 0 \\ x + \alpha g \in \Omega}} \frac{f(x + \alpha g) - f(x)}{\alpha} \tag{3.22}$$

is called the *Dini conditional upper derivative* of the function $f$ at the point $x \in \Omega$ in the direction $g$ *with respect to the set* $\Omega$.

If $x + \alpha g \notin \Omega$ for sufficiently small $\alpha > 0$ then put $f_D^\uparrow(x, g; \Omega) = -\infty$.
The value

$$f_H^\uparrow(x, g; \Omega) = \limsup_{\substack{[\alpha, g'] \to [+0, g] \\ x + \alpha g' \in \Omega}} \frac{f(x + \alpha g') - f(x)}{\alpha} \tag{3.23}$$

is called the *Hadamard conditional upper derivative* of the function $f$ at the point $x \in \Omega$ in the direction $g$ *with respect to the set* $\Omega$.

If $x + \alpha g \notin \Omega$ for sufficiently small $\alpha > 0$ and $g'$ sufficiently close to $g$, then put $f_H^\uparrow(x, g; \Omega) = -\infty$.

**Theorem 3.2.3.** *For a point* $x^* \in \Omega$ *to be a global or local maximizer of the function* $f$ *on the set* $\Omega$ *it is necessary that*

$$f_D^\uparrow(x^*, g; \Omega) \leq 0 \quad \forall g \in X, \tag{3.24}$$

$$f_H^\uparrow(x^*, g; \Omega) \leq 0 \quad \forall g \in X. \tag{3.25}$$

*If the set* $S$ *is compact then the condition*

$$f_H^\uparrow(x^*, g; \Omega) < 0 \quad \forall g \neq 0 = 0_X \tag{3.26}$$

*is sufficient for* $x^*$ *to be a strict local maximizer of the function* $f$ *on the set* $\Omega$.

*Remark 3.2.2.* If the set $\Omega$ is convex then the conditions (3.24) and (3.25) are, respectively, of the form

$$f_D^\uparrow(x^*, x - x^*) \leq 0 \quad \forall x \in \Omega, \tag{3.27}$$

and

$$f_H^\uparrow(x^*, x - x^*) \leq 0 \quad \forall x \in \Omega. \tag{3.28}$$

Definition $f_D^\uparrow(x, g)$, $f_H^\uparrow(x, g)$ see in Section 2.2.

## 3.3  Penalty Functions

### 3.3.1  Definition and Properties of Penalty Functions

Let $[X,\rho]$ be a metric space, $\Omega \subset X$ be a nonempty set of this space. Let a functional $f : X \to \mathbb{R}$ be given. Assume that the set $\Omega$ is defined in the form

$$\Omega = \{x \in X \mid \varphi(x) = 0\}, \tag{3.29}$$

where

$$\varphi : X \to \mathbb{R}, \qquad \varphi(x) \geq 0 \quad \forall x \in X. \tag{3.30}$$

Fix $\lambda \geq 0$ and introduce the function

$$F_\lambda(x) = f(x) + \lambda\varphi(x). \tag{3.31}$$

The function $F_\lambda(x)$ is called a *penalty function* (for the given $f$ and $\varphi$), the number $\lambda$ will be referred to as a *penalty parameter*. Put

$$F_\lambda^* = \inf_{x \in X} F_\lambda(x). \tag{3.32}$$

One has

$$F_\lambda^* := \inf_{x \in X} F_\lambda(x) \leq \inf_{x \in \Omega} F_\lambda(x) = \inf_{x \in \Omega} f(x) =: f_\Omega^* \quad \forall \lambda \geq 0. \tag{3.33}$$

First consider the case where for any $\lambda \geq \lambda_0$ (with $\lambda_0 \geq 0$) there exists a point $x_\lambda \in X$ such that

$$F_\lambda(x_\lambda) = F_\lambda^* \tag{3.34}$$

(i.e., $x_\lambda$ is a minimizer of the function $F_\lambda(x)$ on $X$).

Let us indicate some useful properties of the function

$$h(\lambda) := \inf_{x \in X} F_\lambda(x) = F_\lambda^*. \tag{3.35}$$

1. *The function $h(\lambda)$ is nondecreasing, that is if $\lambda_2 > \lambda_1$ then $h(\lambda_2) \geq h(\lambda_1)$.*
   Indeed, for any $x \in X$ and $\lambda_2 > \lambda_1$ one has

$$F_{\lambda_2}(x) = f(x) + \lambda_2\varphi(x) = f(x) + \lambda_1\varphi(x) + (\lambda_2 - \lambda_1)\varphi(x)$$

$$= F_{\lambda_1}(x) + (\lambda_2 - \lambda_1)\varphi(x) \geq F_{\lambda_1}(x).$$

Therefore

$$h(\lambda_2) = \inf_{x \in X} F_{\lambda_2}(x) \geq \inf_{x \in X} F_{\lambda_1}(x) = h(\lambda_1).$$

2. *If* $\lambda_2 > \lambda_1$, *then*

$$f(x_{\lambda_2}) \geq f(x_{\lambda_1}),\tag{3.36}$$

$$\varphi(x_{\lambda_2}) \leq \varphi(x_{\lambda_1}).\tag{3.37}$$

Indeed, the definitions $x_{\lambda_2}$ and $x_{\lambda_1}$ imply

$$F_{\lambda_1}(x_{\lambda_1}) = f(x_{\lambda_1}) + \lambda_1\varphi(x_{\lambda_1}) \leq F_{\lambda_1}(x_{\lambda_2}) = f(x_{\lambda_2}) + \lambda_1\varphi(x_{\lambda_2}),\tag{3.38}$$

$$F_{\lambda_2}(x_{\lambda_2}) = f(x_{\lambda_2}) + \lambda_2\varphi(x_{\lambda_2}) \leq F_{\lambda_2}(x_{\lambda_1}) = f(x_{\lambda_1}) + \lambda_2\varphi(x_{\lambda_1}).\tag{3.39}$$

The relation (3.39) yields

$$-f(x_{\lambda_1}) - \lambda_2\varphi(x_{\lambda_1}) \leq -f(x_{\lambda_2}) - \lambda_2\varphi(x_{\lambda_2}).\tag{3.40}$$

Now from (3.38)

$$\left(1 - \frac{\lambda_1}{\lambda_2}\right)f(x_{\lambda_1}) \leq \left(1 - \frac{\lambda_1}{\lambda_2}\right)f(x_{\lambda_2}).\tag{3.41}$$

Since $\lambda_1 < \lambda_2$ then (3.41) implies $f(x_{\lambda_1}) \leq f(x_{\lambda_2})$. Using this inequality and (3.39), we get $\varphi(x_{\lambda_2}) \leq \varphi(x_{\lambda_1})$.

3. *If* $\lambda_2 > \lambda_1 > 0$ *and* $f(x_{\lambda_2}) = f(x_{\lambda_1})$ *then*

$$\varphi(x_{\lambda_2}) = \varphi(x_{\lambda_1}),\tag{3.42}$$

*and, conversely, if* $\lambda_2 > \lambda_1 > 0$ *and* $\varphi(x_{\lambda_2}) = \varphi(x_{\lambda_1})$, *then*

$$f(x_{\lambda_2}) = f(x_{\lambda_1}).\tag{3.43}$$

In fact, if $f(x_{\lambda_2}) = f(x_{\lambda_1})$, then (3.38) yields $\varphi(x_{\lambda_1}) \leq \varphi(x_{\lambda_2})$, and (3.39) $-\varphi(x_{\lambda_2}) \leq \varphi(x_{\lambda_1})$. This inequality implies (3.42).

If $\varphi(x_{\lambda_2}) = \varphi(x_{\lambda_1})$, then (3.38) yields $f(x_{\lambda_1}) \leq f(x_{\lambda_2})$, and (3.39) $- f(x_{\lambda_2}) \leq f(x_{\lambda_1})$, i.e., (3.43) holds.

4. *If for some* $\lambda_1, \lambda_2 \in R_+ = \{x \in \mathbb{R} \mid x \geq 0\}$, *such that* $\lambda_2 > \lambda_1$, *it turns out that*

$$h(\lambda_2) = h(\lambda_1),\tag{3.44}$$

*then*

$$\varphi(x_{\lambda_2}) = 0.\tag{3.45}$$

Indeed, assume the opposite, i.e. let $\varphi(x_{\lambda_2}) > 0$. Then

$$h(\lambda_2) = F_{\lambda_2}^* = F_{\lambda_2}(x_{\lambda_2}) = f(x_{\lambda_2}) + \lambda_2\varphi(x_{\lambda_2})$$
$$= F_{\lambda_1}(x_{\lambda_2}) + (\lambda_2 - \lambda_1)\varphi(x_{\lambda_2}) > F_{\lambda_1}(x_{\lambda_2}).$$

The last inequality holds since $\lambda_2 > \lambda_1$, $\varphi(x_{\lambda_2}) > 0$. Hence,

$$h(\lambda_2) > F_{\lambda_1}(x_{\lambda_2}) \geq \inf_{x \in X} F_{\lambda_1}(x) = F_{\lambda_1}^* = h(\lambda_1),$$

that is, $h(\lambda_2) > h(\lambda_1)$, which contradicts (3.44).

*Remark 3.3.1.* For the point $x_{\lambda_1}$, it may be $\varphi(x_{\lambda_1}) > 0$ (see Example 3.7.1 below).

5. *If for some $\lambda_1$, $\lambda_2 \geq 0$, where $\lambda_2 > \lambda_1$, one has*

$$h(\lambda_2) = h(\lambda_1), \tag{3.46}$$

*then*

$$h(\lambda) = h(\lambda_1) \quad \forall \lambda > \lambda_1. \tag{3.47}$$

Indeed, if $\lambda_2 > \lambda > \lambda_1$, then (3.35) yields

$$h(\lambda_2) \geq h(\lambda) \geq h(\lambda_1),$$

and the relation (3.47) follows from (3.46).

If $\lambda > \lambda_2$, then (3.45) implies

$$h(\lambda) \geq h(\lambda_2) = F_{\lambda_2}^* = F_{\lambda_2}(x_{\lambda_2}) = f(x_{\lambda_2}) + \lambda_2 \varphi(x_{\lambda_2}) = f(x_{\lambda_2}). \tag{3.48}$$

However,

$$h(\lambda) = F_\lambda^* = \inf_{x \in X} F_\lambda(x) \leq F_\lambda(x_{\lambda_2}) = f(x_{\lambda_2}) = F_{\lambda_2}(x_{\lambda_2}) = h(\lambda_2).$$

Therefore, the relations (3.46) and (3.48) yield

$$h(\lambda) = h(\lambda_2) = h(\lambda_1).$$

6. *If, for some $\lambda \geq 0$, $\varphi(x_\lambda) = 0$ (i.e. $x_\lambda \in \Omega$) and*

$$\inf_{x \in X} F_\lambda(x) = F_\lambda(x_\lambda) = f(x_\lambda),$$

*then the point $x_\lambda$ is a minimizer of the function $f$ on $\Omega$:*

$$f(x_\lambda) = \inf_{x \in \Omega} f(x) = f_\Omega^*.$$

Indeed, assume the contrary. Then an $\overline{x} \in \Omega$ exists such that $f(\overline{x}) < f(x_\lambda)$. Since $\varphi(\overline{x}) = 0$, one has

$$F_\lambda(\overline{x}) = f(\overline{x}) < f(x_\lambda) = F_\lambda(x_\lambda),$$

which contradicts the fact that the point $x_\lambda$ is a minimizer of the function $F_\lambda(x)$ on $X$.

**Corollary 3.3.1.** *If $\varphi(x_{\lambda_0}) = 0$ for some $\lambda_0 \geq 0$ then the point $x_{\lambda_0}$ is a minimizer of the function $F_\lambda(x)$ on $X$ for all $\lambda > \lambda_0$.*

**Corollary 3.3.2.** *If $F_{\lambda_1}^* = F_{\lambda_2}^*$, where $\lambda_2 > \lambda_1 \geq 0$, then the point $x_{\lambda_2}$ is a minimizer of the function $f$ on $\Omega$.*

*Indeed, (3.45) yields $\varphi(x_{\lambda_2}) = 0$, i.e. $x_{\lambda_2} \in \Omega$, and then Property 6 implies that $x_{\lambda_2}$ is a minimizer of $f$ on $\Omega$. Note again that the point $x_{\lambda_1}$ does not necessarily belong to the set $\Omega$ and, hence, is not a minimizer of the function $f$ on $\Omega$ (see Remark 3.3.1).*

7. If

$$F_{\lambda_1}(x_{\lambda_1}) =: F_{\lambda_1}^* = F_{\lambda_2}^* := F_{\lambda_2}(x_{\lambda_2}),$$

*where $\lambda_2 > \lambda_1 \geq 0$, then the point $x_{\lambda_2}$ is a minimizer of the function $F_{\lambda_1}(x)$ on $X$ (the point $x_{\lambda_1}$ may not belong to the set $\Omega$ while the point $x_{\lambda_2}$ belongs to the set $\Omega$). In other words, the set of (unconstrained) minimizers of the function $F_{\lambda_1}(x)$ contains a point belonging to the set $\Omega$.*

*Really,*

$$F_{\lambda_1}^* = F_{\lambda_2}^* = F_{\lambda_2}(x_{\lambda_2}) = f(x_{\lambda_2}) + \lambda_2 \varphi(x_{\lambda_2}). \tag{3.49}$$

*But $\varphi(x_{\lambda_2}) = 0$, therefore*

$$\begin{aligned} F_{\lambda_1}(x_{\lambda_2}) &= f(x_{\lambda_2}) + \lambda_1 \varphi(x_{\lambda_2}) \\ &= f(x_{\lambda_2}) + \lambda_2 \varphi(x_{\lambda_2}) = F_{\lambda_2}(x_{\lambda_2}) = F_{\lambda_2}^*. \end{aligned} \tag{3.50}$$

*The relations (3.48) and (3.49) yield*

$$F_{\lambda_1}^* = \inf_{x \in X} F_{\lambda_1}(x) \leq F_{\lambda_1}(x_{\lambda_2}) = F_{\lambda_2}^* = F_{\lambda_1}^*. \tag{3.51}$$

*Hence, $F_{\lambda_1}^* = F_{\lambda_1}(x_{\lambda_2})$, which means that $x_{\lambda_2}$ is a minimizer of $F_{\lambda_1}(x)$ on $X$.*

*Put*

$$\mathcal{M}_\lambda = \{x \in X \mid F_\lambda(x) = F_\lambda^* = h(\lambda)\}. \tag{3.52}$$

*If $h(\lambda_2) = h(\lambda_1)$ for some $\lambda_1, \lambda_2 \in \mathbb{R}_+$ such that $\lambda_2 > \lambda_1$, then*

$$\mathcal{M}_\lambda = \mathcal{M}_{\lambda_2} \quad \forall \lambda \in (\lambda_1, \infty), \tag{3.53}$$

$$\mathcal{M}_\lambda \subset \Omega, \quad \mathcal{M}_\lambda \subset \mathcal{M}_{\lambda_1} \quad \forall \lambda \in (\lambda_1, \infty). \tag{3.54}$$

*Furthermore,*

$$\mathcal{M}_\lambda = \mathcal{M}_\Omega^* = \{x \in \Omega \mid f(x) = f_\Omega^*\}. \tag{3.55}$$

**Lemma 3.3.1.** *If*

$$\inf_{x \in X} f(x) = f^* > -\infty, \tag{3.56}$$

*then*

$$\varphi(x_\lambda) \underset{\lambda \to \infty}{\longrightarrow} 0. \tag{3.57}$$

*Proof.* Assume that (3.57) is not valid. Then there exist a sequence $\{\lambda_k\}$ and a number $a > 0$, such that $\lambda_k \to \infty, \varphi(x_{\lambda_k}) \geq a$. We have

$$F_{\lambda_k}(x_{\lambda_k}) = f(x_{\lambda_k}) + \lambda_k \varphi(x_{\lambda_k}) \geq f^* + \lambda_k a \underset{k \to \infty}{\longrightarrow} +\infty. \tag{3.58}$$

By the assumption, $\Omega \neq \emptyset$, therefore there exists $x_0 \in \Omega$. Since $F_{\lambda_k}(x_0) = f(x_0)$, then

$$F_{\lambda_k}(x_{\lambda_k}) \leq f(x_0), \tag{3.59}$$

which contradicts (3.58).           $\square$

Put

$$\Omega_\delta = \{x \in X \mid \varphi(x) < \delta\}, \quad r_\delta = \sup_{y \in \Omega_\delta} \rho(y, \Omega).$$

**Lemma 3.3.2.** *Let*

$$\inf_{x \in X} f(x) = f^* > -\infty, \tag{3.60}$$

$$r_\delta \underset{\delta \downarrow 0}{\to} 0, \tag{3.61}$$

*and the function $f$ be uniformly continuous on $\Omega_{\delta_0}$ for some $\delta_0 > 0$. Then*

$$\lim_{\lambda \to \infty} F_\lambda^* = f_\Omega^*. \tag{3.62}$$

*Proof.* Assume the contrary. Then there exist a number $a > 0$ and sequences $\{\lambda_k\}$ and $\{x_k\}$ such that $\lambda_k \longrightarrow \infty, x_k \in X$,

$$f(x_k) + \lambda_k \varphi(x_k) \leq f_\Omega^* - a. \tag{3.63}$$

It follows from (3.61) that

$$\varphi(x_k) \to 0. \tag{3.64}$$

Choose a sequence $\{\varepsilon_k\}$ such that $\varepsilon_k \downarrow 0$. For every $x_k$, find a $y_k \in \Omega$ such that

$$\rho(y_k, x_k) \leq \rho(x_k, \Omega) + \varepsilon_k \leq r_{\delta_k} + \varepsilon_k, \tag{3.65}$$

where $\delta_k = \varphi(x_k) + \varepsilon_k$. The relation (3.64) means that $\delta_k \to 0$, and then (3.61) yields $r_{\delta_k} \longrightarrow 0$. Then from (3.64) and (3.65) one concludes that $\rho(y_k, x_k) \longrightarrow 0$. Now it follows from (3.63) and the uniform continuity of $f$ on $\Omega_{\delta_0}$ that, for $k$ sufficiently large,

$$f(y_k) = f(x_k) + [f(y_k) - f(x_k)] < f_\Omega^* - \frac{a}{2},$$

which contradicts the definition $f_\Omega^*$, since $y_k \in \Omega$. □

Thus, in the hypotheses of Lemma 3.3.2., for $\lambda \to \infty$ the values of unconstrained infima of the penalty functions $F_\lambda$ tend to the infimal value of the function $f$ on the set $\Omega$. Under quite natural assumptions, the unconstrained minimizers of the functions $F_\lambda$ converge (as $\lambda \to \infty$) to a minimizer of the function $f$ on the set $\Omega$. It turns out that, under rather mild assumptions, for sufficiently large $\lambda$'s, an unconstrained minimizer of the function $F_\lambda$ coincides with a minimizer of the function $f$ on $\Omega$. This case will be discussed in the sequel.

*Remark 3.3.1.* In the hypotheses of Lemma 3.3.2, it is clear from Lemmas 3.3.1 and 3.3.2 how to construct a $\varphi$-minimizing sequence for the function $f$ on $\Omega$. Choose sequences $\{\lambda_k\}$ and $\{\varepsilon_k\}$ such that $\lambda_k \to \infty$, $\varepsilon_k \downarrow 0$. For every $\lambda_k$ let us find $x_k$ such that $F_{\lambda_k}(x_k) \leq F_{\lambda_k}^* + \varepsilon_k$. Due to (3.57), $\varphi(x_k) \longrightarrow 0$. Now (3.62) yields

$$f(x_k) + \lambda_k \varphi(x_k) \longrightarrow f_\Omega^*. \tag{3.66}$$

Note that $f(x_k) \longrightarrow f_\Omega^*$. Indeed, assuming the contrary and arguing as in the proof of the Lemma, one easily gets a contradiction.

Thus, $\varphi(x_k) \longrightarrow 0$, $f(x_k) \longrightarrow f_\Omega^*$, i.e. the sequence $\{x_k\}$ is $\varphi$-minimizing.

## 3.3.2 Indicator Functions

The function

$$\varphi(x) = \begin{cases} +\infty, & x \notin \Omega, \\ 0, & x \in \Omega, \end{cases} \tag{3.67}$$

is called the *indicator function of the set* $\Omega$.

If the condition (3.56) holds and $\Omega \neq \emptyset$, then it is easy to see that for any $\lambda > 0$

$$\inf_{x \in X} F_\lambda(x) = \inf_{x \in \Omega} f(x). \tag{3.68}$$

The function $\varphi(x)$, defined by (3.67), and the function $\varphi(x)$, given by (3.6), defined the same set $\Omega$. (The function (3.6) can also be considered as an indicator function of the set $\Omega$.)

Thus, let $\Omega$ be defined by (3.4), with $\varphi$ described by (3.6). Let $F_\lambda(x) = f(x) + \lambda \varphi(x)$ and let the condition (3.56) be valid.

**Lemma 3.3.3.** *For sufficiently large* $\lambda > 0$

$$\inf_{x \in X} F_\lambda(x) = \inf_{x \in \Omega} f(x). \tag{3.69}$$

*Proof.* Take $\lambda_0 = f_\Omega^* - f^* + a$, where $a > 0$.

Let $\{x_k\}$ be a minimizing sequence for the function $F_{\lambda_0}(x)$, i.e.

$$\lim_{k\to\infty} F_{\lambda_0}(x_k) = \inf_{x\in X} F_{\lambda_0}(x). \tag{3.70}$$

Let us show that $\varphi(x_k) = 0$ for sufficiently large $k$'s. Assume the opposite. Then a subsequence $\{x_{k_s}\}$ exists such that $\varphi(x_{k_s}) = 1$ (remind that $\varphi$ is given by the formula (3.6) and therefore, if $\varphi(x_{k_s}) \neq 0$, then $\varphi(x_{k_s}) = 1$). Since $\{x_{k_s}\}$ is a subsequence of the sequence $\{x_k\}$, then (see (3.70))

$$F_{\lambda_0}(x_{k_s}) \underset{k_s\to\infty}{\longrightarrow} F_{\lambda_0}^*. \tag{3.71}$$

Hence,

$$F_{\lambda_0}(x_{k_s}) = f(x_{k_s}) + \lambda_0 \geq f^* + \lambda_0. \tag{3.72}$$

On the other hand, for any $\overline{x} \in \Omega$,

$$F_{\lambda_0}(\overline{x}) = f(\overline{x}). \tag{3.73}$$

The relations (3.73), (3.71) and (3.72) imply

$$f^* + \lambda_0 \leq F_{\lambda_0}^* \leq f(\overline{x}) \quad \forall \overline{x} \in \Omega. \tag{3.74}$$

If $\{\overline{x}_k\}$ is a minimizing sequence for the function $f$ on the set $\Omega$, i.e.

$$\overline{x}_k \in \Omega \quad \forall k, \quad f(\overline{x}_k) \underset{k\to\infty}{\longrightarrow} f_\Omega^*, \tag{3.75}$$

then (3.74) and (3.75) yield

$$f^* + \lambda_0 \leq f_\Omega^*. \tag{3.76}$$

But for $\lambda_0 > f_\Omega^* - f^*$ the inequality (3.76) is impossible (we have chosen $\lambda_0 = f_\Omega^* - f^* + a$, where $a > 0$).

Hence, (3.69) holds for an arbitrary $\lambda > \lambda^* = f_\Omega^* - f^*$.  □

*Remark 3.3.2.* Thus, if the indicator function (3.67) or the function (3.6) is taken as the function $\varphi$ in (3.4), then there exists $\lambda^* < \infty$ (for the function (3.67) $\lambda^* = 0$, and for the function (3.6) $\lambda^* = f_\Omega^* - f^*$) such that for $\lambda > \lambda^*$ the problem of minimizing the function $f(x)$ on the set $\Omega$ is equivalent to the problem of minimizing the function $F_\lambda(x)$ on the entire space $X$, i.e. the constrained optimization problem is reduced to the unconstrained optimization one. Unfortunately, usually the function $\varphi$ (described by (3.6) or (3.67)) is not known, and, besides, both these functions are essentially discontinuous. Therefore in practice the both mentioned functions are of a little value.

Nevertheless, the above described procedure of reducing the constrained minimization problem to an unconstrained one can be applied to other functions $\varphi$ describing the set $\Omega$.

## 3.4  Exact Penalty Functions and a Global Minimum

### 3.4.1  Global Minimizers

Let $[X, \rho]$ be a metric space, $\Omega \subset X$ be a set described by (3.4), where $\varphi$ satisfies (3.5). For $\lambda \geq 0$ consider the function $F_\lambda(x)$, defined by (3.31). Put (see (2.2))

$$\varphi^{\downarrow}(x) = \liminf_{y \to x} \frac{\varphi(y) - \varphi(x)}{\rho(x, y)}. \tag{3.77}$$

**Theorem 3.4.1.** *Let(3.56) be satisfied and the following conditions hold:*
*1) there exists a $\lambda_0 < \infty$ such that for any $\lambda \geq \lambda_0$ one can find an $x_\lambda \in X$, such that*

$$F_\lambda(x_\lambda) = F_\lambda^* = \inf_{x \in X} F_\lambda(x); \tag{3.78}$$

*2) there exist $\delta > 0$ and $a > 0$ such that*

$$\varphi^{\downarrow}(x) \leq -a < 0 \quad \forall x \in \Omega_\delta \setminus \Omega, \tag{3.79}$$

*where*

$$\Omega_\delta = \{x \in X \mid \varphi(x) < \delta\}, \tag{3.80}$$

*3) the function $f$ is Lipschitz on $\Omega_\delta \setminus \Omega$, i.e., for some $L < \infty$*

$$|f(x_1) - f(x_2)| \leq L\rho(x_1, x_2) \quad \forall x_1, x_2 \in \Omega_\delta \setminus \Omega. \tag{3.81}$$

*Then a $\lambda^* \geq \lambda_0$ exists such that for all $\lambda > \lambda^*$*

$$\varphi(x_\lambda) = 0, \quad f(x_\lambda) = f_\Omega^* = \inf_{x \in \Omega} f(x),$$

*that is, the point $x_\lambda$ is a solution of the problem (3.1).*

*Proof.* First let us show that there exists $\lambda^* \geq \lambda_0$ such that for all $\lambda > \lambda^*$

$$\varphi(x_\lambda) = 0.$$

Assume the contrary. Then one can find a sequence $\{\lambda_s\}$ such that

$$\lambda_s \xrightarrow[s\to\infty]{} \infty; \qquad \varphi(\lambda_s) > 0 \quad \forall s.$$

By Lemma 3.3.1, $\varphi(x_\lambda) \xrightarrow[\lambda\to\infty]{} 0$.

It means that a $\overline{\lambda} \geq \lambda_0$ exists such that $x_\lambda \in \Omega_\delta \setminus \Omega$ for $\lambda \geq \overline{\lambda}$. Fix any $\lambda_s > \overline{\lambda}$ (from our sequence). Due to (3.79), one can find a sequence $\{y_{\lambda_s k}\}$ such that $y_{\lambda_s k} \in X$,

$$\varphi^{\downarrow}(x_{\lambda_s}) = \lim_{k\to\infty} \frac{\varphi(y_{\lambda_s k}) - \varphi(x_{\lambda_s})}{\rho(y_{\lambda_s k}, x_{\lambda_s})} \leq -a, \quad \rho(y_{\lambda_s k}, x_{\lambda_s}) \xrightarrow[k\to\infty]{} 0.$$

Without loss of generality, one may assume that

$$\varphi(y_{\lambda_s k}) - \varphi(x_{\lambda_s}) \leq -\frac{a}{2}\rho(y_{\lambda_s k}, x_{\lambda_s}) \quad \forall k. \tag{3.82}$$

Fix some $k_s$ satisfying (3.82). Then (3.81) and (3.82) yield

$$F_{\lambda_s}(y_{\lambda_s k_s}) - F_{\lambda_s}(x_{\lambda_s}) = f(y_{\lambda_s k_s}) - f(x_{\lambda_s}) + \lambda_s(\varphi(y_{\lambda_s k_s}) - \varphi(x_{\lambda_s}))$$

$$\leq L\rho(y_{\lambda_s k_s}, x_{\lambda_s}) - \frac{a}{2}\lambda_s \rho(x_{\lambda_s}, y_{\lambda_s k_s})$$

$$= \rho(y_{\lambda_s k_s}, x_{\lambda_s})[L - \frac{a}{2}\lambda_s].$$

For

$$\lambda_s > \max\left\{\overline{\lambda}, \frac{2L}{a}\right\} = \lambda^* \tag{3.83}$$

we get $F_{\lambda_s}(y_{\lambda_s k_s}) < F_{\lambda_s}(x_{\lambda_s})$, which contradicts (3.78). Hence, $\varphi(x_\lambda) = 0$ for $\lambda > \lambda^*$. It remains to show that

$$f(x_\lambda) = f_\Omega^* = \inf_{x\in\Omega} f(x).$$

Assume the opposite. Then one can find an $\overline{x} \in \Omega$ such that $f(\overline{x}) < f(x_\lambda)$. We have

$$F_\lambda(\overline{x}) = f(\overline{x}) + \lambda\varphi(\overline{x}) = f(\overline{x}) < f(x_\lambda) = f(x_\lambda) + \lambda\varphi(x_\lambda) = F_\lambda(x_\lambda),$$

which contradicts (3.78). □

**Definition.** A number $\lambda^* \geq 0$ is called an *exact penalty constant* (for the function $f$ on the set $\Omega$) if

$$F_{\lambda^*}^* := \inf_{x\in X} F_{\lambda^*}(x) = \inf_{x\in\Omega} f(x) =: f_\Omega^*. \tag{3.84}$$

It is easy to see that if (3.84) holds then

$$F_\lambda^* := \inf_{x \in X} F_\lambda(x) = \inf_{x \in \Omega} f(x) =: f_\Omega^* \quad \forall \lambda > \lambda^*. \tag{3.85}$$

The lower bounds in (3.84) and (3.85) should not be necessarily attainable.

*Remark 3.4.1.* Thus, if, for a given function $f$ and a given set $\Omega \subset X$, there exists an exact penalty constant $\lambda^*$, then for $\lambda > \lambda^*$ the constrained optimization problem (3.1) is equivalent to the unconstrained optimization problem of minimizing $F_\lambda(x)$ on the entire space. The exact penalty constant is defined by the formula (3.83) where the value $\bar{\lambda}$ is not known in advance. However, if in (3.79) $\delta = +\infty$ (i.e., (3.79) holds for all $x \in X \setminus \Omega$), then $\lambda^* = \frac{2L}{a}$.

*Remark 3.4.2.* the function $\varphi$, defined by (3.6), does not satisfy the condition (3.79). Indeed,

$$\Omega_\delta = \begin{cases} \Omega, & \delta \in [0,1), \\ X, & x \geq 1. \end{cases}$$

For $\delta > 1$ and $x \in \Omega_\delta \setminus \Omega = X \setminus \Omega$, one has $\varphi^\downarrow(x) = 0$. Nevertheless, as it was shown in Lemma 3.3.3, for $\lambda > f_\Omega^* - f^*$ the function $F_\lambda(x)$ is an exact penalty function. Remind that the conditions of Theorem 3.4.1 are only sufficient.

*Remark 3.4.3.* Since

$$F_\lambda(x_\lambda) = \inf_{x \in X} F_\lambda(x) \leq \inf_{x \in \Omega} F_\lambda(x) = \inf_{x \in \Omega} f(x) = f_\Omega^*,$$

then (3.57) implies that, for any $\gamma > 0$ and for sufficiently large $\lambda$, one has $\rho(x_\lambda, M^*) < \gamma$, where

$$M^* = \arg\min_{x \in \Omega} f(x), \quad \rho(x, M) = \inf_{y \in M} \rho(x, y).$$

Therefore the conditions of Theorem 3.4.1 can be weakened:

**Theorem 3.4.2.** *Let (3.56) hold. Assume that*
*1) there exists a $\lambda_0 < \infty$ such that, for every $\lambda \geq \lambda_0$, one can find $x_\lambda \in X$ such that*

$$F_\lambda(x_\lambda) = F_\lambda^* = \inf_{z \in X} F_\lambda(x);$$

*2) there exist $\delta > 0$, $a > 0$ and $\gamma > 0$ such that*

$$\varphi^\downarrow(x) \leq -a \quad \forall x \in B_\gamma(M^*) \cap (\Omega_\delta \setminus \Omega),$$

*where*

$$B_\gamma(M^*) = \{x \in X | \rho(x, M^*) < \gamma\}, \quad \Omega_\delta = \{x \in X | \varphi(x) \leq \delta\};$$

3) the function $f$ is Lipschitz on $T = B_\gamma(M^*) \cap (\Omega_\delta \setminus \Omega)$, i.e., for some $L < \infty$,

$$|f(x_1) - f(x_2)| \le L\rho(x_1, x_2) \quad \forall x_1, x_2 \in T.$$

Then there exists $\lambda^* \ge \lambda^0$ such that

$$\varphi(x_\lambda) = 0 \quad \forall \lambda > \lambda^*.$$

Remark 3.4.4. In condition 3) of the Theorem it is enough to assume the Lipschitzness of the function $f$ on the set $C = (\Omega_\delta \setminus \Omega) \cap D(x_0)$ where

$$D(x_0) = \{x \in X | f(x) \le f(x_0)\}, \quad f(x_0) > f_\Omega^*.$$

Indeed, (3.33) implies

$$f(x_\lambda) \le F_\lambda(x_\lambda) = F_\lambda^* \le f_\Omega^* < f(x_0), \tag{3.86}$$

i.e., $x_\lambda \in \text{int } C$. Now in the proof of Theorem 3.4.1 one may assume that $y_{\lambda_s k_s} \in C \quad \forall s$.


## 3.5  Exact Penalty Functions and Local Minima

### 3.5.1  Local Minimizers

**Definition.** A local minimizer of the function $F_\lambda$ in the absence of constraints is called an *unconstrained local minimizer of the function $F_\lambda$*.

**Theorem 3.5.1.** *In the assumptions of Theorem 3.4.1, there exists a $\lambda^* < \infty$, such that, for $\lambda > \lambda^*$, all local minimizers of the function $F_\lambda(x)$, belonging to the set $\Omega_\delta$, are also local minimizers of the function $f$ on $\Omega$.*

*Proof.* First of all, let us show that a $\lambda^* < \infty$ exists such that, for every $\lambda > \lambda^*$, the following property holds: if $x_\lambda \in \Omega_\delta$ and $x_\lambda$ is a local minimizer of the function $F_\lambda(x)$ then $\varphi(x_\lambda) = 0$ (i.e., $x_\lambda \in \Omega$).

One has $F_\lambda(x_\lambda) = f(x_\lambda) + \lambda\varphi(x_\lambda)$. If $\varphi(x_\lambda) > 0$ (that is, $x_\lambda \notin \Omega$) then, due to (3.79), there exists a sequence of points $\{y_{\lambda k}\}$ (depending, of course, on $\lambda$) such that

$$\rho(y_{\lambda k}, x_\lambda) \xrightarrow[k \to \infty]{} 0, \quad \frac{\varphi(y_{\lambda k}) - \varphi(x_\lambda)}{\rho(y_{\lambda k}, x_\lambda)} \le -\frac{a}{2}. \tag{3.87}$$

It follows from (3.81) and (3.87) that

$$F_\lambda(y_{\lambda k}) - F_\lambda(x_\lambda) = f(y_\lambda) - f(x_\lambda) + \lambda(\varphi(y_k) - \varphi(x_\lambda))$$
$$\le L\rho(x_\lambda, y_{\lambda k}) - \frac{a}{2}\lambda\rho(x_\lambda, y_{\lambda k}) = \rho(x_\lambda, y_{\lambda k})\left[L - \frac{a}{2}\lambda\right].$$

As in the proof of Theorem 3.4.1, for $\lambda > \lambda^* = \max\{\lambda_1, \frac{2L}{a}\}$ (see (3.83)) one gets $F_\lambda(y_{\lambda k}) < F_\lambda(x_\lambda)$, which contradicts the assumption that $x_\lambda$ is a local minimizer of the function $F_\lambda(x)$.

Thus, for $\lambda > \lambda^*$, one has $\varphi(x_\lambda) = 0$.

Now let us show that for $\lambda > \lambda^*$ the point $x_\lambda$ is also a local minimizer of the function $f$ on the set $\Omega$.

Indeed, since $x_\lambda$ is a local minimizer of the function $F_\lambda$ then there exists an $\varepsilon > 0$ such that

$$F_\lambda(x_\lambda) \leq F_\lambda(x) \quad \forall x \in X : \rho(x, x_\lambda) < \varepsilon.$$

Then, for $x \in \Omega$ and such that $\rho(x, x_\lambda) < \varepsilon$, one gets

$$F_\lambda(x_\lambda) = f(x_\lambda) + \lambda\varphi(x_\lambda) = f(x_\lambda) \leq F_\lambda(x) = f(x) + \lambda\varphi(x) = f(x),$$

i.e.

$$f(x_\lambda) \leq f(x) \quad \forall x \in \Omega : \rho(x, x_\lambda) < \varepsilon.$$

It means that $x_\lambda$ is a local minimizer of the function $f$ on the set $\Omega$. $\square$

It turns out that the opposite statement is also valid:

*If $x_0 \in \Omega$ is a local minimizer of $f$ on $\Omega$ then there exists a $\lambda^* < \infty$ such that, for $\lambda > \lambda^*$, the point $x_0$ is a local minimizer of the function $F_\lambda(x)$ on $X$.*

To get this result, first prove the following proposition:

**Lemma 3.5.1.** *Let a function $f : X \to \mathbb{R}$ be Lipschitz with a Lipschitz constant $L < \infty$ in a neighborhood $B_\delta(x_0)$ of the point $x_0$. Then there exists a function $\tilde{f} : X \to \mathbb{R}$ such that*

*a) $\tilde{f}$ is Lipschitz on $X$ with a Lipschitz constant $2L$;*

*b) $\tilde{f}(x) = f(x) \quad \forall x \in B_\delta(x_0) = \{x \in X \mid \rho(x, x_0) \leq \delta\}$;*

*c) $\lim\limits_{\rho(x,x_0) \to \infty} \tilde{f}(x) = +\infty$.*

*Proof.* In the space $X$ let us define the function

$$\tilde{f}(x) = \inf_{u \in B_\delta(x_0)} \{f(u) + 2L\rho(x, u)\}. \tag{3.88}$$

Show that the function $\tilde{f}$ satisfies the conditions a) – c). First, prove the property b). Let $x \in B_\delta(x_0)$. The definition (3.88) yields

$$\tilde{f}(x) \leq f(x) + 2L\rho(x, \lambda) = f(x). \tag{3.89}$$

Let $\{u_k\}$ be a sequence of elements such that

$$u_k \in B_\delta(x_0), \quad \lim_{k \to \infty} [f(u_k) + 2L\rho(x, u_k)] = \tilde{f}(x). \tag{3.90}$$

The Lipschitzness of $f$ implies

$$f(u_k) - f(x) \geq -L\rho(x, u_k). \tag{3.91}$$

Now one concludes from (3.90) that

$$\tilde{f}(x) \geq \lim_{k \to \infty} [f(x) + L\rho(x, u_k)] \geq f(x). \tag{3.92}$$

The property b) follows from (3.92) and (3.89).

Let $x \in X$. From (3.88) one has

$$\tilde{f}(x) \geq \inf_{u \in B_\delta(x_0)} f(u) + \inf_{u \in B_\delta(x_0)} 2L\rho(x, u). \tag{3.93}$$

The function $f$ is Lipschitz on $B_\delta(x_0)$, therefore

$$f(u) \geq f(x_0) - L\rho(x_0, u). \tag{3.94}$$

Since $\rho(x_0, u) \leq \delta \quad \forall u \in B_\delta(x_0)$, then (3.94) yields

$$\inf_{u \in B_\delta(x_0)} f(u) \geq f(x_0) - L\delta. \tag{3.95}$$

On the other hand, the triangle inequality implies

$$\rho(x, u) \geq \rho(x, x_0) - \rho(u, x_0) \geq \rho(x, x_0) - \delta.$$

Hence, from (3.94) and (3.93) one concludes that

$$\tilde{f}(x) \geq f(x_0) - L\delta + 2L(\rho(x, x_0) - \delta) = f(x_0) + 2L\rho(x, x_0) - 3L\delta. \tag{3.96}$$

The property c) follows from (3.96).

It remains to show the Lipschitzness of the function $\tilde{f}$ on $X$. Let $x_1, x_2 \in X$. Choose a sequence $\{\delta_k\}$ such that $\delta_k \downarrow 0$. For every $k$, there exists a $u_k \in B_\delta(x_0)$ such that

$$\tilde{f}(x_2) = f(u_k) + 2L\rho(x_2, u_k) + \mu_k, \tag{3.97}$$

where $|\mu_k| < \delta_k$. Then

$$\tilde{f}(x_1) - \tilde{f}(x_2) = \inf_{u \in B_\delta(x_0)} [f(u) + 2L\rho(x_1, u)] - [f(u_k) + 2L\rho(x_2, u_k) + \mu_k]$$

$$\leq f(u_k) + 2L\rho(x_1, u_k) - [f(u_k) + 2L\rho(x_2, u_k) + \mu_k]$$

$$= 2L[\rho(x_1, u_k) - \rho(x_2, u_k)] - \mu_k. \tag{3.98}$$

Since $\rho$ is a metric, then $\rho(x_1, u_k) \le \rho(x_1, x_2) + \rho(x_2, u_k)$. It follows from (3.97) and (3.98) that

$$\widetilde{f}(x_1) - \widetilde{f}(x_2) \le 2L\rho(x_1, x_2) + \delta_k.$$

Passing to the limit as $k \to \infty$, one gets

$$\widetilde{f}(x_1) - \widetilde{f}(x_2) \le 2L\rho(x_1, x_2). \tag{3.99}$$

Permuting $x_1$ and $x_2$ and arguing analogously, we have

$$\widetilde{f}(x_2) - \widetilde{f}(x_1) \le 2L\rho(x_1, x_2). \tag{3.100}$$

The inequalities (3.99) and (3.100) imply the Lipschitzness of the function $\widetilde{f}$ on the entire space $X$ with a Lipschitz constant $2L$. This completes the proof of the Lemma.    □

**Theorem 3.5.2.** *Let the conditions 2) and 3) of Theorem 3.4.1 hold. If $x_0 \in \Omega$ is a local minimizer of the function $f$ on $\Omega$, then there exists a $\lambda^* < \infty$ such that, for $\lambda > \lambda^*$, the point $x_0$ is a local (unconstrained) minimizer of the function $F_\lambda(x)$.*

*Proof.* Since $x_0 \in \Omega$ is a local minimizer of $f$ on $\Omega$ then a $\delta > 0$ exists such that

$$f(x_0) \le f(x) \quad \forall x \in \Omega \cap B_\varepsilon(x_0), \tag{3.101}$$

where

$$B_\varepsilon(x_0) = \{x \in X | \rho(x, x_0) \le \varepsilon\}.$$

Using Lemma 3.5.1, construct a function $\widetilde{f}$ possessing the properties a) – c) stated in Lemma 3.5.1.

Choose $x \notin B_\varepsilon(x_0)$. Let $\{u_k\}$ be a sequence of points such that

$$u_k \in B_\varepsilon(x_0), \quad \widetilde{f}(x) = f(u_k) + 2L\rho(x, u_k) + \mu_k, \tag{3.102}$$

where

$$|\mu_k| \downarrow 0. \tag{3.103}$$

Since $u_k \in B_\varepsilon(x_0)$, then $f(u_k) \ge \inf\limits_{z \in B_\varepsilon(x_0)} f(z) = f_\varepsilon^*$. This yields

$$\rho(x, u_k) \ge \rho(x, x_0) - \rho(x_0, u_k) \ge \rho(x, x_0) - \varepsilon. \tag{3.104}$$

The property b) from Lemma 3.5.1 implies $\quad \widetilde{f}(z) = f(z) \quad \forall z \in B_\varepsilon(x_0)$. Hence,

$$\inf_{z \in B_\varepsilon(x_0)} f(z) = \inf_{z \in B_\varepsilon(x_0)} \widetilde{f}(z). \tag{3.105}$$

Since $x \notin B_\varepsilon(x_0)$, then from (3.102) and (3.104)

$$\tilde{f}(x) \geq f_\varepsilon^* + 2L(\rho(x, x_0) - \varepsilon) + \mu_k. \tag{3.106}$$

But $\rho(x, x_0) - \varepsilon > 0$, therefore from (3.103), (3.105) and (3.106)

$$\tilde{f}(x) > \inf_{z \in B_\varepsilon(x_0)} f(z) = f_\varepsilon^* = \inf_{z \in B_\varepsilon(x_0)} \tilde{f}(z) \quad \forall x \notin B_\varepsilon(x_0). \tag{3.107}$$

Thus,

$$\inf_{z \in X} \tilde{f}(z) = \inf_{z \in B_\varepsilon(x)} \tilde{f}(z) = \inf_{z \in B_\varepsilon(x)} f(z) = f(x_0).$$

This implies that $x_0$ is a global minimizer of the function $\tilde{f}$ on the set $\Omega$.

For the function $\tilde{f}$ all the assumptions of Theorem 3.4.1 hold, therefore one can find a $\lambda^*$ (depending on $x_0$, since the function $\tilde{f}$ depends on $x_0$), such that, for $\lambda > \lambda^*$, the point $x_0$ is a global (unconstrained) minimizer of the function $\tilde{F}_\lambda(x) = \tilde{f}(x) + \lambda\varphi(x)$, that is,

$$\tilde{F}_\lambda(x_0) = f(x_0) \leq \tilde{F}_\lambda(x) \quad \forall x \in X. \tag{3.108}$$

Since $\tilde{f}(x) = f(x) \quad \forall x \in B_\varepsilon(x_0)$ then (3.108) yields

$$f(x_0) = f(x_0) + \lambda\varphi(x_0) = F_\lambda(x_0)$$
$$\leq \tilde{f}(x) + \lambda\varphi(x) = f(x) + \lambda\varphi(x) = F_\lambda(x) \quad \forall x \in B_\varepsilon(x_0).$$

This implies that $x_0$ is a local minimizer of the function $F_\lambda(x)$.     □

*Remark 3.5.1.* Condition 2) of Theorem 3.4.1, which must be satisfied in Theorem 3.5.2, can be replaced by a weaker condition:

**Theorem 3.5.3.** *Let $x_0 \in \Omega$ be a local minimizer of the function $f$ on the set $\Omega$. Assume that*
*1) There exist $\varepsilon_1 > 0$ and $\lambda_1$ such that, for any $\lambda > \lambda_1$, one can find $x_\lambda \in B_{\varepsilon_1}(x_0)$ satisfying the inequality*

$$F_\lambda(x_\lambda) \leq F_\lambda(x) \quad \forall x \ B_{\varepsilon_1}(x_0)$$

*(i.e., $x_\lambda$ is a minimizer of the function $F_\lambda(x)$ on the set $B_{\varepsilon_1}(x_0)$);*
*2) there exist $\delta > 0$, $\delta_1 > 0$ and $a > 0$ such that*

$$\varphi^\downarrow(x) \leq -a < 0 \quad \forall x \in D = (B_{\delta_1}(x_0) \cap \Omega_\delta) \setminus \Omega; \tag{3.109}$$

*3) the function $f$ is locally Lipschitz on $D$, i.e., for some $L < \infty$,*

$$|f(x_1) - f(x_2)| \leq L\rho(x_1, x_2) \quad \forall x_1, x_2 \in D. \tag{3.110}$$

*Then there exists a $\lambda^* < \infty$ such that, for $\lambda > \lambda^*$, the point $x_0$ is a local minimizer of the function $F_\lambda(x)$.*

*Proof.* Let (3.109) hold and let $\varepsilon > 0$ be a number acting in the proof of Theorem 3.5.2. Without loss of generality, we may assume that $\varepsilon < \varepsilon_1$ (see condition 1) of Theorem 3.5.2). Introduce the function

$$\tilde{\varphi}(x) = \varphi(x) + \left[\max\left\{0, \rho(x, x_0) - \frac{\varepsilon}{2}\right\}\right]^2. \tag{3.111}$$

Then

$$\tilde{\varphi}(x) \begin{cases} = \varphi(x), & x \in B_{\varepsilon/2}(x_0), \\ > \varphi(x), & x \notin B_{\varepsilon/2}(x_0), \\ > \varphi(x) + \frac{\varepsilon^2}{4}, & x \notin B_\varepsilon(x). \end{cases} \tag{3.112}$$

Therefore

$$\tilde{\Omega} = \{x \in X | \tilde{\varphi}(x) = 0\} \subset B_{\varepsilon/2}(x_0). \tag{3.113}$$

Due to (3.112),

$$\tilde{\Omega} = \Omega \cap B_{\varepsilon/2}(x_0), \tag{3.114}$$

$$\tilde{\Omega}_{\varepsilon^2/4} = \left\{x \in X | \tilde{\varphi}(x) \le \frac{\varepsilon^2}{4}\right\} \subset B_\varepsilon(x_0). \tag{3.115}$$

Take $x \in \tilde{\Omega}_{\frac{\varepsilon^2}{4}} \setminus \tilde{\Omega}$. If $x \in B_{\varepsilon/2}(x_0)$, then

$$\tilde{\varphi}(x) = \varphi(x). \tag{3.116}$$

For

$$\varepsilon/2 < \delta_1 \quad \text{or} \quad \frac{\varepsilon^2}{4} < \delta, \tag{3.117}$$

due to (3.116),

$$\tilde{\varphi}^\downarrow(x) = \varphi^\downarrow(x) \le -a \quad \forall x \in B_{\varepsilon/2}(x_0). \tag{3.118}$$

If $x \in B_\varepsilon(x_0) \setminus B_{\varepsilon/2}(x_0)$, then

$$\tilde{\varphi}(x) = \varphi(x) + (\rho(x, x_0) - \varepsilon/2)^2. \tag{3.119}$$

For

$$\varepsilon^2/4 < \delta \tag{3.120}$$

one has $\varphi(x) < \delta$, and for

$$\varepsilon < \delta_1 \tag{3.121}$$

we have (due to (3.112)) $x \in (B_{\delta_1}(x_0) \cap \Omega_\delta) \setminus \Omega$.
Therefore (see (3.109))

$$\varphi^\downarrow(x) \le -a. \tag{3.122}$$

Let us find

$$\widetilde{\varphi}^{\downarrow}(x) = \liminf_{z \to x} \frac{\widetilde{\varphi}(z) - \widetilde{\varphi}(x)}{\rho(z, x)}$$

$$= \liminf_{z \to x} \left[ \frac{\varphi(z) - \varphi(x)}{\rho(z, x)} + \frac{\left(\rho(z, x_0) - \frac{\varepsilon}{2}\right)^2 - \left(\rho(x, x_0) - \frac{\varepsilon}{2}\right)^2}{\rho(z, x)} \right]$$

$$= \liminf_{z \to x} \left[ \frac{\varphi(z) - \varphi(x)}{\rho(z, x)} \right.$$

$$\left. + \frac{(\rho(z, x_0) - \rho(x, x_0))(\rho(z, x_0) + \rho(x, x_0) - \varepsilon)}{\rho(z, x)} \right].$$

$$\tag{3.123}$$

From the triangle axiom (see the properties of a metric)

$$\rho(z, x_0) - \rho(x, x_0) \le \rho(z, x) + \rho(x, x_0) - \rho(x, x_0) = \rho(z, x). \tag{3.124}$$

Since $x \in B_\varepsilon(x_0)$, then $\rho(z, x_0) \le \varepsilon$, $\rho(x, x_0) \le \varepsilon$, and it follows from (3.123) and (3.124) that $\widetilde{\varphi}^{\downarrow}(x) \le \varphi^{\downarrow}(x) + \varepsilon$. For

$$\varepsilon < \frac{a}{2} \tag{3.125}$$

one gets (2. (3.122))

$$\widetilde{\varphi}^{\downarrow}(x) \le -\frac{a}{2} \quad \forall x \in B_\varepsilon(x_0) \setminus B_{\frac{\varepsilon}{2}}(x_0). \tag{3.126}$$

Choose (see (3.117), (3.120), (3.121), (3.125)) $\varepsilon < \min\{2\delta_1, 2\sqrt{\delta}, \frac{a}{2}\}$ and put $\delta_0 = \frac{\varepsilon^2}{4}$. Then (3.119) and (3.126) yield

$$\widetilde{\varphi}^{\downarrow}(x) \le -\frac{a}{2} \quad \forall x \in \widetilde{\Omega}_{\delta_0} \setminus \widetilde{\Omega}. \tag{3.127}$$

(3.101) and (3.114) imply that $x_0$ is a global minimizer of $f$ on the set $\widetilde{\Omega}$. Thus, for the function $f$ and the set $\widetilde{\Omega} = \Omega \cap B_{\varepsilon/2}(x_0)$, defined by (3.113), where the function $\widetilde{\varphi}$ is given by the formula (3.111), all assumptions of Theorem 3.4.1 are satisfied and therefore there exists a $\lambda^* < \infty$ such that, for $\lambda > \lambda^* \ge \lambda_1$ (see condition 1) of the Theorem), the problem of minimizing the function $f$ on the set $\widetilde{\Omega}$ is equivalent to the problem of minimizing on $X$ the function $\widetilde{F}_\lambda(x) = f(x) + \lambda\widetilde{\varphi}(x)$, and $x_0$ is a global minimizer of $\widetilde{F}_\lambda(x)$, that is,

$$\widetilde{F}_\lambda(x_0) \le \widetilde{F}_\lambda(x) \quad \forall x \in X. \tag{3.128}$$

It follows from (3.112) that $\widetilde{\varphi}(x) = \varphi(x) \quad \forall x \in B_{\varepsilon/2}(x_0)$. This relation and (3.128) yield

$$\widetilde{F}_\lambda(x_0) = F_\lambda(x_0) \leq \widetilde{F}_\lambda(x) = F_\lambda(x) \quad \forall x \in B_{\varepsilon/2}(x_0),$$

which means that $x_0$ is a local (unconstrained) minimizer of $F_\lambda$.                    $\square$

*Remark 3.5.2.* The constant $\lambda^*$, which existence is proved in Theorem 3.5.3, depends on the point $x_0$.

## 3.6 Exact Penalty Functions and Stationary Points

### 3.6.1 Inf-Stationary Points

**Theorem 3.6.1.** *Assume that for $\delta > 0$, $a > 0$ and $L < \infty$ the conditions (3.79) and (3.81) hold. Then there exists a $\lambda^* < \infty$ such that, for $\lambda > \lambda^*$, any inf-stationary point of the function $F_\lambda(x)$ on the space $X$, belonging to the set $\Omega_\delta$, is also an inf-stationary point of the function $f$ on the set $\Omega$.*

*Proof.* Remind that (see Subsection 2.1.1) a point $x_\lambda \in X$ is called an inf-stationary point of the function $F_\lambda$ on $X$ if

$$F_\lambda^\downarrow(x_\lambda) = \liminf_{x \to x_\lambda} \frac{F_\lambda(x) - F_\lambda(x_\lambda)}{\rho(x, x_\lambda)} \geq 0. \tag{3.129}$$

First of all, let us show that, for $\lambda$ sufficiently large, for all inf-stationary points of $F_\lambda$, belonging to the set $\Omega_\delta$, one has

$$\varphi(x_\lambda) = 0. \tag{3.130}$$

Assume the contrary, let $x_\lambda \in \Omega_\delta$ be an inf-stationary point and let $\varphi(x_\lambda) > 0$. Then $x_\lambda \in \Omega_\varepsilon \setminus \Omega$, and, by assumption (see (3.79)), $\varphi^\downarrow(x_\lambda) \leq -a < 0$. Therefore there exists a sequence $\{y_{\lambda k}\}$ such that

$$y_{\lambda k} \in X, \quad \rho(y_{\lambda k}, x_\lambda) \underset{k \to \infty}{\longrightarrow} 0; \quad \frac{\varphi(y_{\lambda k}) - \varphi(x_\lambda)}{\rho(y_\lambda, x_\lambda)} \leq -\frac{a}{2} \quad \forall k. \tag{3.131}$$

Now,

$$\frac{F_\lambda(y_{\lambda k}) - F_\lambda(x_\lambda)}{\rho(y_{\lambda k}, x_\lambda)} = \frac{f(y_{\lambda k}) - f(x_\lambda) + \lambda[\varphi(y_{\lambda k}) - \varphi(x_\lambda)]}{\rho(y_{\lambda k}, x_\lambda)}.$$

(3.81) and (3.131) yield

$$\frac{F_\lambda(y_{\lambda k}) - F_\lambda(x_\lambda)}{\rho(y_{\lambda k}, x_\lambda)} \leq L - \frac{a}{2}\lambda.$$

For $\lambda > \frac{2L}{a}$ one has

$$F_\lambda^\downarrow(x_\lambda) \leq \liminf_{k \to \infty} \frac{F_\lambda(y_{\lambda k}) - F_\lambda(x_\lambda)}{\rho(y_{\lambda k}, x_\lambda)} < L - \frac{a}{2}\lambda < 0,$$

which contradicts (3.129).

This contradiction implies that, for $\lambda$ sufficiently large (namely, for $\lambda > \frac{2L}{a}$), the relation (3.130) holds for all inf-stationary points, belonging to the set $\Omega_\delta$.

Since $\varphi(x) = \varphi(x_\lambda) = 0$ for all $x \in \Omega$ and $x_\lambda \in \Omega$, then

$$F_\lambda(x) - F_\lambda(x_\lambda) = f(x) - f(x_\lambda) + \lambda(\varphi(x) - \varphi(x_\lambda)) = f(x) - f(x_\lambda).$$

Therefore from (3.129)

$$f^\downarrow(x_\lambda, \Omega) = \liminf_{\substack{x \in \Omega \\ x \to x_\lambda}} \frac{f(x) - f(x_\lambda)}{\rho(x, x_\lambda)} = \liminf_{\substack{x \in \Omega \\ x \to x_\lambda}} \frac{F_\lambda(x) - F_\lambda(x_\lambda)}{\rho(x, x_\lambda)}$$

$$\geq \liminf_{\substack{x \in X \\ x \to x_\lambda}} \frac{F_\lambda(x) - F_\lambda(x_\lambda)}{\rho(x, x_\lambda)} \geq 0,$$

that is,

$$f^\downarrow(x_\lambda, \Omega) \geq 0. \tag{3.132}$$

The inequality (3.132) means that the point $x_\lambda$ is an inf-stationary point of the function $f$ on the set $\Omega$. $\qquad \square$

*Remark 3.6.1.* Theorems 3.4.1, 3.5.1 and 3.5.2 imply that, in the assumptions of Theorem 3.4.1, for sufficiently large $\lambda$, all global and local minimizers of the functions $F_\lambda(x)$ on $X$ and the function $f$ on $\Omega$, belonging to the set $\Omega_\delta$, coincide.

Usually, the majority of existing numerical methods provide only an inf-stationary point. Theorem 3.6.1 shows that, for sufficiently large $\lambda$, all inf-stationary points of the function $F_\lambda(x)$ on $X$, belonging to the set $\Omega_\delta$, are also inf-stationary points of the function $f$ on the set $\Omega$. The converse property, generally speaking, is not true, but we are not interested in inf-stationary points of the function $f$ on $\Omega$, which are not local or global minimizers.

## 3.6.2   Local Minimizers of the Penalty Function

Let us state one more condition under which the constrained optimization problem is reduced to an unconstrained one.

**Theorem 3.6.2.** *Let $x_0 \in \Omega$ be a local minimizer of the function $f$ on the set $\Omega$ and let there exists a $> 0$ and $\delta > 0$ such that*

$$\varphi(x) \geq a\rho(x, \Omega) \quad \forall x \in B_\delta(x_0). \tag{3.133}$$

*If the function $f$ is Lipschitz on $B_\delta(x_0)$ with a Lipschitz constant $L$ then there exists a $\lambda^* < \infty$, such that, for $\lambda > \lambda^*$, the point $x_0$ is a local minimizer of the function $F_\lambda(x) = f(x) + \lambda\varphi(x)$ on the set $X$.*

*Proof.* Since $x_0$ is a local minimizer of $f$ on $\Omega$, then without loss of generality one can assume that

$$f(x_0) \leq f(x) \quad \forall x \in B_\delta(x_0) \cap \Omega. \tag{3.134}$$

Choose $x \in B_{\varepsilon/2}(x_0)$. Let $\{x_k\}$ be a sequence of points satisfying the relations

$$x_k \in \Omega, \quad \rho(x_k, x) \to \inf_{y \in \Omega} \rho(y, x) = \rho(x, \Omega). \tag{3.135}$$

Note that $x_k \in B_\delta(x_0)$ (since $x \in B_{\delta/2}(x_0)$). We have

$$\begin{aligned}
F_\lambda(x_0) - F_\lambda(x) &= f(x_0) - f(x) - \lambda\varphi(x) \\
&= f(x_0) - f(x_k) + f(x_k) - f(x) - \lambda\varphi(x). \quad (3.136)
\end{aligned}$$

Since $x_k \in \Omega \cap B_\delta(x_0)$, then (3.134) implies $f(x_0) - f(x_k) \leq 0$, and therefore the relations (3.136), (3.133) and Lipschitzness of $f$ imply

$$F_\lambda(x_0) - F_\lambda(x) \leq L\rho(x_k, x) - \lambda a\rho(x, \Omega). \tag{3.137}$$

Passing to the limit in (3.137), one gets from (3.135)

$$F_\lambda(x_0) - F_\lambda(x) \leq \rho(x, \Omega)[L - a\lambda].$$

Hence, for $\lambda > \frac{L}{a}$,

$$F_\lambda(x_0) - F_\lambda(x) \leq 0 \quad \forall x \in B_{\delta/2}(x_0).$$

This means that $x_0$ is a local (unconstrained) minimizer of the function $F_\lambda(x)$ for $\lambda > \frac{L}{a}$. $\qquad\square$

**Corollary 3.6.1.** *Since $\Omega = \{x \in X \mid \rho(x, \Omega) = 0\}$, then, as $\varphi(x)$, it is possible to take the function $\rho(x, \Omega)$, hence, the condition (3.133) holds as an equality for $\alpha = 1$.*

### 3.6.3 Relationship Between Sufficient Conditions

Let us find a relation between the conditions (3.79) and (3.133). In this subsection it is assumed that $X$ is a complete metric space with a metric $\rho$, the function $\varphi : X \to \mathbb{R}_+ = [0, \infty)$ is continuous,

$$\Omega = \{x \in X \mid \varphi(x) = 0\} \neq \emptyset.$$

**Lemma 3.6.1.** *If, for some $\varepsilon > 0$ and $a > 0$,*

$$\varphi^\downarrow(x) \leq -a \quad \forall x \in \Omega_\varepsilon \setminus \Omega, \tag{3.138}$$

*where $\Omega_\varepsilon = \{x \in X \mid \varphi(x) < \varepsilon\}$, then*

$$\varphi(x) \geq a\rho(x, \Omega) \quad \forall x \in \Omega_\varepsilon. \tag{3.139}$$

*Here, as above,*

$$\rho(x, \Omega) = \inf_{y \in \Omega} \rho(x, y). \tag{3.140}$$

*Proof.* Let $x \in \Omega_\varepsilon$. If $x \in \Omega$, then $\varphi(x) = 0$, $\rho(x, \Omega) = 0$, and (3.139) holds. Consider the case $x \notin \Omega$. Fix an arbitrary $\beta > 0$ such that $a' = a - 2\beta > 0$, and a sequence $\{\gamma_k\}$ such that $\gamma_k \downarrow 0$. Let us construct a sequence $\{x_k\}$ as follows. Put $x_0 = x$. Let $x_k \in \Omega_\varepsilon$ have already been found. If $\varphi(x_k) = 0$ (i.e., $x_k \in \Omega$), then put $x^* = x_k$. If $\varphi(x_k) > 0$, then(3.138) yields

$$\liminf_{y \to x_k} \frac{\varphi(y) - \varphi(x_k)}{\rho(x_k, y)} = -a_k \leq -a.$$

Put

$$G_k = \left\{ y \in X \mid \frac{\varphi(y) - \varphi(x_k)}{\rho(x_k, y)} \leq -a' \right\}, \tag{3.141}$$

$$\rho_k^* = \sup_{y \in G_k} \rho(x_k, y). \tag{3.142}$$

Clearly, $G_k \neq \emptyset$. Find $y_k \in G_k$ such that $\rho(x_k, y_k) = \rho_k^* - \delta_k$, where $\delta_k < \gamma_k$. Put $x_{k+1} = y_k$. Clearly, that $\varphi(x_{k+1}) < \varphi(x_k)$, i.e., $x_{k+1} \in \Omega_\varepsilon$. Then (3.141) yields

$$\varphi(x_{k+1}) - \varphi(x_k) \leq -a'\rho(x_k, x_{k+1}) \leq -a'(\rho_k^* - \gamma_k). \tag{3.143}$$

Proceeding analogously, construct a sequence $\{x_k\}$. If this sequence is finite then the last obtained point $x^* \in \Omega$.

Now consider the case where this sequence is infinite. Since

$$\varphi(x_{k+1}) < \varphi(x_k), \quad \varphi(x_k) > 0, \tag{3.144}$$

then there exists a sequence

$$\lim_{k \to \infty} \varphi(x_k) = \varphi^* \geq 0. \tag{3.145}$$

The sequence $\{x_k\}$ is fundamental, i.e.

$$\rho(x_m, x_{m+n}) \underset{m,n \to \infty}{\longrightarrow} 0. \tag{3.146}$$

Indeed, (3.143) implies

$$\varphi(x_k) - \varphi(x_0) = (\varphi(x_k) - \varphi(x_{k-1})) + \cdots + (\varphi(x_1) - \varphi(x_0))$$
$$\leq -a'[\rho(x_k, x_{k-1}) + \cdots + \rho(x_1, x_0)]. \tag{3.147}$$

Therefore (3.145) means that the series $\sum_{k=1}^{\infty} \beta_k$, where $\beta_k = \rho(x_k, x_{k+1})$, converges, and then

$$c_m = \sum_{k=m}^{m+n} \beta_k \longrightarrow 0. \tag{3.148}$$

On the other hand, $\rho(x_m, x_{m+n}) \leq$

$$\leq \rho(x_m, x_{m+1}) + \cdots + \rho(x_{m+n-1}, x_{m+n}) \leq \sum_{k=m}^{m+n} \beta_k = c_m. \tag{3.149}$$

The relations (3.148) and (3.149) imply (3.146).

Thus, the sequence $\{x_k\}$ is fundamental, and since $X$ is a complete metric space, then there exists the limit

$$x^* = \lim_{k \to \infty} x_k. \tag{3.150}$$

Clearly, $x^* \in \Omega_\varepsilon$. Let us show that $\varphi(x^*) = \varphi^* = 0$, that is, $x^* \in \Omega$. Assume the opposite, let $\varphi(x^*) > 0$. It follows from (3.138) that there exists $y^* \in X$ such that

$$\varphi(y^*) - \varphi(x^*) \leq -(a - \beta)\rho(x^*, y^*) = -(a' + \beta)\rho(x^*, y^*), \quad \rho(x^*, y^*) > 0.$$

Therefore
$$\varphi(y^*) \le \varphi(x^*) - (a' + \beta)\rho(x^*, y^*) < \varphi(x^*).$$

One has
$$\varphi(y^*) - \varphi(x_k) = [\varphi(y^*) - \varphi(x^*)] + [\varphi(x^*) - \varphi(x_k)]$$
$$\le -(a' + \beta)\rho(x^*, y^*) + \varphi(x^*) - \varphi(x_k).$$

Since $\rho(x_k, y^*) \le \rho(x_k, x^*) + \rho(x^*, y^*)$, then
$$\rho(x^*, y^*) \ge \rho(x_k, x^*) - \rho(x_k, x^*).$$

Hence, $\varphi(y^*) - \varphi(x_k) \le$
$$\le \varphi(x^*) - \varphi(x_k) - (a' + \beta)[\rho(x_k, y^*) - \rho(x_k, x^*)] = z_k - (a' + \beta)\rho(x_k, y^*),$$

where    $z_k = (a' + \beta)\rho(x_k, x^*) + \varphi(x^*) - \varphi(x_k).$

The relations (3.145) and (3.150) imply that $z_k \to 0$. For sufficiently large $k$

$$\rho(x_k, y^*) \ge \frac{3}{4}\rho(x^*, y^*) > 0, \quad \varphi(y^*) - \varphi(x_k) \le -a'\rho(x_k, y^*), \qquad (3.151)$$

i.e. $y^* \in G_k$. Without loss of generality, one can assume that

$$\gamma_k < \frac{1}{4}\rho(x^*, y^*). \qquad (3.152)$$

One has
$$\rho_k^* \ge \rho(x_k, y^*) \ge \frac{3}{4}\rho(x^*, y^*). \qquad (3.153)$$

Since $\rho(x_{k+1}, x_k) = \rho_k^* - \delta_k > \rho_k^* - \gamma_k$, then (3.152) and (3.153) yield

$$\beta_k = \rho(x_{k+1}, x_k) > \frac{3}{4}\rho(x^*, y^*) - \frac{1}{4}\rho(x^*, y^*) = \frac{1}{2}\rho(x^*, y^*),$$

which contradicts (3.148). This contradiction means that

$$\varphi(x^*) = 0, \qquad (3.154)$$

that is $x^* \in \Omega$. The relation (3.147) implies

$$\varphi(x_k) - \varphi(x_0) \le -a' \sum_{i=0}^{k-1} \rho(x_{i+1}, x_i) \le -a'\rho(x_k, x_0).$$

Passing to the limit as $k \to \infty$, one gets from (3.154)

$$\varphi(x^*) - \varphi(x_0) = -\varphi(x_0) \leq -a'\rho(x^*, x_0),$$

that is,

$$\varphi(x_0) \geq a'\rho(x^*, x_0). \tag{3.155}$$

But $\rho(x^*, x_0) \geq \inf_{y \in \Omega} \rho(y, x_0) = \rho(x_0, \Omega)$, therefore (3.155) yields

$$\varphi(x_0) \geq a'\rho(x_0, \Omega). \tag{3.156}$$

Since $a' = a - 2\beta$ and $\beta > 0$ is arbitrary, then (3.156) implies $\varphi(x_0) \geq a\rho(x^*, x_0)$. To conclude the proof, observe that $x_0 = x$. □


## 3.7 Exact Penalty Functions and Minimizing Sequences

Up to now, we considered the case where, for $\lambda$ sufficiently large, there exists a minimizer $x_\lambda \in X$ of the penalty function $F_\lambda(x)$ on $X$. Now, consider the general case.


### 3.7.1 The Smallest Exact Penalty Constant

Remind that $\lambda \geq 0$ is called an exact penalty constant (for a function $f$ on a set $\Omega$), if

$$F_\lambda^* := \inf_{x \in X} F_\lambda(x) = \inf_{x \in \Omega} f(x) =: f_\Omega^*. \tag{3.157}$$

The function $F_\lambda(x) = f(x) + \lambda\varphi(x)$ is then called an exact penalty function.

**Lemma 3.7.1.** *If $\lambda_0 \geq 0$ is an exact constant for the function $f$ on the set $\Omega$, then any $\lambda > \lambda_0$ is also an exact penalty constant for $f$. Any minimizing sequence for the function $F_\lambda$ on $X$ is a $\varphi$-minimizing sequence for the function $f$ on the set $\Omega$.*

*Proof.* Let $\lambda > \lambda_0$. Since $\varphi(x) \geq 0$, $\lambda - \lambda_0 > 0$, then

$$F_\lambda(x) = f(x) + \lambda\varphi(x) = F_{\lambda_0}(x) + (\lambda - \lambda_0)\varphi(x) \geq F_{\lambda_0}(x).$$

Therefore

$$F_\lambda^* := \inf_{x \in X} F_\lambda(x) \geq \inf_{x \in X} F_{\lambda_0}(x) =: F_{\lambda_0}^*, \tag{3.158}$$

that is, the function $F_\lambda^*$ is nondecreasing. If $\lambda_0$ is an exact penalty constant then (3.157) yields

$$F_\lambda^* \geq F_{\lambda_0}^* = f_\Omega^*. \tag{3.159}$$

But

$$F_\lambda^* := \inf_{x \in X} F_\lambda(x) \le \inf_{x \in \Omega} F_\lambda(x) = \inf_{x \in \Omega} f(x) =: f_\Omega^* \quad \forall \lambda \ge 0. \qquad (3.160)$$

Hence, (3.159) implies $F_\lambda^* = f_\Omega^*$. Thus, for $\lambda > \lambda_0$

$$F_\lambda^* = F_{\lambda_0}^* = f_\Omega^*. \qquad (3.161)$$

Let $\{x_k\}$ be a minimizing sequence for the function $F_\lambda$ on $X$, that is, (due to (3.161))

$$F_\lambda(x_k) = f(x_k) + \lambda\varphi(x_k) \underset{k \to \infty}{\longrightarrow} F_\lambda^* = f_\Omega^*. \qquad (3.162)$$

First show that

$$\varphi(x_k) \underset{k \to \infty}{\longrightarrow} 0. \qquad (3.163)$$

Assume the opposite. Then there exist a subsequence $\{x_{k_s}\}$ and $a > 0$ such that $\varphi(x_{k_s}) > a \quad \forall s; \quad k_s \underset{s \to \infty}{\longrightarrow} \infty$,

$$F_\lambda(x_{k_s}) = F_{\lambda_0}(x_{k_s}) + (\lambda - \lambda_0)\varphi(x_{k_s})) \ge F_{\lambda_0}(x_{k_s}) + (\lambda - \lambda_0)a.$$

Since $F_\lambda(x_{k_s}) \longrightarrow F_\lambda^*$, then, for sufficiently large $k_s$, (see (3.161))

$$F_{\lambda_0}(x_{k_s}) < F_\lambda^* = F_{\lambda_0}^*,$$

which contradicts the definition of $F_{\lambda_0}^*$.

Hence, (3.163) holds, and since in this case (3.162) implies $f(x_k) \to f_\Omega^*$, then $\{x_k\}$ is a $\varphi$-minimizing sequence for the function $f$ on the set $\Omega$.   $\square$

*Remark 3.7.1.* Thus, if, for some $\lambda_0 \ge 0$, the function $F_{\lambda_0}$ is an exact penalty function for the function $f$ on the set $\Omega$, then for $\lambda > \lambda_0$ the function $F_\lambda$ is also an exact penalty function. Any minimizing sequence for the function $F_\lambda$ on $X$ is a $\varphi$-minimizing sequence for the function $f$ on $\Omega$. For the exact penalty constant $\lambda_0$ this property does not necessarily hold. This is clear from the following example.

*Example 3.7.1.* Let $X = R$, $\Omega = [0,1]$, $f(x) = (x+1)^2$. Clearly,

$$\min_{x \in \Omega} f(x) = f(0) = 1 = f_\Omega^*, \quad \min_{x \in R} f(x) = f(-1) = 0 = f^*.$$

Describe $\Omega$ by means of the indicator function

$$\Omega = \{x \in X \mid \varphi(x) = 0\},$$

where

$$\varphi(x) = \begin{cases} 0, & x \in \Omega, \\ 1, & x \neq \Omega. \end{cases}$$

Then

$$F_\lambda(x) = \begin{cases} f(x), & x \in \Omega, \\ f(x) + \lambda, & x \neq \Omega. \end{cases}$$

It is clear that

$$\min_{x \in R}[f(x) + \lambda] = f(-1) + \lambda = \lambda.$$

For $\lambda \in [0, 1)$

$$\min_{x \in R} F_\lambda(x) = F_\lambda(-1) = \lambda < 1 = f_\Omega^*.$$

For $\lambda = 1$

$$\min_{x \in R} F_\lambda(x) = F_\lambda(-1) = F_\lambda(0) = \lambda = 1 = f_\Omega^*.$$

For $\lambda > 1$

$$\min_{x \in R} F_\lambda(x) = F_\lambda(0) = f_\Omega^*.$$

Thus, $\lambda_0 = 1$ is an exact penalty constant (by definition), but not every minimizer of the function $F_{\lambda_0}$ on $X$ is a minimizer of the function $f$ on $\Omega$ (in this case the point $x_0 = -1$ is not a minimizer of the function $f$ on $\Omega$, since does not belong to the set $\Omega$). However, for $\lambda > \lambda_0$, any minimizer of the function $F_\lambda$ on $X$ is a minimizer of the function $f$ on $\Omega$. □

If $\lambda_0$ is an exact penalty constant, but not every minimizer of the function $F_{\lambda_0}$ on $X$ is a minimizer of the function $f$ on $\Omega$, it means that $\lambda_0$ is the *smallest* exact penalty constant. On the other hand, if $\lambda$ is an exact penalty constant then every $\varphi$-minimizing sequence for the function $f$ on $\Omega$ is a minimizing sequence for the function $F_\lambda$ on $X$, that is, if $\lambda_0$ is the smallest exact penalty function then among minimizing sequences for the function $F_{\lambda_0}$ on $X$ is a $\varphi$-minimizing sequence for the function $f$ on $\Omega$.

## 3.7.2   Sufficient Conditions for the Existence of an Exact Penalty Function

**Lemma 3.7.2.** *Let the assumptions of Lemma 3.3.2 hold. If, for some $\lambda \geq 0$ and some minimizing sequence $\{x_k\}$ of the function $F_\lambda$ on $X$, one has $\varphi(x_k) \xrightarrow[k \to \infty]{} 0$, then $F_\lambda(x)$ is an exact penalty function.*

*Proof.* Let

$$F_\lambda(x_k) = f(x_k) + \lambda\varphi(x_k) \xrightarrow[k \to \infty]{} F_\lambda^* := \inf_{x \in X} F_\lambda(x). \qquad (3.164)$$

Let us show that $F_\lambda^* = f_\Omega^*$. Assume the contrary. Then (see (3.160)) $F_\lambda^* < f_\Omega^*$. The relation (3.164) implies that there exists $a > 0$ such that, for $k$ sufficiently large, $f(x_k) \leq f_\Omega^* - a$. Since all assumptions of Lemma 3.3.2 hold, then, for sufficiently large $k$, one can find a point $\bar{x}_k \in \Omega$ such that $f(\bar{x}_k) < f_\Omega^* - \frac{1}{2}a$, which contradicts the definition of $f_\Omega^*$. $\qquad\square$

**Lemma 3.7.3.** *If, for some $x_\lambda \in X$,*

$$F_\lambda(x_\lambda) = F_\lambda^*, \quad \varphi(x_\lambda) = 0, \tag{3.165}$$

*then $F_\lambda(x)$ is an exact penalty function.*

*Proof.* The relation (3.165) yields $F_\lambda(x_\lambda) = f(x_\lambda) = F_\lambda^*$. It remains to show that $f(x_\lambda) = f_\Omega^*$. Assume the contrary. Then there exists a point $\bar{x} \in \Omega$ such that $f(\bar{x}) < f(x_\lambda)$. One has

$$F_\lambda(\bar{x}) = f(\bar{x}) + \lambda\varphi(\bar{x}) = f(\bar{x}) < f(x_\lambda) = F_\lambda^*,$$

which contradicts the definition of $F_\lambda^*$.

*Remark 3.7.2.* In Lemma 3.7.3, it is not required to satisfy the assumptions of Lemma 3.7.2.

**Theorem 3.7.1.** *Let the following conditions hold:*
*1) the function $\varphi$ is continuous on $X$,*
*2) the relation*

$$\inf_{x \in X} f(x) = f^* > -\infty \tag{3.166}$$

*is valid,*
*3) there exist $\delta > 0$ and $a > 0$ such that*

$$\varphi(x) \geq a\rho(x, \Omega) \quad \forall x \in \Omega_\delta \setminus \Omega, \tag{3.167}$$

*where $\Omega_\delta = \{x \in X \mid \varphi(x) < \delta\}$,*
*4) the function $f$ is locally Lipschitz on $\Omega_\delta \setminus \Omega$ with a Lipschitz constant $L$.*
   *Then a $\lambda^* \in [0, \infty)$ exists such that, for $\lambda > \lambda^*$, the function $F_\lambda(x)$ is an exact penalty function.*

*Proof.* Assume that the proposition is not correct. Then for any $\lambda \geq 0$

$$F_\lambda^* := \inf_{x \in X} F_\lambda(x) < \inf_{x \in \Omega} f(x) =: f_\Omega^*. \tag{3.168}$$

Choose an arbitrary $\lambda \geq 0$. Let $\{x_k\} = \{x_{\lambda k}\}$ be a minimizing sequence for the function $F_\lambda$ on $X$. By Lemma 3.7.3, there exists $a_\lambda > 0$ such that, for sufficiently large $k$,

$$\varphi(x_k) \geq a_\lambda. \tag{3.169}$$

Without loss of generality one can assume that (3.169) holds for all $k$. It is also possible to assume that (see Lemma 3.3.1) $\varphi(x_k) < \delta$, that is,

$$x_k \in \Omega_\delta \setminus \Omega \quad \forall k.$$

(3.167) yields

$$\rho_k := \rho(x_k, \Omega) \leq \frac{\varphi(x_k)}{a}. \tag{3.170}$$

By the continuity of the function $\varphi$, there exists $y_k \in \mathcal{B}_{\rho_k}(x_k)$, where

$$\mathcal{B}_{\rho_k}(x_k) = \{x \in X \mid \rho(x_k, x) \leq \rho_k\}, \tag{3.171}$$

such that

$$\varphi(y_k) = \frac{1}{2}\varphi(x_k). \tag{3.172}$$

One gets

$$F_\lambda(y_k) - F_\lambda(x_k) = f_\lambda(y_k) - f_\lambda(x_k) + \lambda[\varphi(y_k) - \varphi(x_k)].$$

The Lipschitzness of the function $f$ and relation (3.172) imply that

$$F_\lambda(y_k) - F_\lambda(x_k) \leq L\rho(y_k, x_k) - \frac{1}{2}\varphi(x_k).$$

It follows from (3.171) that

$$F_\lambda(y_k) - F_\lambda(x_k) \leq L\rho_k - \frac{1}{2}\lambda\varphi(x_k).$$

Now (3.170) yields

$$F_\lambda(y_k) - F_\lambda(x_k) \leq \varphi(x_k)\left[\frac{L}{a} - \frac{\lambda}{2}\right].$$

For $\lambda > \frac{2L}{a}$, it follows from (3.169) that

$$F_\lambda(y_k) - F_\lambda(x_k) \leq a_\lambda\left[\frac{L}{a} - \frac{\lambda}{2}\right] := -b < 0.$$

The sequence $\{x_k\}$ is minimizing for the function $F_\lambda$ on $X$, therefore, for sufficiently large $k$,

$$F_\lambda(y_k) \leq F_\lambda^* - \frac{b}{2} < F_\lambda^*,$$

which contradicts the definition of $F_\lambda^*$.                                                    $\square$

## 3.8 Exact Smooth Penalty Functions

Let $X$ be a normed space with the norm $\|x\|$, $\Omega \subset X$ be a nonempty set and let $f : X \to \mathbb{R}$ be a given functional. Assume that the set $\Omega$ can be represented in the form

$$\Omega = \{x \in \mathbb{R}^n \mid \varphi(x) = 0\}, \tag{3.173}$$

where

$$\varphi(x) \geq 0 \quad \forall x \in X. \tag{3.174}$$

Construct the penalty function $F_\lambda(x) = f(x) + \lambda\varphi(x)$. In the previous part, conditions for the existence of an exact penalty constant were discussed. Let us now discuss the possibility of existence of an exact penalty constant in the case where $\varphi$ is a differentiable function. Assume that the function $\varphi$ is Gâteaux differentiable (see Section 2.4). Then (see formula (2.97))

$$\varphi_D^\downarrow(x_0, g) = \varphi_D'(x_0, g) = \varphi_G'(x_0)(g), \tag{3.175}$$

where $\varphi_G'(x_0)(g)$ is a linear functional. Assume that, for some $\lambda_0 \geq 0$, the function $F_{\lambda_0}(x)$ is an exact penalty function, i.e.,

$$F_{\lambda_0}^* := \inf_{x \in X} F_{\lambda_0}(x) = \inf_{x \in \Omega} f(x) =: f_\Omega^*$$

and there exists a point $x_0 \in \Omega$ such that $F_{\lambda_0}(x_0) = F_{\lambda_0}^*$. Then, by the necessary condition for a minimum of the function $F_{\lambda_0}(x)$ on $X$, the following relation holds (see formula (2.43))

$$F_{\lambda_0 D}^\downarrow(x_0, g) \geq 0 \quad \forall g \in X, \tag{3.176}$$

where

$$F_{\lambda_0 D}^\downarrow(x_0, g) = \liminf_{\alpha \downarrow 0} \frac{F_{\lambda_0}(x_0 + \alpha g) - F_{\lambda_0}(x_0)}{\alpha}.$$

The relation (3.175) yields

$$F_{\lambda_0 D}^\downarrow(x_0, g) = f_D^\downarrow(x_0, g) + \lambda\varphi_D'(x_0, g). \tag{3.177}$$

Since $x_0 \in \Omega$, then $\varphi(x_0) = 0$, and (3.174) implies that $x_0$ is a global minimizer of the function $\varphi(x)$ on $X$. By the necessary condition for a minimum (see Theorem 2.4.3) $\varphi_G'(x_0)(g) = 0 \quad \forall g \in X$. Therefore from (3.176) and (3.177) one concludes that

$$f_D^\downarrow(x_0, g) \geq 0 \quad \forall g \in X, \tag{3.178}$$

which means that $x_0$ is a $D$-inf-stationary point of the function $f$ on $X$. Therefore, if a minimizer of $f$ on $\Omega$ is not a $D$-inf-stationary point of the function $f$ on $X$, then there exists no exact smooth penalty function.

Analogous arguments are valid also in the case where the function $\varphi$ is Fréchet differentiable. Namely, if a minimizer of $f$ on $\Omega$ is not an $H$-inf-stationary point of $f$ on $X$, then there exists no exact smooth penalty function.

## 3.9 Minimization in the Finite-Dimensional Space

### 3.9.1 A Necessary Condition for a Minimum of a Smooth Function on a Convex Set

Let a function $f : \mathbb{R}^n \to \mathbb{R}$ be continuously differentiable on $\mathbb{R}^n$, $\Omega \subset \mathbb{R}^n$ be a convex set. Remind that $\Omega$ is called *a convex set*, if for any $x_1, x_2 \in \Omega$ and $\alpha \in [0,1]$ one has $x_\alpha = \alpha x_1 + (1 - \alpha)x_2 \in \Omega$. Consider the problem of minimizing the function $f$ on the set $\Omega$.

**Theorem 3.9.1.** *For a function $f$ to attain its minimal value on the set $\Omega$ at a point $x^* \in \Omega$ it is necessary that*

$$(f'(x^*), x - x^*) \geq 0 \quad \forall x \in \Omega. \tag{3.179}$$

*Proof.* Let $x^* \in \Omega$ be a minimizer of the function $f$ on the set $\Omega$, i.e.

$$f(x^*) \leq f(x) \quad \forall x \in \Omega. \tag{3.180}$$

Assume that condition (3.179) does not hold. Then there exists $\overline{x} \in \Omega$ such that

$$(f'(x^*), \overline{x} - x^*) = -a < 0. \tag{3.181}$$

Put $x_\alpha = \alpha \overline{x} + (1 - \alpha)x^* = x^* + \alpha(\overline{x} - x^*)$. For $\alpha \in [0,1]$ one has $x_\alpha \in \Omega$ (due to the convexity of the set $\Omega$). Then

$$
\begin{aligned}
f(x_\alpha) &= f(x^* + \alpha(\overline{x} - x^*)) \\
&= f(x^*) + \alpha(f'(x^*), \overline{x} - x^*) + o(\alpha) = f(x^*) - \alpha a + o(\alpha),
\end{aligned}
$$

where $\dfrac{o(\alpha)}{\alpha} \xrightarrow[\alpha \downarrow 0]{} 0$. The relation (3.181) implies that, for $\alpha$ sufficiently small and positive, $x_\alpha \in \Omega$, $f(x_\alpha) < f(x^*)$, which contradicts (3.180).  □

A function $f : \mathbb{R}^n \to \mathbb{R}$ is called *convex* on a convex set $\Omega \subset \mathbb{R}^n$, if

$$f(\alpha x_1 + (1-\alpha)x_2) \leq \alpha f(x_1) + (1-\alpha)f(x_2) \quad \forall x_1, x_2 \in \Omega, \; \forall \alpha \in [0,1]. \quad (3.182)$$

**Exercise 3.9.1.** Prove that if, in the conditions of Theorem 3.9.1, the function $f$ is convex on $\Omega$ then the relation (3.179) is a sufficient condition for the point $x^*$ to be a minimizer of $f$ on $\Omega$.

### 3.9.2 A Necessary Condition for a Minimum of a Max-Type Function

Let

$$f(x) = \max_{i \in I} f_i(x), \quad (3.183)$$

where $I = 1 : N$, functions $f_i : \mathbb{R}^n \to \mathbb{R}$ are continuously differentiable on $\mathbb{R}^n$.

**Theorem 3.9.2.** *For a point $x^* \in \mathbb{R}^n$ to be a minimizer of the function $f$ on $\mathbb{R}^n$, it is necessary that*

$$0_n \in L(x^*) = co\{f_i'(x^*) \mid i \in R(x^*)\}, \quad (3.184)$$

*where*

$$R(x) = \{i \in I \mid f_i(x) = f(x)\}. \quad (3.185)$$

*Proof.* Let $x^* \in \mathbb{R}^n$ be a minimizer of $f$ on the entire space $\mathbb{R}^n$, i.e.

$$f(x^*) \leq f(x) \quad \forall x \in \mathbb{R}^n. \quad (3.186)$$

Assume that condition (3.184) does not hold, that is

$$0_n \notin L(x^*). \quad (3.187)$$

Find

$$\min_{z \in L(x^*)} ||z||^2 = ||z^*||^2.$$

Due to (3.187), $||z^*||^2 = a^2 > 0$. By the necessary condition for a minimum of the function $h(z) = ||z||^2$ on the convex set $L(x^*)$ (see (3.179)) one has

$$(z^*, z - z^*) \geq 0 \quad \forall z \in L(x^*). \quad (3.188)$$

Since $f_i'(x^*) \in L(x^*) \quad \forall i \in R(x^*)$, then (3.188) yields

$$(f_i'(x^*), z^*) \geq ||z^*||^2 = a^2 \quad \forall i \in R(x^*). \quad (3.189)$$

Put $x_\alpha = x^* - \alpha z^*$, where $\alpha > 0$. The relations (3.189) and (3.185) imply

$$f_i(x_\alpha) = f_i(x^*) - \alpha(f_i'(x^*), z^*) + o_i(\alpha) \leq f(x^*) - \alpha a^2 + o_i(\alpha) \quad \forall i \in R(x^*),$$

where $\frac{o_i(\alpha)}{\alpha} \xrightarrow[\alpha \downarrow 0]{} 0$. There exists an $\alpha_1 > 0$ such that

$$f_i(x_\alpha) < f_i(x^*) - \alpha \frac{a^2}{2} \quad \forall \alpha \in (0, \alpha_1], \ \forall i \in R(x^*). \tag{3.190}$$

Since $f_i(x^*) < f(x^*)$ for $i \notin R(x^*)$, then there exists an $\alpha_2 > 0$ such that

$$f_i(x_\alpha) < f(x^*) \quad \forall \alpha \in (0, \alpha_2], \ \forall i \notin R(x^*). \tag{3.191}$$

It follows from (3.190) and (3.191) that

$$f_i(x_\alpha) < f(x^*) \quad \forall \alpha \in (0, \alpha_0], \ \forall i \in I, \tag{3.192}$$

where $\alpha_0 = \min\{\alpha_1, \alpha_2\}$. The relation (3.192) implies $f(x_\alpha) < f(x^*)$ $\forall \alpha \in (0, \alpha_0]$, which contradicts (3.186). $\qquad\square$

A point $x^* \in \mathbb{R}^n$ satisfying (3.184) is called an inf-*stationary* point of the max-function $f$ (defined by (3.183)) on $\mathbb{R}^n$.

### 3.9.3 Mathematical Programming Problems

Let a function $f : \mathbb{R}^n \to \mathbb{R}$ be continuously differentiable on $\mathbb{R}^n$,

$$\Omega = \{x \in \mathbb{R}^n \mid h_i(x) \leq 0 \quad \forall i \in I\}, \tag{3.193}$$

where $I = 1 : N$, functions $h_i : \mathbb{R}^n \to \mathbb{R}$, $i \in I$, are continuously differentiable on $\mathbb{R}^n$. The set $\Omega$ can be written in the form

$$\Omega = \{x \in \mathbb{R}^n \mid h(x) \leq 0\}, \tag{3.194}$$

where

$$h(x) = \max_{i \in I} h_i(x). \tag{3.195}$$

Assume that $\Omega \neq \emptyset$. The set $\Omega$, defined by the relation (3.193), is not necessarily convex, however, if all the functions $h_i$, $i \in I$, are convex on $\mathbb{R}^n$ (see the definition (3.182)), then the set $\Omega$ is also convex.

**Exercise 3.9.1.** Show the convexity of the set $\Omega$, given by (3.193), if $h_i$ are convex functions.

Consider the following problem:  Find $\inf_{x \in \Omega} f(x) = f^*$.

**Theorem 3.9.3.** *For a point $x^* \in \Omega$ to be a minimizer of the function $f$ on the set $\Omega$, it is necessary that*

$$0_n \in L(x^*) = co\Big\{ \{f'(x^*)\} \cup \{h_i'(x^*) \mid i \in R(x^*)\} \Big\}, \qquad (3.196)$$

*where $R(x^*) = \{i \in I \mid h_i(x^*) = 0\}$.*

*Proof.* Since $x^* \in \Omega$, then $h(x^*) \leq 0$. Note that if $h(x^*) < 0$ then $R(x^*) = \emptyset$ and $L(x^*) = \{f'(x^*)\}$. Let $x^* \in \Omega$ be a minimizer of the function $f$ on the set $\Omega$, i.e. $f(x^*) \leq f(x) \quad \forall x \in \Omega$. Assume that the condition (3.196) does not hold, that is,

$$0_n \notin L(x^*). \qquad (3.197)$$

Find

$$\min_{z \in L(x^*)} ||z|| = ||z^*|| = a. \qquad (3.198)$$

Here $||z|| = \sqrt{\sum_{i \in 1:n} z_i^2}$, $z = (z_1, ..., z_n)$. It follows from (3.197) that $a > 0$. The problem (3.198) is equivalent to the problem of minimizing the function $h(z) = ||z||^2$ on the set $L(x^*)$. By the necessary condition for a minimum of the function $h(z)$ on the convex set $L(x^*)$ (see (3.179)) one has

$$(z^*, z - z^*) \geq 0 \quad \forall z \in L(x^*). \qquad (3.199)$$

Since $f'(x^*) \in L(x^*)$, $h_i'(x^*) \in L(x^*) \quad \forall i \in R(x^*)$, then (3.199) yields

$$(f'(x^*), z^*) \geq ||z^*||^2 = a^2, \qquad (3.200)$$

$$(h_i'(x^*), z^*) \geq ||z^*||^2 = a^2 \quad \forall i \in R(x^*). \qquad (3.201)$$

Take $g = -z^*$ and put $x_\alpha = x^* + \alpha g$, where $\alpha > 0$. Since

$$f(x_\alpha) = f(x^* + \alpha g) = f(x^*) + \alpha(f_i'(x^*), g) + o(\alpha),$$

$$h_i(x_\alpha) = h_i(x^*) + \alpha(h_i'(x^*), g) + o_i(\alpha) \quad \forall i \in R(x^*),$$

$$h_i(x^*) = 0 \quad \forall i \in R(x^*),$$

where $\dfrac{o(\alpha)}{\alpha} \xrightarrow[\alpha \downarrow 0]{} 0$, $\dfrac{o_i(\alpha)}{\alpha} \xrightarrow[\alpha \downarrow 0]{} 0 \quad \forall i \in R(x^*)$, then (3.200) and (3.201) imply

that there exists $\alpha_1 > 0$ such that

$$f(x_\alpha) \leq f(x^*) - \alpha \frac{a^2}{2} \quad \forall \alpha \in (0, \alpha_1], \qquad (3.202)$$

$$h_i(x_\alpha) < -\alpha \frac{a^2}{2} \quad \forall \alpha \in (0, \alpha_1], \ \forall i \in R(x^*). \tag{3.203}$$

For $i \notin R(x^*)$ the inequality $h_i(x^*) < 0$ holds. Therefore there exists $\alpha_2 > 0$ such that

$$h_i(x_\alpha) < 0 \quad \forall \alpha \in (0, \alpha_2], \ \forall i \notin R(x^*). \tag{3.204}$$

It follows from (3.202)–(3.204) that for $\alpha \in (0, \alpha_0]$, with $\alpha_0 = \min\{\alpha_1, \alpha_2\}$, one has

$$f(x_\alpha) \le f(x^*) - \alpha \frac{a^2}{2} < f(x^*), \tag{3.205}$$

$$h_i(x_\alpha) < 0 \quad \forall i \in I. \tag{3.206}$$

The relation (3.206) implies that $x_\alpha \in \Omega \ \forall \alpha \in (0, \alpha_0]$, and, hence, (3.205) yields $f(x_\alpha) < f(x^*)$, which contradicts the assumption that $x^*$ is a minimizer of $f$ on $\Omega$. $\qquad\square$

A point $x^* \in \Omega$, satisfying (3.196), is called an *inf-stationary* point of the function $f$ on the set $\Omega$, defined by (3.194).

*Remark 3.9.1.* It follows from (3.196) that there exist coefficients $\alpha_0 \ge 0$, $\alpha_i \ge 0 (i \in R(x^*))$, such that

$$\alpha_0 + \sum_{i \in R(x^*)} \alpha_i = 1, \quad \alpha_0 f'(x^*) + \sum_{i \in R(x^*)} \alpha_i h_i'(x^*) = 0_n. \tag{3.207}$$

Putting $\alpha_i = 0$ for all $i \notin R(x^*)$, one can rewrite (3.207) in the form

$$\alpha_0 f'(x^*) + \sum_{i \in I} \alpha_i h_i'(x^*) = 0_n. \tag{3.208}$$

Note that

$$\alpha_i h_i(x^*) = 0 \quad \forall i \in I. \tag{3.209}$$

The condition (3.209) is called the *complementarity slackness condition*.

The relation (3.208) means that $x^*$ is a stationary point of the function

$$\mathcal{L}_\alpha(x) = \alpha_0 f(x) + \sum_{i \in I} \alpha_i h_i(x) \tag{3.210}$$

on $\mathbb{R}^n$. Here $\alpha = (\alpha_0, \alpha_1, \ldots, \alpha_N) \in \mathbb{R}^{N+1}$. The function $\mathcal{L}_\alpha(x)$ is continuously differentiable on $\mathbb{R}^n$.

*Remark 3.9.2.* Let $h(x^*) = 0$. If $x^* \in \Omega$ is not an inf-stationary point of the function $h(x)$ (defined by the relation (3.195)), then (see (3.184))

$$0_n \notin \mathrm{co}\{h_i'(x^*) \mid i \in R(x^*)\}. \tag{3.211}$$

The condition (3.211) is called a *regularity condition*. Now it follows from (3.207) that $\alpha_0 > 0$, therefore, putting $\lambda_i = \alpha_i/\alpha_0$, one concludes from (3.208) and (3.209) that there exist coefficients $\lambda_i \geq 0$, such that the relations

$$f'(x^*) + \sum_{i \in I} \lambda_i h_i'(x^*) = 0_n, \qquad (3.212)$$

$$\lambda_i h_i(x^*) = 0 \quad \forall i \in I \qquad (3.213)$$

hold.

Let us introduce the function

$$L_\lambda(x) = L(x, \lambda) = f(x) + \sum_{i \in I} \lambda_i h_i(x),$$

where $\lambda = (\lambda_1, \ldots, \lambda_N)$. The function $L_\lambda(x)$ is continuously differentiable on $\mathbb{R}^n$. The condition (3.212) means that $x^*$ is a stationary point of the function $L_\lambda(x)$ on $\mathbb{R}^n$. The multipliers $\lambda_i$ are called (see Section 3.8) *Lagrange multipliers*, and the function $L_\lambda(x)$ is the *Lagrange function*. Hence, the following condition is proved.

**Theorem 3.9.4.** *If $x^* \in \Omega$ is a minimizer of the function $f$ on the set $\Omega$, given by the relation (3.194), then there exist coefficients $\alpha_i \geq 0, i \in 0 : N$, such that $\alpha_0 + \sum_{i \in N} \alpha_i = 1$ and the point $x^*$ is a stationary point of the function $\mathcal{L}_\alpha(x)$, described by (3.210). The complementarity slackness condition (3.209) also holds.*

*If the regularity condition (3.211) is satisfied then there exist coefficients $\lambda_i \geq 0, i \in I$, such that $x^*$ is a stationary point of the Lagrange function $L_\lambda(x)$. The complementarity slackness condition (3.213) is valid as well.*

**Corollary 3.9.1.** *If $h \in C^1(\mathbb{R}^n)$, $\Omega = \{x \in \mathbb{R}^n \mid h(x) \leq 0\}$ and, for some $x^*$, $h'(x^*) \neq 0_n$, then the condition (3.212) takes the form $f'(x^*) + \lambda h'(x^*) = 0_n$, where $\lambda \geq 0$.*


## Bibliographical Comments

### Section 1.

The discovery of the subdifferential by J. -J. Moreau and R. T. Rockafellar (see [71, 89]) was an important breakthrough.

N. Z. Shor [99] proposed the famous generalized gradient method (GGM). Yu. M. Ermoliev [38] and B. T. Polyak [79] elaborated this method. Generalizations and modifications of the GGM are described in [25, 53, 80, 100].

The problems related to convex functions are studied by *Convex Analysis* (see [25, 53, 85, 91]). The formula (1.31) was independently discovered by J. M. Danskin [8], V. F. Demyanov [15] and I.V.Girsanov [48].

Numerical methods for minimax problems based on the notion of sub-differential (1.32) were proposed by V. F. Demyanov, V. N. Malozemov, B. N. Pshenichnyi in [26, 84].

The problems related to maximum-type functions are treated by *Minimax Theory* (see [9, 26]).

The properties of convex and max-type functions inspired B. N. Pshenichnyi (see [86]) to introduce a new class of nonsmooth functions – the *class of subdifferentiable functions* (B. N. Pshenichnyi called them *quasidifferentiable*).

In 1972 N. Z. Shor introduced the notion of almost-gradient (see [100]).

N. Z. Shor used almost-gradients to construct numerical methods for min-imizing Lipschitz functions (see [101]). The Shor subdifferential is a compact set but not convex. In the convex case $co\ \partial_{sh}f(x) = \partial f(x)$. The set $\partial_{sh}f(x)$ is not convex, therefore it provides more "individual" information than the subdifferential $\partial f(x)$.

In 1973 F. Clarke proposed the notion of generalized gradient (see [6, 7]).

This tool (the Clarke subdifferential) was extremely popular in the 1970's and 80's and is still viewed by many people as the most powerful tool available in Nonsmooth Analysis and Nondifferentiable Optimization (see [57]).

R. T. Rockafellar (see [90]) introduced the generalized directional deriva-tive (1.51).

P. Michel and J. -P. Penot in [68] suggested the generalized derivatives (1.54) and (1.55). If $f$ is also d.d. then $\partial_{mp}f(x)$ in (1.54) and (1.55) is the Clarke subdifferential of the function $h(g) = f'(x, g)$ at the point $g = 0_n$. The set $\partial_{mp}f(x)$ is often called [55] the *small subdifferential*.

Many other subdifferentials (convex as well as nonconvex) were proposed later (see, e.g., [54, 57, 63, 69, 76]). They are useful for the study of some prop-erties of specific classes of nonsmooth functions. Their common property is that the function under consideration is investigated by means of one set (convex or nonconvex). The study of nonsmooth functions by means of sev-eral sets was started by V. F. Demyanov and A. M. Rubinov who in 1979 introduced the so-called quasidifferentiable (q.d.) functions (see [17]). In 1980 B. N. Pschenichny (see [85]) proposed the notions of upper convex and lower concave approximations of the directional derivative. In 1982 A. M. Rubinov (see [18]) described exhaustive families of upper convex and lower concave ap-proximations. This led to the notions of upper and lower exhausters [16, 22], and finally the study of directional derivatives ("usual" and generalized) was reduced to the usage of a family of convex sets (see [105]). Some generalized subdifferentials were investigated via exhausters (see [94, 95]).

Properties of quasidifferentiable functions were studied in [21, 24, 25, 29, 62, 82, 98]. Most problems and results of Classical Differential Calculus may be formulated for nonsmooth functions in terms of quasidifferentials (see, e.g., [25, 29]). For example, a Mean-value theorem is valid [108].

The majority of the tools described in Section 1 provide positively homo-geneous approximations of the increment. The used mappings are, in general, discontinuous as functions of a point. This creates computational problems. Nonhomogeneous approximations (by means of *codifferentials* (see Subsection 1.4.5) were introduced in [28, 29] (see [16, 110]).

**Section 2.** Extremum problems in the smooth finite dimensional case are well studied. Unconstrained optimality conditions in different nonsmooth cases were discussed in [25, 26, 28, 44, 58, 81, 86, 91, 101, 106]. The results of Section 6 are published in [27]. The value $f^{\downarrow}(x)$ was, to our best knowledge, first introduced (under the name "slope") by De Giorgi in [10] (for more details see [59]).

**Section 3.** There exist many approaches to solving constrained optimization problems. For mathematical programming problems a lot of methods and algorithms have been proposed (their description one can find in monographs, text-books and reference-books, see, e.g., [3, 10, 34, 45, 61, 81, 87, 106, 112, 113]).

Necessary optimality conditions are of great importance and therefore attract attention and interest of the researchers. Existence of different (and often equivalent) conditions is useful in solving practical problems: the user can choose a condition which is more tailored for a specific problem. Necessary (and sometimes sufficient) conditions were studied in [25, 26, 28, 34, 44, 49, 53, 72, 81, 86, 91, 102, 106].

The idea of reducing the constrained optimization problem to an unconstrained one was implemented by different tools: by means of imbedding methods (see, for example, [5]), via Lagrange multipliers technique (see [3, 58]). Lagrange multipliers are productive for the theoretical study of the problem (see [58, 66, 92, 105]), however, their practical implementation is often difficult due to the fact that the mentioned multipliers are unknown. This problem is overcome by means of exact penalty functions.

The idea of *exact penalties* was first proposed in 1966 by I. I. Eremin in [35] (see also [36, 111]). In the same 1966 G. Danskin and V. F. Demyanov (see [8, 13, 15]) published the formula for the directional derivative of the max-type function. For some special cases this formula was obtained by I. V. Girsanov and B. N. Pschenichny in [48, 85]. Later on Theory of Exact penalty functions [4, 14, 40, 43, 51, 60] and Minimax theory [9, 26] came into being.

Exact penalties were successfully applied to different problems (see [2, 32, 33, 40, 41, 47, 51, 52, 67, 75, 77, 83, 104, 107, 111]).

The main advantage of the exact penalization approach is clear: the constrained optimization problem is replaced by an unconstrained one. The main disadvantage is that, even if the original function is smooth, the penalty function is essentially nonsmooth. However, the present state of Nonsmooth Analysis and Nondifferentiable optimization makes it possible to solve the new nonsmooth problem quite efficiently (see, e.g., [7, 25, 28, 29, 41, 73, 74, 86, 91, 93, 96]). Surveys of such methods can be found in [4, 31, 43, 60]. Extension of the area of applicability of exact penalization depends on new conditions under which exact penalties can be used. One of such conditions was found by V. V. Fedorov in [41]. A more constructive assumption guaranteeing the condition in [41] was proposed by G. Di Pillo and F. Facchinei in [32]. They used the generalized Clarke derivative [7]. This condition was generalized in [14] and [30] (see also [1]). In these papers the Hadamard upper and lower derivatives were employed. Just this approach is used in Section 3. In the above mentioned publications, the penalty parameter $\lambda$ in penalty functions had a linear form. Nonlinear penalty functions were employed by Yu. G. Evtushenko and A. M. Rubinov in [40, 97].

The condition formulated in Subsection 3.6.2 was stated by V. V. Fedorov (see [41]).

# References

1. Aban'kin A. E., *On exact penalty functions*, Vestnik of St. -Petersburg University. Ser. 1, 1966, Issue 3, 13, 3–8.
2. Bertsekas D., *Necessary and sufficient conditions for a penalty function to be exact*, Mathematical Programming, 1975. Vol. 9, 1, 87–99.
3. Bertsekas D., *Constrained optimization and Lagrange multiplier methods*. New York, Acalemic Press. 1982.
4. Boukary D, Fiacco A. V., *Survey of penalty, exact-penalty and multiplier methods from 1968 to 1993*, Optimization, 1995. Vol. 32, 4, 301–334.
5. Bulatov V. P., *Imbedding methods in optimization problems* (In Russian). Novosibirsk: Nauka Publishers, 1977. 158 p.
6. Clarke F., *Generalized gradients and applications*, Trans. Amer. Math. Soc, 1975, **205**, 2, 247–262.
7. Clarke, F. H., *Optimization and nonsmooth analysis.*, N. Y.: Wiley, 1983, 308 p.
8. Danskin J. M., *The theory of max-min, with applications*, SIAM J. on Applied Math., 1966. Vol. 14, p. 641–664.
9. Danskin J. M., *The theory of Max-min and its Application to weapons Allocation Problems*. Berlin, Springer. 1967.
10. De Giorgi E., Marino A., Tosques M., *Problemi di evoluzione in spazi metricie curve di massima pendenza*, Atti Acad. Naz. Lincei Rend. Cl. Sci. Fis. Mat. Natur., 1980. Vol. 6, 68, 180–187.
11. Demyanov V. F., *Conditional derivatives and exhausters in Nonsmooth Analysis.* shape Doklady of the Russian Academy of Sciences, 1999, **338**, No. 6, 730-733.
12. Demyanov V. F. and Rubinov A. M., *Exhaustive families of approximations revisited.* In: R. P. Gilbert, P. D. Panagiotopoulos, P. M. Pardalos (Eds. ) shape From Convexity to Nonconvexity. Nonconvex Optimization and Its Applications, 2001. **55**, 43–50. (Dordrecht: Kluwer Scientific Publishers).
13. Demyanov V. F., *On the solution of some minimax problems. Part I*, Kibernetika, 1966. 6, 58–66. *Part II*, Kibernetika, 1967, 3, 62–66.
14. Demyanov V. F., *Exact penalty functions in nonsmooth optimization problems.*, Vestnik of St. -Petersburg University. Ser. 1, 1994. Issue 4, 22, 21–27.
15. Demyanov V. F., *On the minimization of a maximal deviation*, Vestnik of Leningrad University, 1966, Ser. 1, 7, 21–28.
16. Demyanov V. F., *Exhausters and convexificators – new tools in Nonsmooth Analysis*, In: [88], 2000, 85–137.
17. Demyanov V. F., Rubinov A. M. On quasidifferentiable functionals, Doklady of USSR Acad. of Sci, 1980, 250, 1, 21–25.
18. Demyanov V. F., Rubinov A. M., *Elements of Quasidifferential Calculus*, In: "Nonsmooth Problems of Optimization Theory and Control". (In Russian). Ed. V. F. Demyanov. – Leningrad: Leningrad University Press, 1982, Ch. 1, 5-127. English revised version see in [88], 1–31.

19. Demyanov V. F., Jeyakumar V., *Hunting for a smaller convex subdifferential*, J. of *Global Optimization*, 1997, 10, 3, 305–326.

20. Demyanov V. F., Di Pillo G., Facchinei F., *Exact penalization via Dini and Hadamard conditional derivatives*, Optimization Methods and Software, 1998, 9, 19–36.

21. Demyanov V. F., Rubinov A. M., *Quasidifferential Calculus.*, New York: Optimization Software, 1986. xi+288 p.

22. Demyanov V. F., *Exhausters of a positively homogeneous function*, Optimization., 1999, 45, P. 13–29.

23. Demyanov V. F., *Convexification of a positively homogeneous function*, Doklady of the Russian Academy of Sciences., 1999, 366, N. 6, P. 734–737.

24. Demyanov, V. F., Dixon, L. C. W., (Eds.): *Quasidifferential calculus. Mathematical Programming Study*, 1985, Vol. 29, 221 p.

25. Demyanov, V. F., Vasiliev, L. V., *Nondifferentiable Optimization*. New York: Springer-Optimization Software, 1986 (Russian Edition, Moscow: Nauka, 1981).

26. Demyanov, V. F., Malozemov, V. N., *Introduction to Minimax*. Moscow: Nauka, 1972, 368 p. (English translation by J. Wiley, 1974, 2nd edition, 1990).

27. Demyanov V. F., *Conditions for an Extremum in Metric Spaces.*, Journal of Global Optimization, 2000. Vol. 17, Nos. 1-4. P. 55–63.

28. Demyanov V. F., Rubinov A. M., *Foundations of nonsmooth analysis. Quasidifferential Calculus.* (In Russian) Moscow: Nauka publishers, 1990. 431 p.

29. Demyanov V. F., Rubinov A. M., *Constructive Nonsmooth Analysis.* Frankfurt a/M. : Verlag Peter Lang, 1995. 416 p.

30. Demyanov V. F., Di Pillo G., Facchinei F., *Exact penalization via Dini and Hadamars conditional derivatives.*, Optimization Methods and Software, 1998. Vol. 9. P. 19–36.

31. Di Pillo G., *Exact penalty methods.*, Ed. E. Spedicato: Algorithms for Continuous Optimization. The State-of-the-Art. NATO ASI Series C, Vol. 434. Dordrecht: Kluwer Academic Publishers, 1994. P. 209–253.

32. Di Pillo G., Facchinei F., *Exact penalty functions for nondifferentiable programming problems.*, Eds. F. H. Clarke, V. F. Demyanov and F. Giannessi: Nonsmooth Optimization and Related Topics. N. Y. : Plenum, 1989. P. 89–107.

33. Di Pillo G., Grippo L., *Exact penalty functions in constrained optimization.*, SIAM J. on Contr. and Optimiz., 1989. Vol. 27 P. 1333–1360.

34. Dubovitski, A. Ya., Milyutin, A. A., *Extremal problems in the presence of constraints*, USSR Comput. Mathematics and Math. Physics, 1965, Vol. 5, No. 3, 395–453.

35. Eremin I. I., *On the penalty method in convex programming*, Abstracts of Intern. Congress of Mathematicians. Section 14. Moscow, 1966.

36. Eremin I. I., *Penalty methods in convex programming* (in Russian), Doklady of the USSR Academy of Sciences, 1967, Vol. 143, N 4, P. 748–751.

37. Eremin, I. I., Astafiev, N. N., *Introduction to Theory of Linear and Convex Programming* (in Russian), Moscow: Nauka, 1976, 192 p.

38. Ermoliev Yu. M., *Methods for solving nonlinear extremum problems*, Kibernetika, 1966, N. 4, 1–17.

39. Ermoliev Yu. M., *Stochastic programming methods.* (In Russian) Moscow: Nauka publishers, 1976. 240 p.

40. Evtushenko Yu. G., Zhadan V. G., *Exact auxilliary functions in optimization problems*, J. of computational mathematics and mathematical physics 1990. Vol. 30, N 1. P. 43–57.

41. Fedorov V. V., *Numerical methods for maximin problems.* (In Russian) Moscow: Nauka publishers, 1979. 278 p.

42. Fiacco A. V., McCormick G. P., *Nonlinear Programming: Sequential Unconstrained Minimization Techniques.* New York: John Wiley and Sons, Inc., 1987.

43. Fletcher R., *Penalty functions, Mathematical programming: the-State-of-the-Art.*, Eds. A. Bachen, M. Grötschel, B. Korte). Berlin: Springer–Verlag, 1983. P. 87–114.

44. Gabasov R., Kirillova F. M., *Optimization methods.* (In Russian) Minsk: Belarus University Press, 1981. 350 p.

45. Galeev E. M., Tikhomirov V. N., *A short course of extremal problems theory.* Moscow, Moscow University Press, 1989. 204 p.

46. Giannessi F., *A Common Understanding or a Common Misunderstanding?*, Numer. Funct. Anal. and Optimiz., 1995. Vol. 16, 9-10, 1359–1363.

47. Giannessi F., Niçolucci F., *Connections between nonlinear and integer programming problems.*, Symposia Mathematica, 1976, Vol. 19, 161–175.

48. Girsanov I. V., *Lectures on the Mathematical Theory of Extremum Problems.*– Moscow: Moscow Univ. Press, 1970, 118 p.

49. Golshtein E. G., *Duality theory in mathematical programming and its applications.* (In Russian) Moscow: Nauka publishers, 1971. 352 p.

50. Gorokhovik, V. V., *ε-quasidifferentiability of real-valued functions and optimality conditions in extremal problems.* In: [24], 203–218.

51. Grossman K., Kaplan A. A., *Nonlinear programming via unconstrained minimization.* (In Russian) Novosibirsk: Nauka publishers, 1981. 184 p.

52. Han S., Mangasarian O., *Exact penalty functions in nonlinear programming.*, Math. Progr., 1979, 17, 251–269.

53. Hiriart-Urruty J. B., Lemarechal C., *Convex analysis and Minimization algorithms I, II.* Berlin: Springer-Verlag, 1993. 420 p., 348 p.

54. Hiriart-Urruty J.-B., *New concepts in nondifferentiable programming.*, Bul. Soc. Math. France, 1979, Mem. 60, 57–85.

55. Ioffe A. D., *A Lagrange multiplier rule with small convex-valued subdifferentials for nonsmooth problems of mathematical programming involving equality and nonfunctional constraints, Mathematical Programming.*, 1993, 58, 137–145.

56. Ioffe A. D., Tikhomirov V. M., *Theory of Extremal Problems.* Moscow: Nauka, 1974, 480 p. (English transl. by North-Holland, 1979).

57. Ioffe A. D., *Nonsmooth analysis: differential calculus of nondifferentiable mappings,* Trans. Amer. Math. Soc, 1981, 266, 1, 1–56.

58. Ioffe, A. D. Tikhomirov, V. M., *Theory of Extremal Problems.*, Moscow: Nauka, 1974, 479 p. (English transl. by North-Holland, Amsterdam, New York, 1979).

59. Ioffe A. D., *Metric regularity and subdifferential calculus,* Uspehi Matem. nauk, 2000, 55, 3, 103–162.

60. Kaplan A. A., Tichatschke R., *Stable methods for ill-posed variational problems.* Berlin: Akademie Verlag, 1994.

61. Karmanov V. G., *Mathematical Programming.* In Russian Moscow: Nauka publishers, 1980, 256 p.

62. Kiwiel K. C., *Exact penalty functions in proximal bundle methods for constrained convex nondifferentiable minimization,* Math. Progr., 1991, 52, 285–302.

63. Kruger A. Ya., *On properties of generalized differentials,* Syberian Math. J., 1985, 26, 54–66.

64. Kusraev A. G., Kutateladze S. S., *Subdifferentials: Theory and Applications,* New York: Kluwer Academic Publishers, 1995.

65. Kutateladze S. S., Rubinov A. M., *Minkowski duality and its applications,* Russian Mathem. Surveys, 1972, 27, 131–191.

66. Luderer B., Rosiger R., Wurker U., *On necessary minimum conditions in quasidifferential calculus: independence on the specific choice of quasidifferential,* Optimization, 1991, 22, 643–660.

67. Mangasarian O. L., *Sufficiency in exact penalty minimization,* SIAM J. on Contr. and Optimiz., 1985. Vol. 23, P. 30–37.

68. Michel P., Penot J.-P., *Calcus sous-différential pour les fonctions lipschitzienness et non-lipschitziennes,* C. R. Acad. Sc. Paris, Ser. I, 1984, 298, 269–272.

69. Mordukhovich B. S., *Nonsmooth analysis with nonconvex generalised differentials and conjugate mappings,* Doklady of BSSR Acad. of Sci, 1984, 28, No. 11, P. 976–979.

70. Mordukhovich B. S., *Approximation methods in problems of optimization and control. (In Russian),* Moscow: Nauka, 1988, 359 p.

71. Moreau J.-J., *Fonctionelles sous-differentiables*, C. R. Acad. Sci. Paris, Ser. A-B, 1963, 257, P. 4117–4119.

72. Neustadt L. W., *Optimization: A theory of necessary conditions*. Princeton, N. J. : Princeton University Press, 1976. xii+424 p.

73. Nurminski E. A., *A quasigradient method for solving the mathematical programming problem*, Kibernetika, 1973, 1, 122–125.

74. Pallaschke D., Rolewicz S., *Foundations of mathematical optimization. Convex analysis without linearity*. Dordrecht: Kluwer Academic Publishers, 1997. xii+582 p.

75. Panin V. M., *Finite penalties methods with linear approximation. Part I*, Kibernetika, 1984, 2, 44–50. *Part II*, Kibernetika, 1984, 4, 73–81.

76. Penot J.-P., *Sous-differentiels de fonctions numeriques non convexes*, C. R. Acad. Sc. Paris, Ser. A, 1974, 278, 1553–1555.

77. Pietrzykowski T., *An exact potential method for constrained maxima*, SIAM J. on Numer. Anal., 1969, 6, 299–304.

78. Polak E., *Computational Methods in Optimization: A Unified Approach*. N. Y. : Academic Press, 1971.

79. Polyak B. T., *A general method for solving extremum problems*, Dokl. AN SSSR, 1967, 174, 1, 33–36.

80. Polyak B. T., *Minimization of nonsmooth functionals*, Zhurn. Vychisl. Matem. i Matem. Fiz, 1969, 9, 3, 509–521.

81. Polyak B. T., *Introduction to Optimization*. (In Russian) Moscow: Nauka publishers, 1983, 384 p.

82. Polyakova, L. N., *Necessary conditions for an extremum of a quasidifferentiable function subject to a quasidifferentiable constraint*, Vestnik of Leningrad Univ., 1982, 7, 75–80.

83. Polyakova L. N., *On an exact penalty method for quasidifferentiable functions*, J. of comptutational mathematics and mathematical physics 1990, 30, 1, 43–57.

84. Pschenichny B. N., Danilin Yu. M., *Numerical methods in extremal problems*. (In Russian) Moscow: Nauka publishers, 1975, 320 p.

85. Pschenichny B. N., *Convex Analysis and Extremal Problems* (in Russian), Moscow: Nauka, 1980, 320 p.

86. Pschenichny B. N., *Necessary conditions for an extremum. 2nd ed.*, Moscow: Nauka, 1982, 144 p. (English translation of the 1st ed. by Marsel Dekker, New York, 1971).

87. Pschenichny B. N., *Method of Linearization.*, Moscow: Nauka, 1983, 136 p.

88. Quasidifferentiability and Related Topics., *Nonconvex Optimization and its Applications*, Vol. 43. Eds. V. Demyanov, A. Rubinov, Dordrecht/Boston/London: Kluwer Academic Publishers, 2000, xix+391 p.

89. Rockafellar R. T., *Duality theorems for convex functions*, Bull. Amer. Math. Soc, 1964, 70, 189–192.

90. Rockafellar R. T., *The theory of subgradients and its application to problems of optimization.*, Berlin: Holdermann, 1981, 107p.

91. Rockafellar, R. T., *Convex Analysis.* Princeton N. J: Princeton Univ. Press, 1970.

92. Rockafellar R. T., *Lagrange multipliers and optimality*, SIAM Review., 35, 2, 183–238.

93. Rockafellar R. T., Wets R. J. B., *Variational Analysis*. Berlin: Springer, 1998. xiii+733 p.

94. Roshchina V., *On the relationship between the Fréchet subdifferential and upper exhausters*, shape International Workshop on Optimization: Theory and Algorithms. 19 - 22 August 2006, Zhangjiajie, Hunan, China.

95. Roshchina V., *Relationships Between Upper Exhausters and the Basic Subdifferential in Variational Analysis*, J. Math. Anal. Appl. (2007), doi: 10. 1016/j. jmaa. 2006. 12. 059.

96. Rubinov A. M., *Abstract convexity and global optimization*. Dordrecht: Kluwer Academic Publishers, 2000. xviii+490 p.

97. Rubinov A. M., Yang X. Q., Glover B. M., *Extended Lagrange and penalty functions in optimization*, J. of Optimiz. Theory and Appl., 2001, 111, 2, 381–405.

98. Shapiro A., *On concepts of directional differentiability*, J. Optim. Theory Appl. 66 (1990), 3, 477–487.

99. Shor N. Z., *The development of numerical methods for nonsmooth optimization in the USSR*, In: History of Mathematical Programming, J. K. Lenstra, A. H. G. Rinnoy Kan, and A. Shrijver, eds., Amsterdam, CWI, North-Holland, 1991, 135–139.

100. Shor N. Z., *On a method for minimizing almost-differentiable functions*, Kibernetika, 1972, 4, 65–70.

101. Shor N. Z., *Methods for minimizing nondifferentiable functions.* (In Russian) Kiev: Naukova dumka publishers, 1979. 199 p.

102. Studniarski M., *Necessary and sufficient conditions for isolated local minimum of nonsmooth functions*, SIAM J. Control and Optimization, 1986, 24, 1044–1049.

103. Sukharev A. G., Timokhov A. V., Fedorov V. V., *A course of optimization methods.* (In Russian) Moscow: Nauka publishers, 1986, 325 p.

104. Tretyakov N. V., *An exact estimates method for convex programming problems*, Economics and mathematical methods, 1972, 8, 5, 40–751.

105. Uderzo A., *Quasi-multiplier rules for quasidifferentiable extremum problems*, Optimization, 2002, 51, 6, 761–795.

106. Vasiliev F. P., *Numerical methods of solving extremal problems.* Moscow, Nauka publishers, 1980. 520 p.

107. Ward D. E., *Exact penalties and sufficient conditions for optimality in nonsmooth optimization*, J. of Optimiz. Theory and Appl., 1988, 57, 485–499.

108. Xia, Z. Q., *On mean value theorems in quasidifferential calculus*, J. Math. Res. and Exposition, Dalian, China, D. I. T., 1987, 4, 681–684.

109. Yudin D. B., Golshtein E. G., *Linear programming. Theory and finite methods.* (In Russian) Moscow: Fizmatgiz publishers, 1963, 775 p.

110. Zaffaroni A., *Continuous approximations, codifferentiable functions and minimization methods*, In: [88], 2000, P. 361–391.

111. Zangwill W. I., *Non-linear programming via penalty functions*, Management Science, 1967, 13, 5, 344–358.

112. Zangwill W. I., *Nonlinear Programming: A Unified Approach.* Prentice Hall, Englewood Cliffs, 1969.

113. Zoutendijk G., *Methods of feasible directions – A study in Linear and Nonlinear Programming.* Amsterdam, Elsevier, 1960.

# The Sequential Quadratic Programming Method

Roger Fletcher

## 1 Introduction

Sequential (or Successive) Quadratic Programming (SQP) is a technique for
the solution of Nonlinear Programming (NLP) problems. It is, as we shall
see, an idealized concept, permitting and indeed necessitating many variations and modifications before becoming available as part of a reliable and
efficient production computer code. In this monograph we trace the evolution of the SQP method through some important special cases of nonlinear
programming, up to the most general form of problem. To fully understand
these developments it is important to have a thorough grasp of the underlying theoretical concepts, particularly in regard to optimality conditions. In
this monograph we include a simple yet rigorous presentation of optimality
conditions, which yet covers most cases of interest.

   A nonlinear programming problem is the minimization of a nonlinear *ob-
jective function* $f(\mathbf{x})$, $\mathbf{x} \in \mathbb{R}^n$, of $n$ variables, subject to equation and/or
inequality *constraints* involving a vector of nonlinear functions $\mathbf{c}(\mathbf{x})$. A basic
statement of the problem, useful for didactic purposes is

$$\underset{\mathbf{x}\in\mathbb{R}^n}{\text{minimize}} \ \ f(\mathbf{x})$$

$$\text{subject to } c_i(\mathbf{x}) \geq 0 \qquad i = 1, 2, \ldots, m. \tag{1.1}$$

In this formulation, equation constraints must be encoded as two opposed in-
equality constraints, that is $c(\mathbf{x}) = 0$ is replaced by $c(\mathbf{x}) \geq 0$ and $-c(\mathbf{x}) \geq 0$,
which is usually not convenient. Thus in practice a more detailed formula-
tion is appropriate, admitting also equations, linear constraints and simple
bounds. One way to do this is to add slack variables to the constraints, which

R. Fletcher (✉)
Department of Mathematics, University of Dundee, Dundee DD1 4HN
e-mail: fletcher@maths.dundee.ac.uk

G. Di Pillo et al. (eds.), *Nonlinear Optimization*, Lecture Notes in
Mathematics 1989, DOI 10.1007/978-3-642-11339-0_3,
© Springer-Verlag Berlin Heidelberg 2010

together with simple bounds on the natural variables, gives rise to

$$\begin{aligned}
& \underset{\mathbf{x} \in \mathrm{IR}^n}{\text{minimize}} \quad f(\mathbf{x}) \\
& \text{subject to } A^T \mathbf{x} = \mathbf{b} \\
& \qquad\qquad \mathbf{c}(\mathbf{x}) = \mathbf{0} \\
& \qquad\qquad \mathbf{l} \le \mathbf{x} \le \mathbf{u}.
\end{aligned} \tag{1.2}$$

in which either $u_i$ or $-l_i$ can be set to a very large number if no bound is present. Alternatively one might have

$$\begin{aligned}
& \underset{\mathbf{x} \in \mathrm{IR}^n}{\text{minimize}} \quad f(\mathbf{x}) \\
& \text{subject to } \mathbf{l} \le \begin{pmatrix} \mathbf{x} \\ A^T \mathbf{x} \\ \mathbf{c}(\mathbf{x}) \end{pmatrix} \le \mathbf{u}
\end{aligned} \tag{1.3}$$

in which the user can specify an equation by having $l_i = u_i$.

There are a number of special cases of NLP which are important in their own right, and for which there are special cases of the SQP method for their solution. These include *systems of nonlinear equations, unconstrained optimization,* and *linearly constrained optimization.* Understanding the theoretical and practical issues associated with these special cases is important when it comes to dealing with the general NLP problem as specified above. A common theme in all these cases is that there are *'linear' problems* which can be solved in a finite number of steps (ignoring the effects of round-off error), *nonlinear problems* which can usually only be solved approximately by iteration, and *Newton methods* for solving nonlinear problems by the successive solution of linear problems, obtained by making Taylor series expansions about a current iterate. This context is underpinned by a famous theorem of Dennis and Moré [10] which states, subject to a regularity condition, that superlinear convergence occurs if and only if an iterative method is asymptotically equivalent to a Newton method. Loosely speaking, this tells us that we can only expect rapid convergence if our iterative method is closely related to a Newton method. The SQP method is one realization of a Newton method.

Generally $\mathbf{x} \in \mathrm{IR}^n$ will denote the variables (unknowns) in the problem, $\mathbf{x}^*$ denotes a (local) solution, $\mathbf{x}^{(k)}$, $k = 1, 2 \ldots$ are iterates in some iterative method, and $\mathbf{g}^{(k)}$ for example denotes the function $\mathbf{g}(\mathbf{x})$ evaluated at $\mathbf{x}^{(k)}$. Likewise $\mathbf{g}^*$ would denote $\mathbf{g}(\mathbf{x})$ evaluated at $\mathbf{x}^*$. It is important to recognise that solutions to problems may not exist, or on the other hand, there may exist multiple or non-unique solutions. Generally algorithms are only able to find local minima, and indeed guaranteeing to find a global (best local) minimizer is impractical for problems of any significant size. A more extensive and detailed coverage of topics in this monograph can be found for example in Fletcher [13].

## 2  Newton Methods and Local Optimality

In this and subsequent sections we trace the development of Newton methods from the simplest case of nonlinear equations, through to the general case of nonlinear programming with equations and inequalities.

### 2.1  Systems of $n$ Simultaneous Equations in $n$ Unknowns

In this case the 'linear' problem referred to above is the well known system of linear equations

$$A^T \mathbf{x} = \mathbf{b} \tag{2.1}$$

in which $A$ is a given $n \times n$ matrix of coefficients and $\mathbf{b}$ is a given vector of right hand sides. For all except the very largest problems, the system is readily solved by computing factors $PA^T = LU$ using elimination and partial pivoting. If $A$ is a very sparse matrix, techniques are available to enable large systems to be solved. A less well known approach is to compute implicit factors $LPA^T = U$ (see for example Fletcher [14]) which can be advantageous in certain contexts. The system is *regular* when $A$ is nonsingular in which case a unique solution exists. Otherwise there may be non-unique solutions, or more usually no solutions.

The corresponding nonlinear problem is the system of nonlinear equations

$$\mathbf{r}(\mathbf{x}) = \mathbf{0} \tag{2.2}$$

in which $\mathbf{r}(\mathbf{x})$ is a given vector of $n$ nonlinear functions. We assume that $\mathbf{r}(\mathbf{x})$ is continuously differentiable and denote the $n \times n$ *Jacobian matrix* by $A = \nabla \mathbf{r}^T$, that is $A$ is the matrix whose columns are the gradients of the component functions in $\mathbf{r}$. If $\mathbf{r} = A^T \mathbf{x} - \mathbf{b}$ is in fact linear then its Jacobian matrix is $A$, which accounts for the use of $A^T$ rather than the more usual $A$ in (2.1). A Taylor series about the current iterate $\mathbf{x}^{(k)}$ gives

$$\mathbf{r}(\mathbf{x}^{(k)} + \mathbf{d}) = \mathbf{r}^{(k)} + A^{(k)T}\mathbf{d} + o(\|\mathbf{d}\|). \tag{2.3}$$

Truncating the negligible term in $\mathbf{d}$ and setting the left hand side to zero yields the system of linear equations

$$A^{(k)T}\mathbf{d} = -\mathbf{r}^{(k)}. \tag{2.4}$$

This system forms the basis of the *Newton-Raphson (NR) method* in which (2.4) is solved for a displacement $\mathbf{d} = \mathbf{d}^{(k)}$, and $\mathbf{x}^{(k+1)} = \mathbf{x}^{(k)} + \mathbf{d}^{(k)}$ becomes the next iterate.

A solution $\mathbf{x}^*$ is said to be *regular* iff $A^*$ is nonsingular. Assuming that $A$ is Lipschitz continuous in a neighbourhood of $\mathbf{x}^*$, then the Newton-Raphson method converges locally (that is if some $\mathbf{x}^{(k)}$ is sufficiently close to $\mathbf{x}^*$) and the order of convergence is second order, that is

$$\|\mathbf{x}^{(k+1)} - \mathbf{x}^*\| = O(\|\mathbf{x}^{(k)} - \mathbf{x}^*\|^2).$$

We prove these properties in the next subsection. This property is very desirable, indicating as it does that the number of significant figures of accuracy doubles on each iteration, in the limit. Moreover a theorem of Dennis and Moré [10] indicates that if $\mathbf{x}^*$ is regular then any sequence $\{\mathbf{x}^{(k)}\}$ converging to $\mathbf{x}^*$ exhibits superlinear convergence (that is $\|\mathbf{x}^{(k+1)} - \mathbf{x}^*\| = o(\|\mathbf{x}^{(k)} - \mathbf{x}^*\|)$ if and only if the displacements converge asymptotically to those of the Newton-Raphson method, in the sense that

$$\mathbf{d}^{(k)} = \mathbf{d}_{NR}^{(k)} + o(\|\mathbf{d}_{NR}^{(k)}\|).$$

We shall see in what follows that many successful methods of optimization (Newton methods) are derived from the Newton-Raphson method, including the SQP method, which indicates the fundamental importance of the Dennis and Moré result.

An important class of Newton methods are the so-called *quasi-Newton methods* in which $A^{(k)}$ is approximated by a matrix $B^{(k)}$. Initially $B^{(k)}$ is arbitrary, and is *updated* after each iteration to take advantage of information gained about how $\mathbf{r}(\mathbf{x})$ behaves.

Unfortunately this favourable theoretical profile of the NR method is only valid locally in a neighbourhood of a regular solution. If $\mathbf{x}^{(1)}$ is remote from $\mathbf{x}^*$ then there is no guarantee of convergence, and indeed $A^{(k)}$ can be singular in which case (2.4) usually has no solution and the method aborts. Modifications of the NR method to promote global convergence (that is convergence when $\mathbf{x}^{(1)}$ is remote from $\mathbf{x}^*$) are a subject of much active research interest, some of which is described later in this monograph.

## 2.2  Local Convergence of the Newton-Raphson Method

Let $\mathbf{x}^*$ be a solution of (2.2). Then the following theorem holds.

**Theorem 2.2.1.** *Assume that $\mathbf{r}(\mathbf{x})$ is continuously differentiable in a neighbourhood of $\mathbf{x}^*$, and that $A^*$ is nonsingular. Then there exists a neighbourhood of $\mathbf{x}^*$ such that if any iterate $\mathbf{x}^{(k)}$ is within this neighbourhood, then the Newton-Raphson method converges to $\mathbf{x}^*$ and the order of convergence is superlinear.*

*Proof.* For $\mathbf{x}$ in a neighbourhood of $\mathbf{x}^*$, denote $\mathbf{r} = \mathbf{r}(\mathbf{x})$, $\mathbf{e} = \mathbf{x} - \mathbf{x}^*$ and $\tilde{A} = \int_0^1 A(\mathbf{x} - \theta\mathbf{e})\,d\theta$. The conditions of the theorem permit an integral form of the Taylor series about $\mathbf{x}$

$$\mathbf{0} = \mathbf{r}^* = \mathbf{r} - \tilde{A}^T \mathbf{e} \tag{2.5}$$

to be used. As $\mathbf{e} \to \mathbf{0}$, so $\tilde{A} \to A$ ($A = A(\mathbf{x})$) and hence

$$\|\mathbf{r} - A^T\mathbf{e}\|/\|\mathbf{e}\| \to 0. \tag{2.6}$$

Because $A^*$ is nonsingular there exists a neighbourhood $\mathcal{N}(\mathbf{x}^*)$ in which both $\|A^{-T}\| \le \beta$ is bounded, and

$$\|\mathbf{r} - A^T\mathbf{e}\| \le \alpha\|\mathbf{e}\|/\beta \tag{2.7}$$

for some fixed $\alpha \in (0, 1)$, by virtue of (2.6).

Let $\mathbf{x}^{(k)} \in \mathcal{N}(\mathbf{x}^*)$ and denote the error $\mathbf{e}^{(k)} = \mathbf{x}^{(k)} - \mathbf{x}^*$. Then

$$\begin{align}
\mathbf{e}^{(k+1)} &= \mathbf{e}^{(k)} + \mathbf{d}^{(k)} \tag{2.8}\\
&= \mathbf{e}^{(k)} - (A^{(k)T})^{-1}\mathbf{r}^{(k)} \tag{2.9}\\
&= -(A^{(k)T})^{-1}(\mathbf{r}^{(k)} - A^{(k)T}\mathbf{e}^{(k)}). \tag{2.10}
\end{align}$$

It follows from (2.7) that $\|\mathbf{e}^{(k+1)}\| \le \alpha\|\mathbf{e}^{(k)}\|$ and hence that $\mathbf{x}^{(k+1)} \in \mathcal{N}(\mathbf{x}^*)$. By induction, $\mathbf{e}^{(k)} \to \mathbf{0}$, and hence $\mathbf{x}^{(k)} \to \mathbf{x}^*$. It also follows from (2.10) and (2.6) that $\|\mathbf{e}^{(k+1)}\|/\|\mathbf{e}^{(k)}\| \to 0$, showing that the order of convergence is superlinear.

$\square$

**Corollary 2.2.1.** *If, in addition, the Jacobian matrix $A(\mathbf{x})$ satisfies a Lipschitz condition, then the order of convergence is second order.*

*Proof.* In this case we can write the Taylor series (2.3) in the form

$$\mathbf{r}(\mathbf{x}^{(k)} + \mathbf{d}) = \mathbf{r}^{(k)} + A^{(k)T}\mathbf{d} + O(\|\mathbf{d}\|^2). \tag{2.11}$$

Now we can replace (2.6) by the stronger result that $\|\mathbf{r} - A^T\mathbf{e}\| = O(\|\mathbf{e}\|^2)$. Following a similar argument to the above, we then deduce that

$$\mathbf{e}^{(k+1)} = O(\|\mathbf{e}^{(k)}\|^2) \tag{2.12}$$

which is the definition of second order convergence. $\square$

## 2.3   Unconstrained Optimization

In this case the 'linear' problem is that of minimizing a quadratic function
of $n$ variables

$$q(\mathbf{x}) = \tfrac{1}{2}\mathbf{x}^T G \mathbf{x} + \mathbf{h}^T \mathbf{x} \qquad (2.13)$$

where the Hessian matrix of second derivatives $G$ is symmetric and positive
definite. The corresponding nonlinear problem is to find a local minimizing
point $\mathbf{x}^*$ of a given non-quadratic function $f(\mathbf{x})$. We refer to the gradient
(column) vector of first partial derivatives of $f(\mathbf{x})$ by $\mathbf{g}(\mathbf{x}) = \boldsymbol{\nabla} f(\mathbf{x})$ and the
Hessian by $G(\mathbf{x}) = \boldsymbol{\nabla}\mathbf{g}(\mathbf{x})^T$. If $\mathbf{x}^*$ is a local minimizer of $f(\mathbf{x})$ then clearly it
is a minimizer along any line

$$\mathbf{x}(\alpha) = \mathbf{x}^* + \alpha\mathbf{s}, \qquad \mathbf{s} \neq \mathbf{0}$$

through $\mathbf{x}^*$. It follows from this and the chain rule that the slope

$$df(\mathbf{x}(\alpha))/d\alpha|_{\alpha=0} = \mathbf{g}^{*T}\mathbf{s} = 0$$

for any $\mathbf{s}$. Consequently a necessary condition for $\mathbf{x}^*$ to be a local minimizer
of $f(\mathbf{x})$ is that $\mathbf{g}^* = \mathbf{0}$. Points $\mathbf{x}$ which satisfy $\mathbf{g}(\mathbf{x}) = \mathbf{0}$ are referred to
as *stationary points*, and include saddle points and maximizers as well as
minimizers. Similarly another necessary condition for a minimizer is that the
second derivative of $f(\mathbf{x}(\alpha))$ at $\alpha = 0$ is non-negative, that is $\mathbf{s}^T G^* \mathbf{s} \geq 0$ for
all $\mathbf{s}$, which is the condition that $G^*$ is positive semi-definite. On the other
hand, if both $\mathbf{g}^* = \mathbf{0}$ and $G^*$ is positive definite, then this is a *sufficient*
condition for $\mathbf{x}^*$ to be a local minimizer.

The stationary point condition $\mathbf{g}(\mathbf{x}) = \mathbf{0}$ is a system of nonlinear equations
that can be solved by the NR method to find a stationary point of $f(\mathbf{x})$. The
Jacobian $\boldsymbol{\nabla}\mathbf{g}(\mathbf{x})^T$ is just the Hessian $G(\mathbf{x})$. For minimization we are interested in the case that $G$ is positive definite. In the case of a quadratic function
(2.13), $\mathbf{g}(\mathbf{x}) = G\mathbf{x}+\mathbf{h}$ and we solve the system $G\mathbf{x} = -\mathbf{h}$. In the non-quadratic
case, the appropriate regularity condition is that $G^*$ is positive definite, and
the NR iteration formula (2.4) becomes $G^{(k)}\mathbf{d} = -\mathbf{g}^{(k)}$. These linear systems are most efficiently solved using Choleski factors $G = LL^T$ when $G$ is
positive definite. We refer to this method as the *Newton method for minimization*. The method inherits all the favourable local properties of the NR
method, described in the previous subsection. Likewise, there is also the possibility that when the initial point is remote from a minimizer, the method
might fail to converge, and so must be modified to promote global convergence. There is an additional issue that not only might $G^{(k)}$ be singular,
whence the method aborts, but also $G^{(k)}$ might become indefinite, remote
from the solution. In this case the local quadratic approximating function
$q(\mathbf{d}) = \tfrac{1}{2}\mathbf{d}^T G^{(k)}\mathbf{d} + \mathbf{g}^{(k)T}\mathbf{d}$ no longer has a minimizer, and the resulting displacement $\mathbf{d}^{(k)}$ obtained by finding the stationary point of $q(\mathbf{d})$ is unlikely

to be useful. Thus the most effective way of promoting global convergence is still a subject of some interest. In this respect, quasi-Newton methods are of particular interest because there exist methods of updating a Hessian approximating matrix $B^{(k)}$ so as to maintain the property that $B^{(k)}$ is symmetric positive definite (see for example [13]).

## 2.4  Optimization with Linear Equality Constraints

This section is mainly important for the techniques it introduces in regard to handling linear constraints, which form a major feature of the Quadratic Programming (QP) method that is the subproblem of the SQP method. The 'linear' problem that we consider is the Equality QP (EQP) problem

$$\underset{\mathbf{x} \in \mathbb{R}^n}{\text{minimize}} \quad q(\mathbf{x}) = \tfrac{1}{2}\mathbf{x}^T G \mathbf{x} + \mathbf{h}^T \mathbf{x}$$
$$\text{subject to } A^T \mathbf{x} = \mathbf{b}. \tag{2.14}$$

In this problem $A$ is an $n \times m$ matrix with $m \leq n$. The constraints are regular iff $\text{rank}(A) = m$, which we assume to be the case. When $m = n$ the solution is simply that given by (2.1) and $q(\mathbf{x})$ plays no part. We shall focus therefore on the case that $m < n$, whence the equations in (2.14) are under-determined.

In this case we can express the general solution of $A^T \mathbf{x} = \mathbf{b}$ as

$$\mathbf{x} = \mathbf{x}^\circ + Z\mathbf{t} \tag{2.15}$$

where $\mathbf{x}^\circ$ is a *particular solution* of $A^T \mathbf{x} = \mathbf{b}$, $Z$ is an $n \times (n - m)$ matrix whose columns are a basis for the null space of $A^T$, that is $\text{null}(A^T) = \{\mathbf{z} \mid A^T \mathbf{z} = \mathbf{0}\}$, and $\mathbf{t} \in \mathbb{R}^{n-m}$ is an arbitrary vector. The $Z\mathbf{t}$ term in (2.15) expresses the non-uniqueness of solutions of $A^T \mathbf{x} = \mathbf{b}$. Neither $\mathbf{x}^\circ$ nor $Z$ are uniquely defined and any valid choice is acceptable, although there are possible considerations relating to ill-conditioning. There are various ways of finding a suitable $\mathbf{x}^\circ$ and $Z$. One is to reduce $A^T$ to upper echelon form, as described in any basic linear algebra text, using pivoting to avoid ill-conditioning. More relevant to QP software is to find any matrix, $V$ say, such that $[A \mid V]$ is nonsingular, and also well-conditioned, insofar as that is possible. Denote $[A \mid V]^{-T} = [Y \mid Z]$ where $Z$ has $n - m$ columns. Then it can readily be verified using $A^T Y = I$, $A^T Z = 0$ and $\text{rank}(Z) = n - m$ that $\mathbf{x}^\circ = Y\mathbf{b}$ is a particular solution and columns of $Z$ are a basis for $\text{null}(A^T)$.

What follows is known as the *Null Space Method* for solving EQP problems. Simply, we substitute the general solution (2.15) into the definition of $q(\mathbf{x})$, giving a *reduced quadratic function*

$$Q(\mathbf{t}) = q(\mathbf{x}^\circ + Z\mathbf{t}) = \tfrac{1}{2}(\mathbf{x}^\circ + Z\mathbf{t})^T G(\mathbf{x}^\circ + Z\mathbf{t}) + \mathbf{h}^T(\mathbf{x}^\circ + Z\mathbf{t}).$$

We find a stationary point of $Q(\mathbf{t})$ by applying the condition $\nabla_{\mathbf{t}} Q(\mathbf{t}) = \mathbf{0}$, giving rise to the system of equations

$$Z^T G Z \mathbf{t} = -Z^T (\mathbf{h} + G \mathbf{x}^\circ) \qquad (2.16)$$

which is solved for $\mathbf{t}$. Then (2.15) defines the solution $\mathbf{x}^*$. The solution is a unique minimizer if and only if $Z^T G Z$, referred to as the *reduced Hessian matrix*, is positive definite.

The Null Space Method extends to solve any linear equality constrained problem (LECP) in which the objective function is non-quadratic, that is

$$\begin{aligned} &\underset{\mathbf{x} \in \mathbb{R}^n}{\text{minimize}} \quad f(\mathbf{x}) \\ &\text{subject to } A^T \mathbf{x} = \mathbf{b}. \end{aligned} \qquad (2.17)$$

Again, we just substitute for $\mathbf{x}$ using (2.15) giving a reduced problem

$$\underset{\mathbf{t} \in \mathbb{R}^{n-m}}{\text{minimize}} F(\mathbf{t}) = f(\mathbf{x}^\circ + Z \mathbf{t}).$$

This is now a non-quadratic unconstrained minimization problem and can be solved by the methods of the previous subsection.

# 3   Optimization with Nonlinear Equations

In this section we consider the Equality constrained NLP problem

$$\text{ENLP} \begin{cases} \underset{\mathbf{x} \in \mathbb{R}^n}{\text{minimize}} \quad f(\mathbf{x}) \\ \text{subject to } \mathbf{c}(\mathbf{x}) = \mathbf{0} \end{cases}$$

where in general $\mathbf{c}(\mathbf{x})$ is a vector of $m$ nonlinear functions. We assume that these functions are continuous and continuously differentiable ($\mathbb{C}^1$) functions of $\mathbf{x}$. In the null space method we looked for a parametrization of the feasible region which allows us to eliminate the linear constraints $A^T \mathbf{x} = \mathbf{b}$, and so solve an unconstrained problem. In this section we seek to do the same when the constraint manifold is nonlinear. The development of this section provides an elegant and concise introduction to the concept of so-called *Lagrange multipliers* and their relation to the reduced optimization problems seen in the previous section. We make use of a locally valid nonlinear transformation, which, although computationally unattractive, does enable us to state necessary optimality conditions, and to derive a Newton method which is ultimately the basis for the SQP method.

## 3.1  Stationary Points and Lagrange Multipliers

In this section we follow the rationale of the null space method and attempt to derive an equivalent reduced unconstrained optimization problem. In this case however it is necessary to make a nonlinear transformation of variables, and there exist exceptional situations in which this is not possible. In order therefore to ensure that our transformation is well defined, local to a solution $\mathbf{x}^*$ of the ENLP, we make the *regularity assumption* that the columns of the Jacobian matrix $A^*$ are linearly independent, or equivalently that $\operatorname{rank}(A^*) = m$.

Existence and some properties of the transformation are a consequence of the *Inverse Function Theorem*, an important result which can be found in texts on real variable calculus. It may be stated as follows. Let $\mathbf{r}(\mathbf{x})$, $\mathbb{R}^n \to \mathbb{R}^n$, be a $\mathbb{C}^1$ nonlinear mapping, and let $\mathbf{x}^*$ be such that $\nabla \mathbf{r}(\mathbf{x}^*)$ is nonsingular. Then open neighbourhoods of $\mathbf{x}^*$ and $\mathbf{r}^*$ $(= \mathbf{r}(\mathbf{x}^*))$ exist within which a $\mathbb{C}^1$ inverse mapping $\mathbf{x}(\mathbf{r})$ is uniquely defined, so that $\mathbf{x}(\mathbf{r}(\mathbf{x})) = \mathbf{x}$ and $\mathbf{r}(\mathbf{x}(\mathbf{r})) = \mathbf{r}$. Moreover, derivatives of the mappings are related by

$$\nabla_{\mathbf{r}}\mathbf{x}^T = (\nabla_{\mathbf{x}}\mathbf{r}^T)^{-1}. \tag{3.1}$$

In the case of the ENLP above, we choose any fixed matrix $V$ such that $[A^* \,|\, V]$ is nonsingular (this is possible by virtue of the regularity assumption), and consider the nonlinear mapping

$$\mathbf{r}(\mathbf{x}) = \begin{pmatrix} \mathbf{c}(\mathbf{x}) \\ V^T(\mathbf{x} - \mathbf{x}^*) \end{pmatrix}, \tag{3.2}$$

noting that $\mathbf{r}(\mathbf{x}^*) = \mathbf{0}$. The Jacobian of the transformation is $\nabla_{\mathbf{x}}\mathbf{r}^T = [A \,|\, V]$ which is nonsingular at $\mathbf{x}^*$. It follows by virtue of the inverse function theorem that a well defined inverse function $\mathbf{x}(\mathbf{r})$ exists in a neighbourhood of $\mathbf{x}^*$. We consider the constrained form of $\mathbf{x}(\mathbf{r})$ in which $\mathbf{r} = \begin{pmatrix} \mathbf{0} \\ \mathbf{t} \end{pmatrix}$. This defines a function $\mathbf{x}(\mathbf{t})$, $\mathbf{t} \in \mathbb{R}^{n-m}$, for which $\mathbf{c}(\mathbf{x}(\mathbf{t})) = \mathbf{0}$, and so provides a parametrization of the feasible region of the ENLP which is valid local to $\mathbf{x}^*$. Moreover, for any $\mathbf{x}$ local to $\mathbf{x}^*$ there corresponds a unique value of $\mathbf{t} = V^T(\mathbf{x} - \mathbf{x}^*)$, and $\mathbf{x} = \mathbf{x}^*$ corresponds to $\mathbf{t} = \mathbf{0}$. It also follows from (3.1) that

$$\nabla_{\mathbf{r}}\mathbf{x}(\mathbf{r})^T = [A \,|\, V]^{-1} = \begin{bmatrix} Y^T \\ Z^T \end{bmatrix}, \tag{3.3}$$

say, so that

$$\nabla_{\mathbf{t}}\mathbf{x}(\mathbf{t})^T = Z^T \qquad \text{and hence} \qquad \partial x_i / \partial t_j = z_{ij}. \tag{3.4}$$

Note in general that $Z$ is no longer a constant matrix, and the expressions are only valid local to $\mathbf{x}^*$, in contrast to the null space method for handling

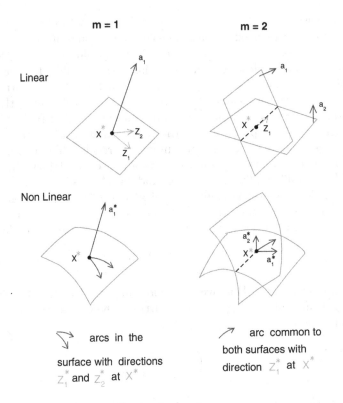

**Fig. 1** Illustrating dimension reduction for $n = 3$

linear constraints. However, in the linear case the reduced problem that is obtained below is identical to that described in Section 2.4.

The process of dimension reduction is illustrated in Figure 1 when the full space is three dimensional. In the linear case the constraints can be represented by planes, and the normal vectors $\mathbf{a}_1$ and $\mathbf{a}_2$ (columns of the matrix $A$) are perpendicular to the planes. When $m = 1$, the null space has dimension two, and is spanned by two independent vectors $\mathbf{z}_1$ and $\mathbf{z}_2$ which are the columns of $Z$. Any point $\mathbf{x}$ in the plane can be represented uniquely by the linear combination $\mathbf{x} = \mathbf{x}^* + t_1\mathbf{z}_1 + t_2\mathbf{z}_2$. When $m = 2$ the feasible set is the intersection of two planes, there is only one basis vector $\mathbf{z}_1$, and the feasible set is just $\mathbf{x} = \mathbf{x}^* + t_1\mathbf{z}_1$. It can be seen in both cases that the vectors $\mathbf{a}_i$ and $\mathbf{z}_j$ are mutually perpendicular for all $i$ and $j$, which expresses the condition $A^T Z = 0$. In the nonlinear case, the planes are replaced by curved surfaces, and feasible lines in the linear case have become feasible arcs in the nonlinear case, whose directions at $\mathbf{x}^*$ are the vectors $\mathbf{z}_i^*$. It can be seen that $A$ and $Z$ are no longer constant in the nonlinear case.

We can now state an equivalent reduced unconstrained optimization problem for the ENLP, that is

$$\underset{\mathbf{t}\in\mathrm{I\!R}^{n-m}}{\text{minimize}} \, F(\mathbf{t}), \tag{3.5}$$

where $F(\mathbf{t}) = f(\mathbf{x}(\mathbf{t}))$. A stationary point of $F(\mathbf{t})$ is defined by the condition that

$$\nabla_\mathbf{t} F(\mathbf{t}) = \mathbf{0}. \tag{3.6}$$

By the chain rule

$$\frac{\partial}{\partial t_j} = \sum_i \frac{\partial x_i}{\partial t_j} \frac{\partial}{\partial x_i} = \sum_i z_{ij} \frac{\partial}{\partial x_i}$$

from (3.4), or

$$\nabla_\mathbf{t} = Z^T \nabla_\mathbf{x}. \tag{3.7}$$

This result shows how derivatives in the reduced space are related to those in the full space. Applying these derivatives to $F(\mathbf{t})$ and $f(\mathbf{x})$ at $\mathbf{x}^*$, it follows from (3.6) that the stationary point condition for the ENLP problem is

$$Z^{*T} \mathbf{g}(\mathbf{x}^*) = \mathbf{0}, \tag{3.8}$$

where $Z^*$ denotes the matrix given by (3.3) when $\mathbf{x} = \mathbf{x}^*$ and $A = A^*$. The vector $Z^T \mathbf{g}$ is referred to as the *reduced gradient* and is zero at a stationary point of the ENLP problem.

There is also an alternative formulation of the stationary point condition that can be deduced. Arising from (3.3) and the regularity assumption $\text{rank}(A^*) = m$, we know that both $A^*$ and $Z^*$ have linearly independent columns. The definition of the inverse in (3.3) implies that $A^{*T} Z^* = 0$ showing that columns of $Z^*$ are in $\text{null}(A^{*T})$. But we might equally write $Z^{*T} A^* = 0$, showing that columns of $A^*$ are in $\text{null}(Z^{*T})$, and indeed provide a basis for $\text{null}(Z^{*T})$ by virtue of linear independence. Now the stationary point condition (3.8) states that $\mathbf{g}^* \in \text{null}(Z^{*T})$. Thus we can express $\mathbf{g}^*$ as a linear combination of basis vectors for $\text{null}(Z^{*T})$, that is

$$\mathbf{g}^* = A^* \boldsymbol{\lambda}^* = \sum_{i=1}^{m} \mathbf{a}_i^* \lambda_i^*. \tag{3.9}$$

The multipliers $\boldsymbol{\lambda}^*$ in the linear combination are referred to as *Lagrange multipliers*. There is one Lagrange multiplier for each constraint.

Equations (3.8) and (3.9) provide alternative and equivalent statements of the stationary point conditions for an ENLP problem. They are often referred to as *first order necessary conditions* for a local solution of a regular ENLP problem. We can also express (3.9) as $\mathbf{g}^* \in \text{range}(A^*)$. These alternative viewpoints of null space and range space formulations pervade much of both the theoretical and computational aspects of quadratic programming and nonlinear programming, as we shall see below.

Satisfying (3.9) and feasibility provides a method for solving the ENLP, that is: find $\mathbf{x}^*, \boldsymbol{\lambda}^*$ to solve the system of $n + m$ equations

$$\mathbf{r}(\mathbf{x}, \boldsymbol{\lambda}) = \begin{pmatrix} \mathbf{g} - A\boldsymbol{\lambda} \\ -\mathbf{c} \end{pmatrix} = \mathbf{0}, \tag{3.10}$$

in the $n + m$ variables $\mathbf{x}, \boldsymbol{\lambda}$, where $\mathbf{g}$, $A$ and $\mathbf{c}$ are functions of $\mathbf{x}$. This is the so-called *method of Lagrange multipliers*. However, the system is generally nonlinear in $\mathbf{x}$ and may not be straightforward to solve, as we have observed above. It also can only be expected to yield stationary points of the ENLP. Equation (3.10) can also be interpreted as defining a stationary point of a *Lagrangian function*

$$\mathcal{L}(\mathbf{x}, \boldsymbol{\lambda}) = f(\mathbf{x}) - \boldsymbol{\lambda}^T \mathbf{c}(\mathbf{x}), \tag{3.11}$$

since $\nabla_{\mathbf{x}} \mathcal{L} = \mathbf{g} - A\boldsymbol{\lambda}$ and $\nabla_{\boldsymbol{\lambda}} \mathcal{L} = -\mathbf{c}$. The Lagrangian function plays a central rôle in both the theoretical and computational aspects of nonlinear programming and the SQP method.

Lagrange multipliers also have a useful interpretation in terms of the sensitivity of the ENLP to perturbations in the constraints. Consider an ENLP problem

$$\begin{array}{c} \underset{\mathbf{x} \in \mathbb{R}^n}{\text{minimize}} \quad f(\mathbf{x}) \\ \text{subject to } \mathbf{c}(\mathbf{x}) = \boldsymbol{\varepsilon} \end{array} \tag{3.12}$$

in which the right hand sides of the constraints have been perturbed by an amount $\boldsymbol{\varepsilon}$. Let $\mathbf{x}(\boldsymbol{\varepsilon})$, $\boldsymbol{\lambda}(\boldsymbol{\varepsilon})$ be the solution and multipliers of the perturbed problem, and consider $f(\mathbf{x}(\boldsymbol{\varepsilon}))$. Then

$$df(x(\varepsilon))/d\varepsilon_i = \lambda_i, \tag{3.13}$$

showing that $\lambda_i$ measures the change of $f(\mathbf{x}(\boldsymbol{\varepsilon}))$ with respect to a change $\varepsilon_i$ in constraint $i$, to first order. To prove this result, let the Lagrangian of the perturbed ENLP problem be

$$\mathcal{L}(\mathbf{x}, \boldsymbol{\lambda}, \boldsymbol{\varepsilon}) = f(\mathbf{x}) - \boldsymbol{\lambda}^T (\mathbf{c}(\mathbf{x}) - \boldsymbol{\varepsilon}),$$

and observe that $\mathcal{L}(\mathbf{x}(\boldsymbol{\varepsilon}), \boldsymbol{\lambda}(\boldsymbol{\varepsilon}), \boldsymbol{\varepsilon}) = f(\mathbf{x}(\boldsymbol{\varepsilon}))$. Then the chain rule gives

$$\frac{df}{d\varepsilon_i} = \frac{d\mathcal{L}}{d\varepsilon_i} = \frac{\partial \mathbf{x}^T}{\partial \varepsilon_i} \nabla_{\mathbf{x}} \mathcal{L} + \frac{\partial \boldsymbol{\lambda}^T}{\partial \varepsilon_i} \nabla_{\boldsymbol{\lambda}} \mathcal{L} + \frac{\partial \mathcal{L}}{\partial \varepsilon_i} = \lambda_i$$

by virtue of the stationarity of the Lagrangian function with respect to $\mathbf{x}$ and $\boldsymbol{\lambda}$.

## 3.2  Second Order Conditions for the ENLP Problem

Let $\mathbf{x}^*$ solve the ENLP and let $\text{rank}(A^*) = m$ (regularity). Then we have seen that (3.5) is an equivalent reduced unconstrained minimization problem. Thus, from Section 2.2, a second order necessary condition is that the Hessian matrix $\nabla_t^2 F(\mathbf{t})$ is positive semi-definite. To relate this to the ENLP, we use equation (3.7) which relates derivatives in the reduced and full systems. Thus

$$\nabla_t^2 F(\mathbf{t}) = \boldsymbol{\nabla}_t(\boldsymbol{\nabla}_t F(\mathbf{t}))^T = Z^T \boldsymbol{\nabla}_{\mathbf{x}}(\mathbf{g}^T Z).$$

When the constraints are linear, such as in (2.17), $Z$ is a constant matrix so we can differentiate further to get

$$Z^T \boldsymbol{\nabla}_{\mathbf{x}}(\mathbf{g}^T Z) = Z^T(\boldsymbol{\nabla}_{\mathbf{x}}\mathbf{g}^T)Z = Z^T G Z$$

where $G$ is the Hessian matrix of $f(\mathbf{x})$. Thus the second order necessary condition in this case is that the *reduced Hessian matrix* $Z^T G^* Z$ is positive semi-definite. Moreover, $Z^T \mathbf{g}^* = \mathbf{0}$ and $Z^T G^* Z$ being positive definite are sufficient conditions for $\mathbf{x}^*$ to solve (3.5).

For an ENLP with nonlinear constraints, $Z$ depends on $\mathbf{x}$ and we can no longer assume that derivatives of $Z$ with respect to $\mathbf{x}$ are zero. To make progress, we observe that the ENLP is equivalent to the problem

$$\begin{array}{c} \underset{\mathbf{x}\in\mathbb{R}^n}{\text{minimize}} \quad \mathcal{L}(\mathbf{x},\boldsymbol{\lambda}^*) \\[4pt] \text{subject to } \mathbf{c}(\mathbf{x}) = \mathbf{0} \end{array} \tag{3.14}$$

since $f(\mathbf{x}) = \mathcal{L}(\mathbf{x},\boldsymbol{\lambda})$ when $\mathbf{c}(\mathbf{x}) = \mathbf{0}$. We now define $\mathbf{x}(\mathbf{t})$ as in (3.2), and consider the problem of minimizing a reduced function $F(\mathbf{t}) = \mathcal{L}(\mathbf{x}(\mathbf{t}),\boldsymbol{\lambda}^*)$. Then $\boldsymbol{\nabla}_t F = \mathbf{0}$ becomes $Z^T \boldsymbol{\nabla}_{\mathbf{x}}\mathcal{L} = \mathbf{0}$, or $Z^T(\mathbf{g} - A\boldsymbol{\lambda}^*) = \mathbf{0}$. At $\mathbf{x}^*$, it follows that $Z^{*T}\mathbf{g}^* = \mathbf{0}$ which is the first order necessary condition. For second derivatives,

$$\nabla_t^2 F(\mathbf{t}) = \boldsymbol{\nabla}_t(\boldsymbol{\nabla}_t F(\mathbf{t}))^T = Z^T \boldsymbol{\nabla}_{\mathbf{x}}((\mathbf{g} - A\boldsymbol{\lambda}^*)^T Z).$$

At $\mathbf{x}^*$, derivatives of $Z$ are multiplied by $\mathbf{g}^* - A^*\boldsymbol{\lambda}^*$ which is zero, so we have

$$\nabla_t^2 F(\mathbf{t}^*) = Z^T \boldsymbol{\nabla}_{\mathbf{x}}((\mathbf{g} - A\boldsymbol{\lambda}^*)^T)Z|_{\mathbf{x}^*} = Z^{*T} W^* Z^*,$$

where

$$W(\mathbf{x},\boldsymbol{\lambda}) = \nabla_{\mathbf{x}}^2 \mathcal{L}(\mathbf{x},\boldsymbol{\lambda}) = \nabla^2 f(\mathbf{x}) - \sum_{i=1}^m \lambda_i \nabla^2 c_i(\mathbf{x}) \tag{3.15}$$

is the Hessian with respect to $\mathbf{x}$ of the Lagrangian function, and $W^* = W(\mathbf{x}^*,\boldsymbol{\lambda}^*)$. Thus the second order necessary condition for the regular ENLP problem is that the reduced Hessian of the Lagrangian function is positive semi-definite. As above, a sufficient condition is that the reduced Hessian is positive definite and the reduced gradient is zero.

## 3.3 The SQP Method for the ENLP Problem

We have seen in (3.10) that a stationary point of a regular ENLP can be found by solving a system of nonlinear equations. Applying the Newton-Raphson method to these equations enables us to derive a Newton type method with rapid local convergence properties. First however we consider solving (3.10) in the case of an EQP problem (2.14). In this case, $\mathbf{g} = G\mathbf{x}+\mathbf{h}$ and $\mathbf{c} = A^T\mathbf{x}-\mathbf{b}$, so (3.10) can be written as the system of $n + m$ linear equations in $n + m$ unknowns

$$\begin{bmatrix} G & -A \\ -A^T & 0 \end{bmatrix} \begin{pmatrix} \mathbf{x} \\ \boldsymbol{\lambda} \end{pmatrix} = - \begin{pmatrix} \mathbf{h} \\ \mathbf{b} \end{pmatrix}. \tag{3.16}$$

Although symmetric, the coefficient matrix in (3.16) is indefinite so cannot be solved by using Choleski factors. The Null Space Method (2.16) is essentially one way of solving (3.16), based on eliminating the constraints $A^T\mathbf{x} = \mathbf{b}$. When $G$ is positive definite, and particularly when $G$ permits sparse Choleski factors $G = LL^T$ to be obtained, it can be more effective to use the first block equation to eliminate $\mathbf{x} = G^{-1}(A\boldsymbol{\lambda} - \mathbf{h})$. Then the system

$$A^T G^{-1} A \boldsymbol{\lambda} = \mathbf{b} + A^T G^{-1} \mathbf{h} \tag{3.17}$$

is used to determine $\boldsymbol{\lambda}$, and hence implicitly $\mathbf{x}$. Of course, operations with $G^{-1}$ are carried out by making triangular solves with the Choleski factor $L$. This method might be regarded as a *Range Space Method for EQP*, in contrast to the Null Space Method described earlier.

We now proceed to consider the ENLP problem in the general case of nonlinear constraints. In this case we attempt to solve the equations $\mathbf{r}(\mathbf{x}, \boldsymbol{\lambda}) = \mathbf{0}$ in (3.10) by the Newton-Raphson method. First we need the Jacobian matrix of this system, which is the $(n + m) \times (n + m)$ matrix

$$\begin{pmatrix} \boldsymbol{\nabla}_{\mathbf{x}} \\ \boldsymbol{\nabla}_{\boldsymbol{\lambda}} \end{pmatrix} \mathbf{r}^T = \begin{bmatrix} W & -A \\ -A^T & 0 \end{bmatrix} \tag{3.18}$$

where $W$ as defined in (3.15) is the Hessian with respect to $\mathbf{x}$ of the Lagrangian function. The current iterate in the NR method is the pair of vectors $\mathbf{x}^{(k)}$, $\boldsymbol{\lambda}^{(k)}$, and the iteration formula, generalising (2.4), is

$$\begin{bmatrix} W^{(k)} & -A^{(k)} \\ -A^{(k)T} & 0 \end{bmatrix} \begin{pmatrix} \mathbf{d}^{(k)} \\ \boldsymbol{\delta}^{(k)} \end{pmatrix} = -\mathbf{r}^{(k)} = \begin{pmatrix} A^{(k)}\boldsymbol{\lambda}^{(k)} - \mathbf{g}^{(k)} \\ \mathbf{c}^{(k)} \end{pmatrix} \tag{3.19}$$

where superscript $(k)$ denotes quantities calculated from $\mathbf{x}^{(k)}$ and $\boldsymbol{\lambda}^{(k)}$. Then updated values for the next iteration are defined by $\mathbf{x}^{(k+1)} = \mathbf{x}^{(k)} + \mathbf{d}^{(k)}$ and $\boldsymbol{\lambda}^{(k+1)} = \boldsymbol{\lambda}^{(k)} + \boldsymbol{\delta}^{(k)}$. These formulae may be rearranged by moving the $A^{(k)}\boldsymbol{\lambda}^{(k)}$ term to the left hand side of (3.19), giving

$$
\begin{bmatrix} W^{(k)} & -A^{(k)} \\ -A^{(k)T} & 0 \end{bmatrix} \begin{pmatrix} \mathbf{d}^{(k)} \\ \boldsymbol{\lambda}^{(k+1)} \end{pmatrix} = \begin{pmatrix} -\mathbf{g}^{(k)} \\ \mathbf{c}^{(k)} \end{pmatrix}. \tag{3.20}
$$

This then is a Newton iteration formula for finding a stationary point of an EQP problem. For rapid local convergence to a stationary point $\mathbf{x}^*$ with multipliers $\boldsymbol{\lambda}^*$, we require that the Jacobian matrix

$$
\begin{bmatrix} W^* & -A^* \\ -A^{*T} & 0 \end{bmatrix} \tag{3.21}
$$

is nonsingular.

So where does the SQP method come in? There are two important observations to make. First, if the constraints in the ENLP problem are regular $(\text{rank}(A^*) = m)$, and the ENLP satisfies second order sufficiency conditions $(Z^{*T}W^*Z^*$ is positive definite), then it is a nice exercise in linear algebra to show that the matrix (3.21) is nonsingular (see [13]). Thus the local rapid convergence of (3.20) is assured. Moreover, it also follows that $\text{rank}(A^{(k)}) = m$ and $Z^{(k)T}W^{(k)}Z^{(k)}$ is positive definite in a neighbourhood of $\mathbf{x}^*, \boldsymbol{\lambda}^*$. Under these conditions, the EQP problem

$$
\text{EQP}^{(k)} \begin{cases} \underset{\mathbf{d} \in \mathbb{R}^n}{\text{minimize}} & \frac{1}{2}\mathbf{d}^T W^{(k)}\mathbf{d} + \mathbf{d}^T \mathbf{g}^{(k)} + f^{(k)} \\ \text{subject to} & \mathbf{c}^{(k)} + A^{(k)T}\mathbf{d} = 0 \end{cases}
$$

is regular and has a unique local minimizer, which can be found by solving the stationary point condition (see (3.16)), which for $\text{EQP}^{(k)}$ is none other than (3.20). Thus, for finding a local minimizer of an ENLP problem, it is better to replace the iteration formula (3.20) by one based on solving $\text{EQP}^{(k)}$ for a correction $\mathbf{d}^{(k)} = \mathbf{d}$ and multiplier vector $\boldsymbol{\lambda}^{(k+1)}$. This correctly accounts for the second order condition required by a local minimizer to an ENLP problem. In particular, any solution of (3.20) which corresponds to a saddle point or maximizer of $\text{EQP}^{(k)}$ is not accepted. ($\text{EQP}^{(k)}$ is unbounded in this situation.) Also $\text{EQP}^{(k)}$ has a nice interpretation: the constraints are linear Taylor series approximations about $\mathbf{x}^{(k)}$ to those in the ENLP problem, and the objective function is a quadratic Taylor series approximation about $\mathbf{x}^{(k)}$ to the objective function in the ENLP, plus terms in $W^{(k)}$ that account for constraint curvature. The objective function can equally be viewed as a quadratic approximation to the Lagrangian function. (In fact the term $f^{(k)}$ in the objective function of $\text{EQP}^{(k)}$ is redundant, but is included so as to make these nice observations.)

To summarize, solving an ENLP in this way may be interpreted as a Sequential EQP method (SEQP) with the following basic structure

```
initialize  x⁽¹⁾, λ⁽¹⁾
for  k = 1, 2,... until converged
    solve  EQP⁽ᵏ⁾  giving  d⁽ᵏ⁾  and  multipliers  λ⁽ᵏ⁺¹⁾
    set   x⁽ᵏ⁺¹⁾ = x⁽ᵏ⁾ + d⁽ᵏ⁾
end
```

As with other Newton methods, the method may not converge globally, and $EQP^{(k)}$ may have no solution, for example it may be unbounded, or possibly infeasible if $A^{(k)}$ is rank deficient. It is therefore essential that extra features are included in any practical implementation. We return to this subject later in the monograph.

# 4  Inequality Constraints and Nonlinear Programming

In this section we examine the extra complication caused by having inequality constraints in the formulation of an optimization problem. As above we discuss 'linear' problems which can be solved in a finite number of steps, and nonlinear problems for which iteration is required, leading to a general formulation of the SQP method. We also discuss changes to optimality conditions to accommodate inequality constraints.

## 4.1  Systems of Inequalities

Corresponding to the development of Section 2.1, the 'linear' problem we now consider is that of a system of linear inequalities

$$A^T\mathbf{x} \geq \mathbf{b} \tag{4.1}$$

in which $A$ is a given $n \times m$ matrix of coefficients and $\mathbf{b}$ is a given vector of right hand sides. Usually $m > n$ when there is no objective function present, although in general, $m \leq n$ is also possible. Each inequality $\mathbf{a}_i^T\mathbf{x} \geq b_i$ in (4.1) divides $\mathbb{R}^n$ into two parts, a *feasible* side and an *infeasible* side, with respect to the inequality. Equality holds on the boundary. Any $n$ independent such equations define a point of intersection, referred to as a *vertex* in this context. Usually methods for solving (4.1) attempt to locate a feasible vertex. Each vertex can be found by solving a system of linear equations as in (2.1). There are only a finite number of vertices so the process will eventually find a solution, or establish that none exists. However, there may be as many as $\binom{m}{n}$ vertices, which can be extremely large for problems of any size. Thus it is important to enumerate the vertices in an efficient way. This can be done by a modification known as *Phase I* of the so-called Simplex Method for Linear Programming, which we describe briefly below (see also [13] for example). An important saving in this case is that adjacent vertices differ by only one equation, and this can be used to update matrix factors to gain efficiency.

The corresponding nonlinear problem is the system of nonlinear inequalities

$$\mathbf{r}(\mathbf{x}) \geq \mathbf{0} \tag{4.2}$$

in which $\mathbf{r}(\mathbf{x})$ is a given vector of $m$ nonlinear functions. This problem can be solved by a Newton type algorithm in which a sequence of linearized subproblems are solved, each being obtained by a linear Taylor series approximation to $\mathbf{c}(\mathbf{x})$ about a current iterate $\mathbf{x}^{(k)}$. This just becomes a special case of the SQP method, and we shall defer discussion on it until later.

In cases when no solution exists to (4.1) or (2.1), it is often of interest to find a 'best' solution which minimizes some measure of constraint infeasibility. Exactly what measure to choose is a decision for the user, and is one which has implications for the type of method that is possible.

## 4.2  Optimization with Inequality Constraints

In this case our generic linear problem, which can be solved finitely, is the QP problem

$$\begin{aligned} \underset{\mathbf{x}\in \mathbb{R}^n}{\text{minimize}} \quad & q(\mathbf{x}) = \tfrac{1}{2}\mathbf{x}^T G\mathbf{x} + \mathbf{h}^T\mathbf{x} \\ \text{subject to} \quad & A^T\mathbf{x} \geq \mathbf{b}. \end{aligned} \tag{4.3}$$

In this problem $A$ is an $n \times m$ matrix with no restrictions on the value of $m$. The corresponding nonlinear problem is the NLP problem (1.1). There is also the intermediate stage of a linearly constrained problem (LCP say) in which the objective function is non-quadratic, that is

$$\begin{aligned} \underset{\mathbf{x}\in \mathbb{R}^n}{\text{minimize}} \quad & f(\mathbf{x}) \\ \text{subject to} \quad & A^T\mathbf{x} \geq \mathbf{b}. \end{aligned} \tag{4.4}$$

As in Section 1, a more general formulation of the QP and NLP problems is often appropriate for practical use, but the simplified form is convenient for introducing the main features. An important special case of QP, which can also be solved finitely, is the *Linear Programming* (LP) problem, characterized by $G = 0$ in (4.3).

In this section our main aim is to discuss optimality conditions for an NLP problem. An important concept is that of an *active constraint*. The set of active constraints at a point $\mathbf{x}$ is defined by

$$\mathcal{A}(\mathbf{x}) = \{i \mid c_i(\mathbf{x}) = 0\} \tag{4.5}$$

so that $i \in \mathcal{A}(\mathbf{x})$ indicates that $\mathbf{x}$ is on the boundary of constraint $i$. The set of active constraints at the solution is denoted by $\mathcal{A}^*$. Constraints not in $\mathcal{A}^*$ have no influence on the behaviour of the NLP problem, local to a local solution $\mathbf{x}^*$ of (1.1).

Clearly $\mathbf{x}^*$ solves the following ENLP problem

$$\begin{aligned} \underset{\mathbf{x} \in \mathbb{R}^n}{\text{minimize}} \quad & f(\mathbf{x}) \\ \text{subject to} \quad & c_i(\mathbf{x}) = 0 \ i \in \mathcal{A}^*. \end{aligned} \tag{4.6}$$

Let the gradient vectors $\mathbf{a}_i^* = \nabla c_i^*$, $i \in \mathcal{A}^*$ be linearly independent (regularity). Then from (3.9) in Section 3.1 there exist multipliers $\boldsymbol{\lambda}^*$ such that

$$\mathbf{g}^* = \sum_{i \in \mathcal{A}^*} \mathbf{a}_i^* \lambda_i^* = \sum_{i=1}^{m} \mathbf{a}_i^* \lambda_i^* = A^* \boldsymbol{\lambda}^*, \tag{4.7}$$

denoting $\lambda_i^* = 0$ for inactive constraints $i \notin \mathcal{A}^*$. Moreover, by virtue of regularity, we can perturb the right hand side of (4.6) by a sufficiently small amount $\varepsilon_i > 0$, and $\varepsilon_j = 0$, $j \neq i$, and still retain feasibility in (1.1). Then, if $\lambda_i^* < 0$, it follows from (3.13) that $df/d\varepsilon_i = \lambda_i^* < 0$, which contradicts optimality. Thus

$$\lambda_i^* \geq 0 \tag{4.8}$$

for an *inequality constraint* is also necessary. (The multiplier of an equality constraint can take either sign.) The convention that the multiplier of an inactive constraint is zero may also be expressed as

$$\lambda_i^* c_i^* = 0 \tag{4.9}$$

which is referred to as the *complementarity condition*. If $\lambda_i^* > 0$ for all active inequality constraints, then *strict complementarity* is said to hold. Collectively, feasibility in (1.1), (3.9), (4.8) for an inequality constraint, and (4.9) are known as KT (Kuhn-Tucker) (or KKT (Karush-Kuhn-Tucker)) conditions (Karush [31], Kuhn and Tucker [32]). Subject to a regularity assumption of some kind, they are necessary conditions for a local solution of (1.1). A point $\mathbf{x}^*$ which satisfies KT conditions for some $\boldsymbol{\lambda}^*$ is said to be a *KT point*.

A second order necessary condition that can be deduced from (4.6) is that $Z^{*T} W^* Z^*$ is positive semi-definite, where $Z^*$ is the null-space basis matrix for (4.6), and $W$ is defined in (3.15). A sufficient condition is that $\mathbf{x}^*$ is a KT point, strict complementarity holds, and $Z^{*T} W^* Z^*$ is positive definite.

The regularity assumption used in these notes (that the gradient vectors $\mathbf{a}_i^* = \nabla c_i^*$, $i \in \mathcal{A}^*$ are linearly independent) is known as the Linear Independence Constraint Qualification (LICQ). If LICQ fails at any point, *degeneracy* is said to hold at that point. However KT conditions can hold under weaker conditions, most notably when all the active constraints are linear. In this case the validity of KT conditions is a consequence of the famous *Farkas Lemma* (see [13] for example). Weaker regularity conditions than LICQ have been proposed, from which KT conditions can be deduced, for instance by Kuhn and Tucker [32] (KTCQ), and by Mangasarian and Fromowitz [33] (MFCQ). A sufficient condition for KTCQ to hold is that the active constraints are all linear. For problems in which there are some nonlinear active constraints, LICQ is arguably the only practical way of establishing that a point is regular, and failure of LICQ is rare at local solutions of nonlinear constraint problems.

## 4.3  Quadratic Programming

Before describing the SQP method, it is important to know how to solve a QP problem (4.3) that contains inequality constraints. A method with finite termination is the Active Set Method (ASM), which has features that are favourable in the context of SQP. The method solves (4.3) by solving a sequence of EQP problems, whilst retaining feasibility in (4.3), until the correct active set is determined. The method is described in the case that $G$ is positive definite, all the constraints are inequalities, and there is no degeneracy at the iterates $\mathbf{x}^{(k)}$. The method is initialized by finding a feasible vertex, $\mathbf{x}^{(1)}$ say, as described in Section 4.1. We let $\mathcal{A}$ denote the current set of active constraints at $\mathbf{x}^{(k)}$. The current EQP is defined by

$$
\begin{aligned}
&\underset{\mathbf{x}\in\mathbb{R}^n}{\text{minimize}} \quad \tfrac{1}{2}\mathbf{x}^T G\mathbf{x} + \mathbf{h}^T\mathbf{x} \\
&\text{subject to } \mathbf{a}_i^T\mathbf{x} = b_i \qquad i \in \mathcal{A}.
\end{aligned}
\tag{4.10}
$$

Because $\mathbf{x}^{(1)}$ is a vertex, it is in fact the solution of the current EQP defined by $\mathcal{A}$. The ASM has two major steps.

(i) If $\mathbf{x}^{(k)}$ solves the current EQP, then find the corresponding multipliers $\boldsymbol{\lambda}^{(k)}$. Choose any $i : \lambda_i^{(k)} < 0$ (if none exist, then **finish** with $\mathbf{x}^* = \mathbf{x}^{(k)}$). Otherwise, remove $i$ from $\mathcal{A}$ and goto step (ii).

(ii) Find the solution, $\hat{\mathbf{x}}$ say, of the current EQP. If $\hat{\mathbf{x}}$ is feasible in the QP problem then [set $\mathbf{x}^{(k+1)} = \hat{\mathbf{x}}$, $k = k + 1$ and goto step (i)]. Otherwise, set $\mathbf{x}^{(k+1)}$ as the closest feasible point to $\hat{\mathbf{x}}$ along the line segment from $\mathbf{x}^{(k)}$ to $\hat{\mathbf{x}}$. Add the index of a newly active constraint to $\mathcal{A}$. Set $k = k + 1$. If $|\mathcal{A}| = n$ then goto step (i) else goto step (ii).

The motivation for the algorithm is provided by the observation in Section 4.2 that if there exists $i : \lambda_i^{(k)} < 0$ at the solution of the current EQP, then it is possible to relax constraint $i$ whilst reducing the objective function. If $G = 0$ in (4.3) then we have a Linear Programming (LP) problem, and the ASM is essentially the same as the *Simplex method* for LP, although it is not often explained in this way.

Special linear algebra techniques are required to make the method efficient
in practice. Changes to the current active set involve either adding or sub-
tracting one constraint index. Updates to matrices such as $Z$ and $Z^T G Z$ can
be performed much more quickly than re-evaluating the matrices. For large
problems, it is important to take advantage of sparsity in $A$ and possibly $G$.

There are some complicating factors for the ASM. If the Hessian $G$ is not
positive definite, then it is possible that the EQP obtained by removing $i$
from $\mathcal{A}$ may be unbounded, so that $\hat{\mathbf{x}}$ does not exist. In this case an arbitrary
choice of feasible descent direction is chosen, to make progress. If $G$ has
negative eigenvalues, then the QP problem may have local solutions, and the
ASM does not guarantee to find a global solution. Any solution found by
the ASM will be a KT point of the QP problem, but may not be a local
solution unless strict complementarity holds. A more serious complicating
factor is that of *degeneracy* which refers to the situation where regularity
of the active constraints at the solution of an EQP problem fails to hold.
An example would be where there are more than $n$ active constraints at a
feasible vertex. In this case, deciding whether $\mathbf{x}^{(k)}$ solves the current EQP, or
whether a feasible descent direction exists, is a more complex issue, although
a finite algorithm to decide the issue is possible. Degeneracy is often present
in practical instances of QP problems, and it is important that it is correctly
accounted for in a computer code.

More recently an alternative class of methods has become available for
the solution of LP or QP problems in which $G$ is positive semi-definite.
These *interior point methods* have the advantage that they avoid the worst
case behaviour of ASM and Simplex methods, in which the number of iterations required to locate the solution may grow exponentially with $n$. However,
interior point methods also have some disadvantages in an SQP context.

## 4.4  The SQP Method

We are now in a position to describe the basic SQP method for an NLP (1.1)
with inequality constraints. The method was first suggested in a thesis of
Wilson (1960), [48], and became well known due to the work of Beale [1]. The
idea follows simply from the SEQP method for an ENLP problem, where the
equality constraints $\mathbf{c}(\mathbf{x}) = \mathbf{0}$ are approximated by the linear Taylor series
$\mathbf{c}^{(k)} + A^{(k)T}\mathbf{d} = \mathbf{0}$ in the subproblem EQP$^{(k)}$. In an NLP with inequality
constraints $\mathbf{c}(\mathbf{x}) \geq \mathbf{0}$ we therefore make the same approximation, leading to
a QP subproblem with linear inequality constraints $\mathbf{c}^{(k)} + A^{(k)T}\mathbf{d} \geq \mathbf{0}$, that is

$$\text{QP}^{(k)} \begin{cases} \underset{\mathbf{d} \in \mathbb{R}^n}{\text{minimize}} \quad \tfrac{1}{2}\mathbf{d}^T W^{(k)}\mathbf{d} + \mathbf{d}^T \mathbf{g}^{(k)} \\ \text{subject to } \mathbf{c}^{(k)} + A^{(k)T}\mathbf{d} \geq \mathbf{0}. \end{cases}$$

The basic form of the algorithm therefore is that described at the end of Section 3.3, with the substitution of QP$^{(k)}$ for EQP$^{(k)}$. To view this method as a Newton-type method, we need to assume that strict complementarity $\lambda_i^* > 0$, $i \in \mathcal{A}^*$ holds at a regular solution to (1.1). Then, if $\mathbf{x}^{(k)}$, $\boldsymbol{\lambda}^{(k)}$ is sufficiently close to $\mathbf{x}^*$, $\boldsymbol{\lambda}^*$, it follows that the solution of EQP$^{(k)}$ with active constraints $\mathcal{A}^*$, also satisfies the sufficient conditions for QP$^{(k)}$. Thus we can ignore inactive constraints $i \notin \mathcal{A}^*$, and the SQP method is identical to the SEQP method on the active constraint set $\mathcal{A}^*$. Thus the SQP method inherits the local rapid convergence of a Newton type method under these circumstances.

The progress of the SQP method on the NLP problem

$$\begin{aligned}
&\underset{\mathbf{x}\in\mathbb{R}^2}{\text{minimize}} \quad f(\mathbf{x}) = -x_1 - x_2 \\
&\text{subject to } c_1(\mathbf{x}) = x_2 - x_1^2 \geq 0 \\
&\qquad\qquad\quad c_2(\mathbf{x}) = 1 - x_1^2 - x_2^2 \geq 0
\end{aligned}$$

is illustrated in Table 1, and has some instructive features. Because the initial multiplier estimate is zero, and $f(\mathbf{x})$ is linear, the initial $W^{(1)}$ matrix is zero, and QP$^{(1)}$ is in fact an LP problem. Consequently, $\mathbf{x}^{(1)}$ has to be chosen carefully to avoid an unbounded subproblem (or alternatively one could add simple upper and lower bounds to the NLP problem). The solution of QP$^{(1)}$ delivers some non-zero multipliers for $\boldsymbol{\lambda}^{(2)}$, so that $W^{(2)}$ becomes positive definite. The solution of QP$^{(2)}$ predicts that constraint 1 is inactive, and we see that $\lambda_1^{(3)}$ is zero. This situation persists on all subsequent iterations. For this NLP problem, the active set is $\mathcal{A}^* = \{2\}$, and we see for $k \geq 3$, that the SQP method converges to the solution in the same way as the SEQP method with the single equality constraint $1 - x_1^2 - x_2^2 = 0$. The onset of rapid local convergence, characteristic of a Newton method, can also be observed.

However, the basic method can fail to converge if $\mathbf{x}^{(k)}$ is remote from $\mathbf{x}^*$ (it is not as important to have $\boldsymbol{\lambda}^{(k)}$ close to $\boldsymbol{\lambda}^*$ because if $\mathbf{x}^{(k)}$ is close to $\mathbf{x}^*$, one solution of QP$^{(k)}$ will give an accurate multiplier estimate). It is also possible that QP$^{(k)}$ has no solution, either because it is unbounded, or because the linearized constraints are infeasible.

For these reasons, the SQP method is only the starting point for a fully developed NLP solver, and extra features must be added to promote convergence from remote initial values. This is the subject of subsequent sections of

**Table 1** A numerical example of the SQP method

| $k$ | $x_1^{(k)}$ | $x_2^{(k)}$ | $\lambda_1^{(k)}$ | $\lambda_2^{(k)}$ | $c_1^{(k)}$ | $c_2^{(k)}$ |
|---|---|---|---|---|---|---|
| 1 | $\frac{1}{2}$ | 1 | 0 | 0 | $\frac{3}{4}$ | $-\frac{1}{2}$ |
| 2 | $\frac{11}{12}$ | $\frac{2}{3}$ | $\frac{1}{3}$ | $\frac{2}{3}$ | $-0.173611$ | $-0.284722$ |
| 3 | 0.747120 | 0.686252 | 0 | 0.730415 | 0.128064 | $-0.029130$ |
| 4 | 0.708762 | 0.706789 | 0 | 0.706737 | 0.204445 | $-0.001893$ |
| 5 | 0.707107 | 0.707108 | 0 | 0.707105 | 0.207108 | $-0.28_{10} - 5$ |

this monograph. Nonetheless it has been and still is the method of choice for many researchers. The success of the method is critically dependent on having an efficient, flexible and reliable code for solving the QP subproblem. It is important to be able to take advantage of *warm starts*, that is, initializing the QP solver with the active set from a previous iteration. Also important is the ability to deal with the situation that the matrix $W^{(k)}$ is not positive semi-definite. For both these reasons, an active set method code for solving the QP subproblems is likely to be preferred to an interior point method. However, NLP solvers are still a very active research area, and the situation is not at all clear, especially when dealing with very large scale NLPs.

## 4.5 SLP-EQP Algorithms

An early idea for solving NLP problems is the successive linear programming (SLP) algorithm in which an LP subproblem is solved ($W^{(k)} = 0$ in $\mathrm{QP}^{(k)}$). This is able to take advantage of fast existing software for large scale LP. However, unless the solution of the NLP problem is at a vertex, convergence is slow because of the lack of second derivative information. A more recent development is the SLP-EQP algorithm, introduced by Fletcher and Sainz de la Maza [22], in which the SLP subproblem is used to determine the active set and multipliers, but the resulting step $\mathbf{d}$ is not used. Instead an SEQP calculation using the subproblem $\mathrm{EQP}^{(k)}$ in Section 3.3 is made to determine $\mathbf{d}^{(k)}$. The use of a trust region in the LP subproblem (see below) is an essential feature in the calculation. The method is another example of a Newton-type method and shares the rapid local convergence properties. The idea has proved quite workable, as a recent software product SLIQUE of Byrd, Gould, Nocedal and Waltz [3] demonstrates.

## 4.6 Representing the Lagrangian Hessian $W^{(k)}$

An important issue for the development of an SQP code is how to represent the Hessian matrix $W^{(k)}$ that arises in the SQP subproblem. As defined in $\mathrm{QP}^{(k)}$, it requires evaluation of all the Hessian matrices of $f$ and $c_i$, $i = 1, 2, \ldots, m$ at $\mathbf{x}^{(k)}$, and their combination using the multipliers $\boldsymbol{\lambda}^{(k)}$. Writing code from which to evaluate second derivatives can be quite error prone, and in the past, this option has not always been preferred. However, the use of an exact $W^{(k)}$ matrix has been given new impetus through the availability of easy-to-use automatic differentiation within modelling languages such as AMPL, GAMS and TOMLAB (see Section 7). In large problems the exact $W^{(k)}$ may be a sparse matrix, which provides another reason to consider this option. On the other hand, it may be that the globalization strategy requires $W^{(k)}$ to be positive definite, in which case it will usually not be possible to use $W^{(k)}$ directly (except for certain 'convex' problems, $W$ can be indefinite, even at the solution). However, if $W^{(k)}$ is readily and cheaply available, it is probably best to make use of it in some way, as this can be expected to keep the iteration count low. There are various ideas for modifying $W^{(k)}$ to obtain a positive definite matrix. One is to add a suitable multiple of a unit matrix to $W^{(k)}$. Another is to add outer products using active constraint gradients, as when using an augmented Lagrangian penalty function.

Otherwise, the simplest approach, generically referred to as *quasi-Newton* SQP, is to update a symmetric matrix $B^{(k)}$ which approximates $W^{(k)}$. In this method, $B^{(1)}$ is initialized to some suitable positive definite or semi-definite matrix (often a multiple of the unit matrix), and $B^{(k)}$ is updated after each iteration to build up information about second derivatives. A suitable strategy (e.g. Nocedal and Overton [37]) is usually based on evaluating difference vectors

$$\delta^{(k)} = \mathbf{x}^{(k+1)} - \mathbf{x}^{(k)} \tag{4.11}$$

in $\mathbf{x}$, and

$$\gamma^{(k)} = \nabla_{\mathbf{x}}\mathcal{L}(\mathbf{x}^{(k+1)}, \lambda^{(k)}) - \nabla_{\mathbf{x}}\mathcal{L}(\mathbf{x}^{(k)}, \lambda^{(k)}) \tag{4.12}$$

in the gradient of the Lagrangian function $\mathcal{L}(\mathbf{x}, \lambda^{(k)})$, using the latest available estimate $\lambda^{(k)}$ of the multipliers. Then the updated matrix $B^{(k+1)}$ is chosen to satisfy the *secant condition* $B^{(k+1)}\delta^{(k)} = \gamma^{(k)}$.

There are many ways in which one might proceed. For small problems, where it is required to maintain a positive definite $B^{(k)}$ matrix, the BFGS formula (see [13]) might be used, in which case it is necessary to have $\delta^{(k)T}\gamma^{(k)} > 0$. It is not immediately obvious how best to meet this requirement in an NLP context, although a method suggested by Powell [41] has been used widely with some success. For large problems, some form of *limited memory update* is a practical proposition. The L-BFGS method, Nocedal [36], as implemented by Byrd, Nocedal and Schnabel [4] is attractive, although other ideas have also been tried. Another method which permits low costs is the low rank Hessian approximation $B = UU^T$ (Fletcher [15]), where $U$ has relatively few columns. For ENLP, updating the reduced Hessian matrix $M \approx Z^T W Z$, $B = VMV^T$, using differences in reduced gradients, is appropriate, essentially updating the Hessian of the reduced objective function $F(\mathbf{t})$ in (3.5). However, this idea does not translate easily into the context of NLP with inequality constraints, due to the change in dimension of $m$ when the number of active constraints changes.

An intermediate situation for large scale SQP is to update an approximation which takes the sparsity pattern of $W^{(k)}$ into account, and updates only the non-sparse elements. The LANCELOT project (see Conn, Gould and Toint [8] for many references) makes use of *partially separable* functions in which $B^{(k)}$ is the sum of various low dimensional *element*

*Hessians*, for which the symmetric rank one update is used. Other sparsity respecting updates have also been proposed, for example Toint [45], Fletcher, Grothey and Leyffer [17], but the coding is complex, and there are some difficulties.

Various important conditions exist regarding rapid local convergence, relating to the asymptotic properties of $WZ$ or $Z^T WZ$ (see [13] for references). Significantly, low storage methods like L-BFGS do not satisfy these conditions, and indeed slow convergence is occasionally observed, especially when the true reduced Hessian $Z^{*T}W^*Z^*$ is ill-conditioned. For this reason, obtaining rapid local convergence when the null space dimension is very large is still a topic of research interest. Indeed the entire subject of how best to provide second derivative information in an SQP method is very much an open issue.

# 5  Globalization of NLP Methods

In this section we examine the transition from Newton type methods with rapid local convergence, such as the SQP method, to globally convergent methods suitable for incorporation into production NLP software. By globally convergent, we refer to the ability to converge to local solutions of an NLP problem from globally selected initial iterates which may be remote from any solution. This is not to be confused with the problem of guaranteeing to find global solutions of an NLP problem in the sense of the best local solution, which is computationally impractical for problems of any size (perhaps >40 variables, say), unless the problem has some special convexity properties, which is rarely the case outside of LP and QP. We must also be aware that NLP problems may have no solution, mainly due to the constraints being infeasible (that is, no feasible point exists). In this case the method should ideally be able to indicate that this is the case, and not spend an undue amount of time in searching for a non-existent solution. In practice even to guarantee that no feasible solution exists is an unrealistic aim, akin to that of finding a global minimizer of some measure of constraint infeasibility. What is practical is to locate a point which is *locally infeasible* in the sense that the first order Taylor series approximation to the constraints set is infeasible at that point. Again the main requirement is that the method should be able to converge rapidly to such a point, and exit with a suitable indication of local infeasibility. Another possibility, which can be excluded by bounding the feasible region, is that the NLP is unbounded, that is $f(\mathbf{x})$ is not bounded below on the feasible region, or that there are no KT points in the feasible region. Again the software has to recognize the situation and terminate accordingly.

Ultimately the aim is to be able to effectively solve NLP problems created by scientists, engineers, economists etc., who have a limited background in optimization methods. For this we must develop general purpose software which is

- Efficient
- Reliable
- Well documented
- Thoroughly tested
- Flexible
- Easy to use, and has
- Large scale capacity.

Efficiency and reliability are self evident, as is being well-documented, Being thoroughly tested involves monitoring the behaviour on test problems (e.g. CUTE [2]) which encompass a wide range of possible situations, including problems with no solution as referred to above. Flexibility includes the ability to specify simple bounds, linear constraints, etc., to take account of sparsity in the formulation, and to be able to make warm starts from the solution of a previously solved problem. It can also include the flexibility to decide whether or not to supply second derivatives (indeed some codes have been developed which require no derivative information to be supplied, but these are very limited in the size of problem that can be solved, and the accuracy that can be achieved). By easy to use we envisage issues such as the provision of default parameter values which do not need tuning by the user, and access to the software via modelling languages like AMPL, GAMS or TOMLAB. Large scale capacity is required for the software to make a significant impact: even run of the mill applications of optimization now have 1000's of variables and/or constraints and may be computationally intractable without the use of sparse matrix techniques, or special iterative methods such as Conjugate Gradients.

## 5.1 Penalty and Barrier Functions

From an historical perspective, almost all general purpose NLP solvers until about 1996 aimed to promote global convergence by constructing an auxiliary function from $f(\mathbf{x})$ and $\mathbf{c}(\mathbf{x})$ known variously as a penalty, barrier, or merit function. In the earlier days, the idea was to apply successful existing techniques for unconstrained minimization to the auxiliary function, in such a way as to find the solution of the NLP problem. Later, there came the idea of using the auxiliary function to decide whether or not to accept the step given by the SQP method, hence the term merit function.

For an ENLP problem, an early idea was the Courant [9] penalty function

$$\phi(\mathbf{x}; \sigma) = f(\mathbf{x}) + \tfrac{1}{2}\sigma \mathbf{c}^T \mathbf{c} = f(\mathbf{x}) + \tfrac{1}{2}\sigma \sum_{i=1}^{m} c_i^2 \quad \text{where} \quad \mathbf{c} = \mathbf{c}(\mathbf{x}), \quad (5.1)$$

where $\sigma > 0$ is a parameter. The $\mathbf{c}^T\mathbf{c}$ term 'penalizes' points which violate the constraints, and $\sigma$ determines the strength of the penalty. Let a minimizer, $\mathbf{x}(\sigma)$ say, of $\phi(\mathbf{x};\sigma)$ be found by some technique for unconstrained minimization. Then we choose a sequence of values of $\sigma \to \infty$, for example $\sigma = \{1,\ 10,\ 100,\ \dots\}$ and observe the behaviour of $\mathbf{x}(\sigma)$. Under some assumptions, it can be shown that the NLP solution is given by $\mathbf{x}^* = \lim_{\sigma\to\infty}\mathbf{x}(\sigma)$. A simple modification for the NLP problem is to include terms $(\min(c_i,\ 0))^2$ in the summation for any inequality constraints $c_i(\mathbf{x}) \geq 0$. In practice the methods are slow as compared with SQP techniques, and more seriously, suffer serious effects due to ill-conditioning of the Hessian matrix of $\phi$ as $\sigma \to \infty$.

Another early idea for NLP with inequality constraints was the Frisch [24] *log function*

$$\phi(\mathbf{x};\mu) = f(\mathbf{x}) - \mu \sum_{i=1}^{m} \log_e c_i(\mathbf{x}), \qquad (5.2)$$

where $\mu > 0$ is a parameter. In this case we require $\mathbf{x}$ to lie in the interior of the feasible region where $c_i(\mathbf{x}) > 0$, $i = 1, 2, \dots, m$, so that the log terms are well defined. Then each term $-\log_e c_i(\mathbf{x})$ approaches $+\infty$ as $\mathbf{x}$ approaches the boundary of the feasible region, and creates a 'barrier' which prevents iterates from escaping out of the feasible region. The parameter $\mu$ determines the extent to which the influence of barrier extends into the interior of the feasible region. For any fixed value of $\mu$ we again find $\mathbf{x}(\mu)$ to minimize $\phi(\mathbf{x};\mu)$. Then we choose a sequence of values of $\mu \to 0$, for example $\mu = \{1,\ \frac{1}{10},\ \frac{1}{100},\dots\}$ and observe the behaviour of $\mathbf{x}(\mu)$. Under some assumptions, it can be shown that $\mathbf{x}^* = \lim_{\mu\to 0}\mathbf{x}(\mu)$. Again the methods are slow and suffer from ill-conditioning of the Hessian. Moreover the need to find a strictly feasible starting point is usually a difficult problem in its own right. More recently the log barrier function has been used as a merit function in conjunction with interior point methods.

## 5.2 Multiplier Penalty and Barrier Functions

From around 1969 onwards, a more satisfactory class of auxiliary function came into use, involving an additional parameter $\boldsymbol{\lambda} \in \mathrm{I\!R}^m$ which could be interpreted as an estimate of the Lagrange multiplier vector. For ENLP the Lagrangian function is augmented with a penalty term giving

$$\phi(\mathbf{x};\boldsymbol{\lambda},\sigma) = f(\mathbf{x}) - \boldsymbol{\lambda}^T\mathbf{c} + \tfrac{1}{2}\sigma\mathbf{c}^T\mathbf{c} \qquad \text{where} \qquad \mathbf{c} = \mathbf{c}(\mathbf{x}), \qquad (5.3)$$

sometimes referred to as the *augmented Lagrangian function* (Hestenes [30], Powell [40]). We note that

$$\nabla_{\mathbf{x}}\phi = \mathbf{g} - A\boldsymbol{\lambda} + \sigma A\mathbf{c}, \qquad (5.4)$$

so that $\nabla\phi(\mathbf{x}^*; \boldsymbol{\lambda}^*, \sigma) = \mathbf{0}$. Moreover if the local solution is regular and satisfies second order sufficient conditions, and if $\sigma$ is sufficiently large, then $\nabla^2\phi(\mathbf{x}^*; \boldsymbol{\lambda}^*, \sigma)$ is positive definite (see [13]). Thus, if $\boldsymbol{\lambda}^*$ is known, the solution of the ENLP can be found by minimizing $\phi(\mathbf{x}; \boldsymbol{\lambda}^*, \sigma)$, without the need to drive $\sigma$ to infinity, in contrast to the methods of the previous subsection, and this avoids the worst effects of the ill-conditioning. Unfortunately $\boldsymbol{\lambda}^*$ is not known a-priori, so a sequential minimization technique must still be used. In an outer iteration a sequence of parameters $\boldsymbol{\lambda}^{(k)} \rightarrow \boldsymbol{\lambda}^*$ is chosen, whilst in the inner iteration, a minimizer $\mathbf{x}(\boldsymbol{\lambda}^{(k)}, \sigma)$ of $\phi(\mathbf{x}; \boldsymbol{\lambda}^{(k)}, \sigma)$ is found by means of an unconstrained minimization technique. The behaviour of $\mathbf{x}(\boldsymbol{\lambda}^{(k)}, \sigma)$ is observed as $\boldsymbol{\lambda}^{(k)}$ changes. If the iteration is not converging to $\mathbf{x}^*$, it may be necessary to increase $\sigma$ at some stage. To update $\boldsymbol{\lambda}^{(k)}$, a formula

$$\boldsymbol{\lambda}^{(k+1)} = \boldsymbol{\lambda}^{(k)} - (A^T W A)^{-1} \mathbf{c}^{(k)} \qquad (5.5)$$

derived from the SQP iteration formula (3.20) may be used. For large $\sigma$ this may be approximated by the scheme

$$\boldsymbol{\lambda}^{(k+1)} = \boldsymbol{\lambda}^{(k)} - \sigma \mathbf{c}^{(k)}, \qquad (5.6)$$

see [13].

A multiplier penalty function for the NLP (1.1) with inequality constraints is

$$\phi(\mathbf{x}; \boldsymbol{\lambda}, \sigma) = f(\mathbf{x}) + \sum_i \begin{cases} -\lambda_i c_i + \frac{1}{2}\sigma c_i^2 & \text{if } c_i \leq \lambda_i/\sigma \\ -\frac{1}{2}\lambda_i^2/\sigma & \text{if } c_i \geq \lambda_i/\sigma \end{cases} \qquad (5.7)$$

suggested by Rockafellar [44]. The piecewise term does not cause any discontinuity in first derivatives, and any second derivative discontinuities occur away from the solution. Otherwise its use is similar to that in (5.3) above.

A multiplier based modification of the Frisch barrier function due to Polyak [39] is

$$\phi(\mathbf{x}; \boldsymbol{\lambda}, \mu) = f(\mathbf{x}) - \mu \sum_{i=1}^{m} \lambda_i \log_e(c_i/\mu + 1), \qquad (5.8)$$

in which the boundary occurs where $c_i = -\mu$, which is strictly outside the feasible region. Thus the discontinuity of the Frisch function at the solution $\mathbf{x}^*$ is moved away into the infeasible region. We note that

$$\nabla_{\mathbf{x}}\phi(\mathbf{x}^*; \boldsymbol{\lambda}^*, \mu) = \mathbf{g}^* - \mu \sum_{i=1}^{m} \frac{\lambda_i^* \mathbf{a}_i^*/\mu}{(c_i^*/\mu + 1)} = \mathbf{0}$$

using KT conditions (including complementarity). If the solution $\mathbf{x}^*$ is regular, and sufficient conditions hold, with strict complementarity, then it can also be shown that $\nabla_{\mathbf{x}}^2\phi(\mathbf{x}^*; \boldsymbol{\lambda}^*, \mu)$ is positive definite if $\mu$ is sufficiently small. Thus a suitable fixed value of $\mu > 0$ can be found and the worst effects of ill-conditioning are avoided. Both the Rockafellar and Polyak functions are used in a sequential manner with an outer iteration in which $\boldsymbol{\lambda}^{(k)} \to \boldsymbol{\lambda}^*$.

All the above proposals involve sequential unconstrained minimization, and as such, are inherently less effective than the SQP method, particularly in regard to the rapidity of local convergence. Errors in $\boldsymbol{\lambda}^{(k)}$ induce errors of similar order in $\mathbf{x}(\boldsymbol{\lambda}^{(k)})$, which is not the case for the SQP method. Many other auxiliary functions have been suggested for solving NLP or ENLP problems in ways related to the above. In a later section we shall investigate so-called exact penalty functions which avoid the sequential unconstrained minimization aspect.

A major initiative to provide robust and effective software with large scale capability based on the augmented Lagrangian function was the LANCELOT code of Conn, Gould and Toint (see [8] for references). The code applies to an NLP in the form

$$\begin{aligned} \underset{\mathbf{x} \in \mathbb{R}^n}{\text{minimize}} \quad & f(\mathbf{x}) \\ \text{subject to } & \mathbf{c}(\mathbf{x}) = \mathbf{0} \\ & \mathbf{l} \le \mathbf{x} \le \mathbf{u}. \end{aligned} \qquad (5.9)$$

which treats simple bounds explicitly but assumes that slack variables have been added to any other inequality constraints. In the inner iteration, the augmented Lagrangian function (5.3) is minimized subject to the simple bounds $\mathbf{l} \le \mathbf{x} \le \mathbf{u}$ on the variables. A potential disadvantage of this approach for large scale computation is that the Hessian $\nabla_{\mathbf{x}}^2 \phi(\mathbf{x}; \boldsymbol{\lambda}^{(k)}, \sigma)$ is likely to be much less sparse than the Hessian $W^{(k)}$ in the SQP method. To avoid this difficulty, LANCELOT uses a simple bound minimization technique based on the use of the preconditioned conjugate gradient method, and solves the subproblem to lower accuracy when $\boldsymbol{\lambda}^{(k)}$ is inaccurate. It also uses an innovative idea of building the Hessian from a sum of elementary Hessians through the concept of group partial separability. LANCELOT has been successfully used to solve problems with upwards of $10^4$ variables, particularly those with large dimensional null spaces arising for example from the discretization of a partial differential equation. It is less effective for problems with low dimensional null spaces and does not take advantage of any linear constraints.

## 5.3 Augmented Lagrangians with SQP

A fruitful way to take advantage of linear constraints has been to merge the globalization aspect of the augmented Lagrangian function with the rapid local convergence of the SQP method. This was the motivation of the very successful MINOS code of Murtagh and Saunders [35], which was arguably the first SQP-like NLP solver with large scale capability. In fact MINOS was influenced by a method due to Robinson [43] which is not an SQP method in the strict sense described above, but is closely related to it. We describe Robinson's method in the context of the NLP problem (1.1). The method generates a sequence of major iterates $\mathbf{x}^{(k)}$, $\boldsymbol{\lambda}^{(k)}$ and solves the LCP problem

$$\text{LCP}^{(k)} \left\{ \begin{array}{l} \underset{\mathbf{x} \in \mathbb{R}^n}{\text{minimize}} \ f(\mathbf{x}) - \boldsymbol{\lambda}^{(k)T}(\mathbf{c}(\mathbf{x}) - \mathbf{s}) \\ \text{subject to } \mathbf{s} = \mathbf{c}^{(k)} + A^{(k)T}(\mathbf{x} - \mathbf{x}^{(k)}) \geq \mathbf{0}, \end{array} \right.$$

where $\mathbf{s} = \mathbf{s}(\mathbf{x}, \mathbf{x}^{(k)}) = \mathbf{c}^{(k)} + A^{(k)T}(\mathbf{x} - \mathbf{x}^{(k)})$ is the first order Taylor series approximation to $\mathbf{c}(\mathbf{x})$ about the current point, and the quantity $\mathbf{c}(\mathbf{x}) - \mathbf{s}$ may be thought of as the *deviation from linearity*. The solution and multipliers of $\text{LCP}^{(k)}$ then become the iterates $\mathbf{x}^{(k+1)}$, $\boldsymbol{\lambda}^{(k+1)}$ for the next major iteration. The method differs from SQP, firstly in that $\text{LCP}^{(k)}$ cannot be solved finitely, which is a disadvantage, and secondly that second derivatives are not required, which is an advantage. Robinson intended that $\text{LCP}^{(k)}$ should be solved by a reduced Hessian quasi-Newton method. If a Taylor expansion of the objective function in $\text{LCP}^{(k)}$ about the current point is made, then it agrees with that of $\text{SQP}^{(k)}$ up to and including second order terms. Also the method has the same fixed point property as SQP that if $\mathbf{x}^{(k)}$, $\boldsymbol{\lambda}^{(k)}$ is equal to $\mathbf{x}^*$, $\boldsymbol{\lambda}^*$, then $\mathbf{x}^*$, $\boldsymbol{\lambda}^*$ is the next iterate, and the process terminates. Consequently the method has the same rapid local convergence properties as the SQP method, assuming that the $\text{LCP}^{(k)}$ subproblem is solved sufficiently accurately. However there is no global convergence result available, for example there is no mechanism to force the iterates $\mathbf{x}^{(k)}$ to accumulate at a feasible point.

The MINOS code attempts to mitigate the lack of a global convergence property by augmenting the objective function in $\text{LCP}^{(k)}$ with a squared penalty term. As with LANCELOT, the method is applicable to an NLP in the form (5.9), and the LCP subproblem that is solved on the $k$-th major iteration is

$$\begin{array}{l} \underset{\mathbf{x} \in \mathbb{R}^n}{\text{minimize}} \ f(\mathbf{x}) - \boldsymbol{\lambda}^{(k)T}(\mathbf{c}(\mathbf{x}) - \mathbf{s}) + \tfrac{1}{2}\sigma(\mathbf{c}(\mathbf{x}) - \mathbf{s})^T(\mathbf{c}(\mathbf{x}) - \mathbf{s}) \\ \text{subject to } \mathbf{s} = \mathbf{c}^{(k)} + A^{(k)T}(\mathbf{x} - \mathbf{x}^{(k)}) = \mathbf{0} \\ \qquad\qquad \mathbf{l} \leq \mathbf{x} \leq \mathbf{u}. \end{array}$$

In the original source, MINOS refers to the active set method used to solve this LCP subproblem, and MINOS/AUGMENTED refers to the major iterative procedure for solving (5.9). However it is more usual now to refer to the NLP solver by MINOS. The code has sparse matrix facilities, and also allows 'linear variables' to be designated, so allowing the use of a smaller Hessian approximation. MINOS was probably the first SQP-type code with the capability to solve large scale problems, and as such has been very successful and is still in use.

A development of MINOS is the SNOPT code of Gill, Murray and Saunders [25] which first appeared in about 1992. In place of an LCP subproblem it solves a QP subproblem, but using an approximate Hessian matrix. Slack variables **s** are explicitly included, and each iteration involves the solution of a line search subproblem based on the MINOS augmented Lagrangian

$$\phi(\mathbf{x}, \mathbf{s}, \boldsymbol{\lambda}) = f(\mathbf{x}) - \boldsymbol{\lambda}^{(k)T}(\mathbf{c}(\mathbf{x}) - \mathbf{s}) + \tfrac{1}{2}(\mathbf{c}(\mathbf{x}) - \mathbf{s})^T D(\mathbf{c}(\mathbf{x}) - \mathbf{s}), \quad (5.10)$$

where $D = \operatorname{diag} \sigma_i$ is a diagonal matrix of penalty parameters. The entire triple $\mathbf{x}^{(k)}$, $\mathbf{s}^{(k)}$, $\boldsymbol{\lambda}^{(k)}$ is varied in the line search. Various treatments of the Hessian approximation are possible, depending for example on the size of the problem. Another difference from MINOS is the use of 'elastic mode' (essentially the $l_1$ penalty function of the next subsection) to resolve significant deviations from infeasibility. It is impossible here to do justice to all the features of the code, and the reader is referred to the comprehensive description in [25], although it is quite likely that further development of the code has taken place. For NLP problems in which the null space dimension is not too large, up to 1000 say, SNOPT is currently amongst the best currently available NLP solvers (see Byrd, Gould, Nocedal and Waltz [3]).

## 5.4   The $l_1$ Exact Penalty Function

The penalty and barrier functions in Sections 5.1 and 5.2 are inherently sequential, that is the solution of the NLP problem is obtained by a sequence of unconstrained minimization calculations. It is however possible to construct a so-called *exact penalty function*, that is a penalty function of which $\mathbf{x}^*$, a solution of the NLP problem, is a local minimizer. It is convenient here to consider an NLP problem in the form

$$\begin{aligned} \operatorname*{minimize}_{\mathbf{x} \in \mathbb{R}^n} \quad & f(\mathbf{x}) \\ \text{subject to } & c(\mathbf{x}) \le \mathbf{0}. \end{aligned} \quad (5.11)$$

The most well known exact penalty function (Pietrzykowski [38], see [13] for more references) is the $l_1$ *exact penalty function* ($l_1$EPF)

$$\phi(\mathbf{x}; \sigma) = f(\mathbf{x}) + \sigma \|\mathbf{c}^+(\mathbf{x})\|_1 = f(\mathbf{x}) + \sigma \sum_{i=1}^{m} c_i^+(\mathbf{x}), \quad (5.12)$$

where $c_i^+ = \max(c_i, 0)$ is the amount by which the $i$–th constraint is violated. The parameter $\sigma$ controls the strength of the penalty.

First we consider optimality conditions for a local minimizer of $\phi(\mathbf{x}, \sigma)$. The function is nonsmooth due to the discontinuity in derivative of $\max(c_i, 0)$

at zero, so we cannot refer to the stationary point condition of Section 2.2. In fact we proceed very much as in Section 4.2, by defining the set of active constraints as in (4.5), and the set of *infeasible constraints*

$$\mathcal{I}(\mathbf{x}) = \{i \mid c_i(\mathbf{x}) > 0\}. \tag{5.13}$$

Infeasible constraints at $\mathbf{x}^*$ are denoted by the set $\mathcal{I}^*$ and are assigned a multiplier $\lambda_i^* = \sigma$. Constraints that are neither active nor infeasible at $\mathbf{x}^*$ play no part in the conditions and can be ignored. As before we assign them a multiplier $\lambda_i^* = 0$. We observe that if $\mathbf{x}^*$ minimizes $\phi(\mathbf{x}, \sigma)$, then it must solve the problem

$$\begin{array}{ll} \underset{\mathbf{x} \in \mathbb{R}^n}{\text{minimize}} & f(\mathbf{x}) + \sigma \sum_{i=1}^{m} c_i^+(\mathbf{x}) \\ \text{subject to } c_i(\mathbf{x}) \leq 0 & i \in \mathcal{A}^*. \end{array} \tag{5.14}$$

and hence the problem

$$\begin{array}{ll} \underset{\mathbf{x} \in \mathbb{R}^n}{\text{minimize}} & f(\mathbf{x}) + \sigma \sum_{i \in \mathcal{I}^*} c_i(\mathbf{x}) \\ \text{subject to } c_i(\mathbf{x}) \leq 0 & i \in \mathcal{A}^*. \end{array} \tag{5.15}$$

since the penalty term comes only from the infeasible constraints. Therefore if we make the same regularity assumption that the vectors $\mathbf{a}_i^*$, $i \in \mathcal{A}^*$ are linearly independent, then the KT conditions for this problem are also necessary for a minimizer of (5.12). Moreover, if we perturb the right hand side of a constraint $i \in \mathcal{A}^*$ in (5.15) by a sufficiently small $\varepsilon_i > 0$, $\varepsilon_j = 0$, $j \neq i$, we make constraint $i$ infeasible, but do not change the status of any other constraints. This causes an increase of $\sigma \varepsilon_i$ in the penalty term. Moreover the change in $f(\mathbf{x}) + \sigma \sum_{i \in \mathcal{I}} c_i(\mathbf{x})$ to first order is $-\lambda_i^* \varepsilon_i$. (The negative sign holds because of the sign change in (5.11)). Hence the change in $\phi(\mathbf{x}, \sigma)$ to first order is $\varepsilon_i(\sigma - \lambda_i^*)$. If $\lambda_i^* > \sigma$ then $\phi$ is reduced by the perturbation, which contradicts the optimality of $\mathbf{x}^*$ in the $l_1$ EPF. Thus the condition $\lambda_i^* \leq \sigma$, $i \in \mathcal{A}^*$ is also necessary. This result tells us that unless the penalty parameter $\sigma$ is sufficiently large, a local minimizer will not be created. We can therefore summarize the first order necessary conditions as

$$\mathbf{g}^* + \sigma \sum_{i \in \mathcal{I}^*} \mathbf{a}_i^* + \sum_{i \in \mathcal{A}^*} \mathbf{a}_i^* \lambda_i^* = \mathbf{g}^* + \sum_{i=1}^{m} \mathbf{a}_i^* \lambda_i^* = \mathbf{g}^* + A^* \boldsymbol{\lambda}^* = \mathbf{0} \tag{5.16}$$

$$\left. \begin{array}{l} 0 \leq \lambda_i^* \leq \sigma \\ c_i^* < 0 \Rightarrow \lambda_i^* = 0 \\ c_i^* > 0 \Rightarrow \lambda_i^* = \sigma \end{array} \right\} \quad i = 1, 2, \ldots, m. \tag{5.17}$$

If $\Rightarrow$ in (5.17) can be replaced by $\Leftrightarrow$ then *strict complementarity* is said to hold. Second order necessary conditions are the same as for (5.15), that is $Z^{*T}W^*Z^*$ is positive semi-definite. Sufficient are that first order conditions hold, with strict complementarity, and $Z^{*T}W^*Z^*$ is positive definite.

We see that these conditions are very similar to those for solving the NLP problem (5.11). Consequently there is a strong correlation between local minimizers of the $l_1$EPF and local solutions of the NLP problem, which justifies the practical use of the $l_1$EPF as a means of solving an NLP problem. To be precise, assume that the appropriate second order sufficient conditions and strict complementarity hold. Then if $\mathbf{x}^*$ solves the NLP problem and $\sigma > \|\boldsymbol{\lambda}^*\|_\infty$, it follows that $\mathbf{x}^*$ is a local minimizer of the $l_1$EPF. Conversely, if $\mathbf{x}^*$ is a local minimizer of the $l_1$EPF, and $\mathbf{x}^*$ is feasible in the NLP problem, then $\mathbf{x}^*$ is a local solution of the NLP problem.

## 5.5  SQP with the $l_1$EPF

The SQP method first came into prominent use when used in conjunction with the $l_1$EPF as suggested by Han [29] and Powell [41]. The vector $\mathbf{d}^{(k)}$ generated by the SQP subproblem QP$^{(k)}$ is regarded as a direction of search, and the $l_1$EPF is used as a merit function, so that the next iterate is $\mathbf{x}^{(k+1)} = \mathbf{x}^{(k)} + \alpha^{(k)}\mathbf{d}^{(k)}$, with $\alpha^{(k)}$ being chosen to obtain a sufficient reduction in $\phi(\mathbf{x}, \sigma)$. For this to be possible requires that $\mathbf{d}^{(k)}$ is a descent direction at $\mathbf{x}^{(k)}$ for $\phi(\mathbf{x}, \sigma)$. If the Hessian $W^{(k)}$ (or its approximation) is positive definite, it is possible to ensure that this is the case, if necessary by increasing $\sigma$. Early results with this technique were quite promising, when compared with sequential unconstrained penalty and barrier methods. However the use of a nonsmooth merit function is not without its difficulties. In particular the discontinuities in derivative cause 'curved valleys', with sides whose steepness depends on the size of $\sigma$. If $\sigma$ is large, the requirement to monotonically improve $\phi$ on every iteration can only be achieved by taking correspondingly small steps, leading to slow convergence. Unfortunately, increasing $\sigma$ to obtain descent exacerbates this situation.

A way round this is the Sequential $l_1$ Quadratic Programming (S$l_1$QP) method of Fletcher [11]. The idea (also applicable to an $l_\infty$ exact penalty function) is to solve a subproblem which more closely models the $l_1$EPF, by moving the linearized constraints into the objective function, in an $l_1$ penalty term. Thus the $l_1$QP subproblem is

$$\underset{\mathbf{d}\in\mathbb{R}^n}{\text{minimize}} \quad \mathbf{g}^{(k)T}\mathbf{d} + \tfrac{1}{2}\mathbf{d}^TW^{(k)}\mathbf{d} + \sigma\|(\mathbf{c}^{(k)} + A^{(k)T}\mathbf{d})^+\|_1$$
$$\text{subject to } \|\mathbf{d}\|_\infty \le \rho. \tag{5.18}$$

It is necessary that $\sigma$ is sufficiently large as discussed in Section 5.4, and also below. The restriction on $\mathbf{d}$ is the trust region constraint and $\rho$ is the trust region radius. Solving the subproblem ensures descent, and quite strong results regarding global convergence can be proved by using standard trust region ideas. The use of an $l_\infty$ trust region (a 'box constraint') fits conveniently into a QP type framework.

Even so, there are still some issues to be resolved. Firstly, (5.18) is not a QP in standard form due to the presence of $l_1$ terms in the objective, although it is still a problem that can be solved in a finite number of steps. Ideally a special purpose $l_1$QP solver with sparse matrix capabilities would be used. This would enable an efficient $l_1$ piecewise quadratic line search to be used within the solver. Unfortunately a fully developed code of this type is not easy to come by. The alternative is to transform (5.18) into a regular QP by the addition of extra variables. For example a constraint $\mathbf{l} \le \mathbf{c}(\mathbf{x}) \le \mathbf{u}$ can be written as $\mathbf{l} \le \mathbf{c}(\mathbf{x}) - \mathbf{v} + \mathbf{w} \le \mathbf{u}$ where $\mathbf{v} \ge \mathbf{0}$ and $\mathbf{w} \ge \mathbf{0}$ are auxiliary variables, and a penalty term of the form $\sigma \|\mathbf{v} + \mathbf{w}\|_1$, which is linear in $\mathbf{v}$ and $\mathbf{w}$, would then be appropriate. However $2m$ extra variables need be added, which is cumbersome, and the benefit of the piecewise quadratic line search is not obtained. A related idea is to use the SLP-EQP idea of Fletcher and Sainz de la Maza [22], referred to in Section 4.5. In this case an $l_1$LP subproblem would be used to find an active set and multipliers, followed by an EQP calculation to obtain the step $\mathbf{d}^{(k)}$. As above, the $l_1$LP subproblem can be converted to an LP problem by the addition of extra variables, and this allows fast large scale LP software to be used.

It is also not easy for the user to choose a satisfactory value of $\sigma$. If it is chosen too small, then a local minimizer may not be created, if too large then the difficulties referred to above become apparent. There is also a possibility of the *Maratos effect* [34] occurring, in which, close to the solution, the Newton-type step given by the SQP method increases $\phi$ and cannot be accepted if monotonic improvement in $\phi$ is sought. Thus the expected rapid local convergence is not realised. More recently, ideas for circumventing these difficulties have been suggested, including *second order corrections* [12], the *watchdog technique* [7], and a *non-monotonic line search* [27].

# 6  Filter Methods

Filter methods were introduced in response to the perceived difficulties in using penalty function methods for globalization, that is the difficulty of choosing suitable penalty parameters, the inefficiency of sequential methods, and the slow convergence associated with monotonic minimization methods, particularly in the case of nonsmooth exact penalty functions. Along with this is the observation that the basic SQP method is able to quickly solve a significant proportion of test problems without the need for modifications to induce global convergence. Thus the goal of filter methods is to provide global optimization safeguards that allow the full SQP step to be taken much more often. In this section we describe the main ideas and possible pitfalls, and discuss the way in which a global convergence result for a filter method has been constructed.

A penalty function is an artefact to combine two competing aims in NLP, namely the minimization of $f(\mathbf{x})$ and the need to obtain feasibility with respect to the constraints. The latter aim can equivalently be expressed as the minimization of some measure $h(\mathbf{c}(\mathbf{x}))$ of constraint violation. For example, in the context of (5.11) we could define $h(\mathbf{c}) = \|\mathbf{c}^+\|$ in some convenient norm. Thus, in a filter method, we view NLP as the resolution of two competing aims of minimizing $f(\mathbf{x})$ and $h(\mathbf{c}(\mathbf{x}))$. This is the type of situation addressed by Pareto (multi-objective) optimization, but in our context the minimization of $h(\mathbf{c}(\mathbf{x}))$ has priority, in that it is essential to find a Pareto solution that corresponds to a feasible point. However it is useful to borrow the concept of *domination* from multi-objective optimization. Let $\mathbf{x}^{(k)}$ and $\mathbf{x}^{(l)}$ be two points generated during the progress of some method. We say that $\mathbf{x}^{(k)}$ dominates $\mathbf{x}^{(l)}$ if and only if $h^{(k)} \leq h^{(l)}$ and $f^{(k)} \leq f^{(l)}$. That is to say, there is no reason to prefer $\mathbf{x}^{(l)}$ on the basis of either measure. Now we define a *filter* to be a list of pairs $(h^{(k)}, f^{(k)})$ such that no pair dominates any other. As the algorithm progresses, a filter is built up from all the points that have been sampled by the algorithm. A typical filter is shown in Figure 2, where the shaded region shows the region dominated by the filter entries (the outer vertices of this shaded region). The contours of the $l_1$ exact penalty function would be straight lines with slope $-\sigma$ on this plot, indicating that at least for a single entry, the filter provides a less restrictive acceptance condition than the penalty function.

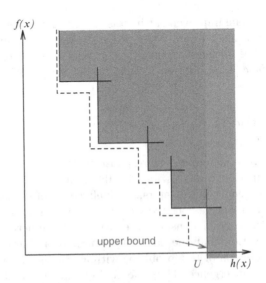

**Fig. 2** A Typical Filter Diagram

## 6.1  SQP Filter Methods

Filter methods were first introduced in the context of trust region SQP methods in 1997 by Fletcher and Leyffer [18], making use of a subproblem

$$QP^{(k)}(\rho) \begin{cases} \underset{\mathbf{d} \in \mathbb{R}^n}{\text{minimize}} & \frac{1}{2}\mathbf{d}^T W^{(k)}\mathbf{d} + \mathbf{d}^T \mathbf{g}^{(k)} \\ \text{subject to} & \mathbf{c}^{(k)} + A^{(k)T}\mathbf{d} \geq \mathbf{0}. \\ & \|\mathbf{d}\|_\infty \leq \rho, \end{cases}$$

obtained by adding a trust region constraint to $QP^{(k)}$. A *trust region* (see [13]) is a heuristic aimed at restricting the step size to lie in a region in which there is 'adequate' agreement between the true functions and their Taylor series approximations. The trust region radius $\rho$ is adjusted during or after each iteration to achieve this aim.

A first attempt at a filter algorithm might go as follows. On iteration $k = 1$ the filter $\mathcal{F}^{(1)}$ is empty. On iteration $k$ we solve $QP^{(k)}(\rho)$ giving a step $\mathbf{d}$ and evaluate $f = f(\mathbf{x}^{(k)} + \mathbf{d})$ and $h = h(\mathbf{c}(\mathbf{x}^{(k)} + \mathbf{d}))$. If the resulting pair $(h, f)$ is acceptable to $\mathcal{F}^{(k)}$ (that is, it is not dominated by any of the entries in $\mathcal{F}^{(k)}$), then we update $\mathbf{x}^{(k)}$ and $\boldsymbol{\lambda}^{(k)}$ as described in Section 4.4. We also update the filter, adding the new pair $(h, f)$ and removing any entries that are dominated by it. Possibly we might also increase the trust region radius. If, on the other hand, the pair is not acceptable, then we reject it, reduce the trust region radius, and re-solve $QP^{(k)}(\rho)$.

There are various ways in which this simple approach can fail, or become unsatisfactory. One is that unacceptably large violations in the constraints may occur. This is readily handled by imposing an upper bound $U$ on constraint violations, and initializing $\mathcal{F}^{(1)}$ to $(U, -\infty)$ (see Figure 2). More serious is the possibility that, if the current point $\mathbf{x}^{(k)}$ is infeasible ($h^{(k)} > 0$), and if $\rho$ is reduced sufficiently in the algorithm, then the constraints of $QP^{(k)}(\rho)$ can become incompatible, and the algorithm stalls. In this case our approach has been to enter a *feasibility restoration* phase, in which a different SQP-like algorithm (see Fletcher and Leyffer [19] for a filter-like approach) is invoked to find a new acceptable point $\mathbf{x}^{(k+1)}$ for which the TR subproblem is solvable. Of course, we cannot predict the effect on $f(\mathbf{x})$ and we must be prepared for it to increase. Another feature that the algorithm lacks is any sense of *sufficient reduction* in either $f$ or $h$, as is used in other convergence proofs. For instance we would not like new pairs to become arbitrarily close to existing filter entries, because this might allow convergence to a non-KT point. With this in mind we strengthen the acceptance condition by adding a *filter envelope* around the current filter entries, so as to extend the set of unacceptable points (see Figure 2). Most recently we use the *sloping envelope* of Chin [5], Chin and Fletcher [6], in which acceptability of a pair $(h, f)$ with respect to a filter $\mathcal{F}$ is defined by

$$h \leq \beta h_i \quad \text{or} \quad f \leq f_i - \gamma h \qquad \forall (h_i, f_i) \in \mathcal{F} \tag{6.1}$$

where $\beta$ and $\gamma$ are constants in $(0, 1)$. Typical values might be $\beta = 0.9$ and $\gamma = 0.01$. (In earlier work we used $f \leq f_i - \gamma h_i$ for the second test, giving a rectangular envelope. However, this allows the possibility that $(h, f)$ dominates $(h_i, f_i)$ but the envelope of $(h, f)$ does *not* dominate the envelope of $(h_i, f_i)$, which is undesirable.)

During testing of the filter algorithm, another less obvious disadvantage became apparent. Say the current filter contains an entry $(0, f_i)$ where $f_i$ is relatively small. If, subsequently, feasibility restoration is invoked, it may be impossible to find an acceptable point which is not dominated by $(0, f_i)$. Most likely, the feasibility restoration phase then converges to a feasible point that is not a KT point. We refer to $(0, f_i)$ as a *blocking entry*. We were faced with two possibilities. One is to allow the removal of blocking entries on emerging from feasibility restoration. This we implemented in the first code, reported by Fletcher and Leyffer [18]. To avoid the possibility of cycling, we reduce the upper bound when a blocking entry is removed. We did not attempt to provide a global convergence proof for this code, although it may well be possible to do so. Subsequently it became clear that other heuristics in the code were redundant and further work resulted in a related filter algorithm (Fletcher, Leyffer and Toint [20]) for which a convergence proof can be given. In this algorithm we resolve the difficulty over blocking by not including all accepted points in the filter. This work is described in the next section. However, the earlier code proved very robust, and has seen widespread use. It shows up quite well on the numbers of function and derivative counts required to solve a problem, in comparison say with SNOPT. Actual computing times are less competitive, probably because the QP solver used by SNOPT is more efficient. It has the same disadvantage as SNOPT that it is inefficient for large null space problems. Otherwise, good results were obtained in comparison with LANCELOT and an implementation of the $l_1$EPF method.

## 6.2  A Filter Convergence Proof

In this section an outline is given of the way in which a global convergence proof has been developed for the trust region filter SQP method. The theory has two aspects: how to force $h^{(k)} \to 0$, and how to minimize $f(\mathbf{x})$ subject to $h(\mathbf{c}(\mathbf{x})) = 0$.

First we review existing trust region SQP convergence theory for *unconstrained* optimization. At any non-KT point $\mathbf{x}^{(k)}$ and radius $\rho$ in $\mathrm{QP}^{(k)}(\rho)$, we define the *predicted reduction*

$$\Delta q = q(\mathbf{0}) - q(\mathbf{d}) = -\mathbf{g}^{(k)T}\mathbf{d} - \tfrac{1}{2}\mathbf{d}^T W^{(k)}\mathbf{d} > 0 \qquad (6.2)$$

and the *actual reduction*

$$\Delta f = f^{(k)} - f(\mathbf{x}^{(k)} + \mathbf{d}). \qquad (6.3)$$

As $\rho \to 0$, and with suitable assumptions, so $\mathbf{d} \to \mathbf{0}$, $\Delta q \sim \rho \|\mathbf{g}^{(k)}\|$ and $\Delta f / \Delta q \to 1$. It follows for sufficiently small $\rho$ that the inequality

$$\Delta f \ge \sigma \Delta q \qquad (6.4)$$

is valid. This is referred to as the *sufficient reduction* condition. In the TR algorithm of Figure 4 (with the feasibility restoration and filter boxes stripped out) we essentially choose $\rho^{(k)}$ as large as possible (within a factor of 2) subject to (6.4) holding. If the gradients $\mathbf{g}^{(k)}$ are accumulating at a value $\mathbf{g}^{\infty} \ne \mathbf{0}$, then $\rho^{(k)}$ is uniformly bounded away from zero, and it follows that $\Delta f^{(k)} \ge \sigma \Delta q^{(k)} \sim \sigma \rho^{(k)}$. Summing over all $k$ shows that $f^{(k)} \to -\infty$ which is a contradiction. Thus the gradients can only accumulate at $\mathbf{0}$ and any accumulation point $\mathbf{x}^{\infty}$ is stationary.

We aim to make use of these ideas in an NLP context. However there is a difficulty when $h^{(k)} > 0$ that $\Delta q < 0$ may be possible. This is illustrated by the left hand diagram of Figure 3. However, with a larger trust region radius, it is possible to have $\Delta q > 0$, as in the right hand diagram. We describe the resulting steps respectively as being either *h-type* or *f-type*, according to whether $\Delta q \le 0$ or not. Of course, if $h^{(k)} = 0$ the resulting step must be f-type. We construct our TR algorithm such that whilst f-type steps are being taken, we make no new entries into the filter, and rely to a large extent on the above convergence theory. Only h-type steps give rise to a filter entry (we include the use of feasibility restoration as an h-type step). The resulting algorithm is detailed in Figure 4. Note that the current pair $(h^{(k)}, f^{(k)})$ is not in the filter, and is only included subsequently if the algorithm takes an h-type step.

We now turn to examine the way in which the slanting envelope (6.1) operates. If an infinite sequence of entries are made, then it is a consequence

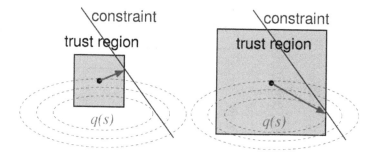

**Fig. 3** Illustration of h-type and f-type steps

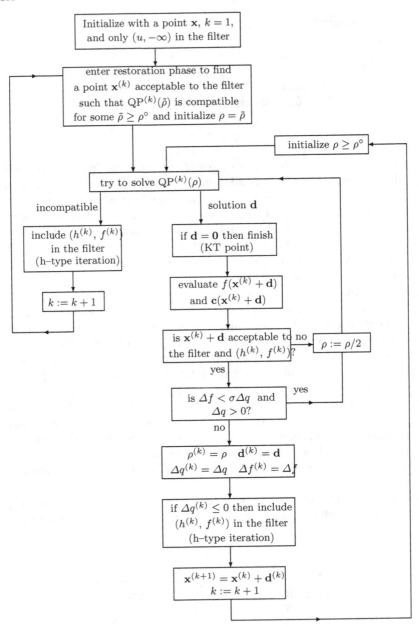

**Fig. 4** A Filter–SQP Algorithm

that $h^{(k)} \to 0$ (otherwise the condition $f \le f_i - \gamma h$ has the effect of forcing $f \to -\infty$: a contradiction). Thus the convergence proof claims that either

1. The restoration phase converges to a locally infeasible point,
2. The algorithm terminates at a KT point, or
3. There exists a feasible accumulation point that is either a KT point, or the Mangasarian-Fromowitz constraint qualification (MFCQ) fails.

The proof proceeds as follows. The first case corresponds to the situation that the local approximation to the constraint set is infeasible, and no further progress can be made. So we only need to examine case 3. If there are an infinite number of h-type iterations, it is possible to find a subsequence on which $h^{(k)} \to 0$ by virtue of (6.1). By taking thinner subsequences if necessary, we examine the behaviour of iterates in the neighbourhood of a feasible non-KT point that satisfies MFCQ (a type of regularity condition). Because MFCQ holds, there exist feasible descent directions and it is shown that the TR algorithm takes f-type steps in the limit, which is a contradiction.

The only other possibility to consider is that there exists some $K$ such that the algorithm takes f-type steps for all $k \ge K$. We can deduce from (6.1) and the fact that $(h^{(k+1)}, f^{(k+1)})$ is always acceptable to $(h^{(k)}, f^{(k)})$, that $h^{(k)} \to 0$. Then an argument similar to that in the unconstrained case contradicts the fact that there are feasible descent directions at any accumulation point. Because MFCQ holds, it follows that the accumulation point is a KT point. In passing, note that the proof of the Corollary to Lemma 1 in [20] contains an error. A corrected version of the paper can be found on my web site.

## 6.3 Other Filter SQP Methods

Another filter SQP algorithm, analysed by Fletcher, Gould, Leyffer, Toint and Wächter [16], decomposes the SQP step into a normal and tangential component. The normal step provides feasibility for the linearized constraints and the tangential step minimizes the quadratic model in the feasible region. Related ideas are discussed by Gonzaga, Karas and Varga [26] and Ribiero, Karas and Gonzaga [42]. Wächter and Biegler describe line search filter methods in [46] and [47]. Chin [5] and Chin and Fletcher [6] consider SLP-EQP trust region filter methods. Gould and Toint [28] present a non-monotone filter SQP method which extends the non-monotonic properties of filter SQP type algorithms. A review of other recent developments of filter methods, outwith SQP, but including interior point methods for NLP, appears in the SIAG/OPT activity group newsletter (March 2007) and can be accessed in [21].

# 7  Modelling Languages and NEOS

Solving complex optimization problems that arise in practice has many difficulties. Writing your own NLP solver is not recommended, as there are many difficulties to be circumvented, even if you have access to a good QP solver. However, access to a fully developed production code, e.g. MINOS, LANCELOT, etc., is not the end of the story. The interface between the requirements of the code, and the features of the problem can be very difficult to set up. It is usually necessary to provide derivatives of any nonlinear functions, which is prone to error. Modelling languages have been designed to allow the user to present the problem to the NLP solver in as friendly a way as possible. The user is able to define constructions in terms of concepts familiar to the problem. Three languages come to mind, AMPL, GAMS and TOMLAB. All are hooked up to a number of well known NLP solvers. TOMLAB is based on MATLAB syntax, the other have their own individual syntax that has to be learned. Unfortunately all are commercial products.

In this review I shall describe AMPL (A Mathematical Programming Language) which I have found very flexible and easy to use. A student edition is freely available which allows problems of up to 300 variables and constraints, and gives access to MINOS and some other solvers. Access to AMPL for larger problems is freely available through the so-called NEOS system, described below in Section 7.5.

## 7.1  The AMPL Language

AMPL is a high level language for describing optimization problems, submitting them to a solver, and manipulating the results. An AMPL program has three main parts, the model, in which the problem constructs are defined, the data which is self evident, and programming in which instructions for activating a solver, and displaying or manipulating the results are carried out. A list of model, data and programming *commands* are prepared in one or more files and presented to the AMPL system. This processes the information, evaluates any derivatives automatically, and presents the problem to a designated solver. Results are then returned to the user. The AMPL system is due to Fourer, Gay and Kernighan, and is described in the AMPL reference manual [23]. The syntax of AMPL is concisely defined in the Appendix of [23], and provides an invaluable aid to debugging an AMPL model. The user is strongly advised to come to terms with the notation that is used. Other more abridged introductions to AMPL are available, as can be found by surfing the web.

The main features of an NLP problem are the *variables*, presented via the keyword var, the *objective*, presented via either minimize or maximize, and *constraints*, presented via the keyword subject to. These constructions are described using entities introduced by the keywords **param** for fixed parameters, and **set** for multivalued set constructions. An example which illustrates all the features is to solve the HS72 problem in CUTE. Here the model is specified in a file **hs72.mod**. Note that upper and lower case letters are different: here we have written user names in upper case, but that is not necessary. All AMPL commands are terminated by a semicolon.

hs72.mod

```
set ROWS = {1..2};
set COLUMNS = {1..4};
param A {ROWS, COLUMNS};
param B {ROWS};
var X {COLUMNS} >= 0.001;
minimize OBJFN: 1 + sum {j in COLUMNS} x[j];
subject to
    CONSTR {i in ROWS}: sum {j in COLUMNS}
        A[i,j]/X[j] <= B[i];
    UBD {j in COLUMNS}: X[j] <= (5-j)*1e5;
```

In this model, the sets are just simple ranges, like 1..4 (i.e. 1 up to 4). We could have shortened the program by deleting the set declaration and replacing ROWS by 1..2 etc., in the rest of the program. But the use of ROWS and COLUMNS is more descriptive. The program defines a vector of variables X, and the data is the matrix A and vector B which are parameters. Simple lower bounds on the variables are specified in the **var** statement, and **sum** is a construction which provides summation. The constraint CONSTR implements the system of inequalities $\sum_{j=1}^{4} a_{i,j}/x_j \le b_i$, $i = 1, 2$. Note that indices are given within square brackets in AMPL. Constructions like j in COLUMNS are referred to as *indexing* in the AMPL syntax. The constraints in UBD define upper bounds on the variables which depend upon j. Note that the objective function and each set of constraint functions must be given a name by the user.

The data of the problem is specified in the file **hs72.dat**. Note the use of tabular presentation for elements of A and B.

hs72.dat

```
param A: 1 2 3 4 :=
    1   4       2.25  1       0.25
    2   0.16    0.36  0.64    0.64;
param B :=
    1   0.0401
    2   0.010085;
```

The next stage is to fire up the AMPL system on the computer. This will result in the user receiving the AMPL prompt **ampl:**. The programming session to solve HS72 would proceed as follows.

An AMPL session

```
ampl: model hs72.mod;
ampl: data hs72.dat;
ampl: let {j in COLUMNS} X[j] := 1;
ampl: solve;
ampl: display OBJFN;
ampl: display X;
```

The **model** and **data** keywords read in the model and data. The keyword **let** allows assignment of initial values to the variables, and the **display** commands initiate output to the terminal. Output from AMPL (not shown here) would be interspersed between the user commands. It is also possible to aggregate data and, if required, programming, into a single **hs72.mod** file. In this case the data must follow the model, and must be preceded by the statement **data;**. One feature to watch out for occurs when revising the model. In this case, repeating the command **model hs72.mod;** will *add* the new text to the database, rather than overwrite it. To remove the previous model from the database, the command **reset;** should first be given.

## 7.2 Networks in AMPL

AMPL is a most convenient system for modelling networks of diverse kinds (road, gas supply, work flow, ...). We illustrate some of the useful AMPL constructions in this section by reference to an *electrical power network* in which the objective is to minimize the power generation required to meet the demand on a given network. The model also illustrates some other useful AMPL features, notably the use of *dependent variables*. A toy system used by electrical engineers is illustrated in Figure 5. It is described in AMPL

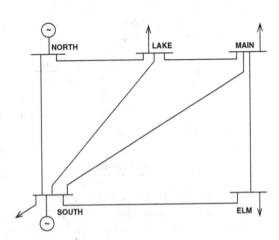

**Fig. 5** A Simple Power Distribution Network

A Power Generation Problem

```
set CONSUMERS;
set GENERATORS;
set NODES = CONSUMERS union GENERATORS;
set POWERLINES within (NODES cross NODES);
param LOADP {NODES};
param R {POWERLINES};
param X {POWERLINES};
param ZSQ {(i,j) in POWERLINES} = R[i,j]^2+X[i,j]^2;
param C {(i,j) in POWERLINES} = R[i,j]/ZSQ[i,j];
param D {(i,j) in POWERLINES} = X[i,j]/ZSQ[i,j];
...

var V {NODES};          # Voltages
var THETA {NODES};        # Phase angles
var PG {GENERATORS} ¿= 0;  # Power generation
...

var PSI {(i,j) in POWERLINES} = THETA[i] - THETA[j];
var P {(i,j) in POWERLINES} = C[i,j]*V[i]^2 +
  V[i]*V[j]*(D[i,j]*sin(PSI[i,j])-C[i,j]*cos(PSI[i,j]));
...

minimize PGEN: sum {i in GENERATORS} PG[i];
...

subject to
  EQUALPC {i in CONSUMERS}:
    sum {(j,i) in POWERLINES} P[j,i]=
    sum {(i,j) in POWERLINES} P[i,j] + LOADP[i];
  EQUALPG {i in GENERATORS}:
    PG[i,j] + sum {(j,i) in POWERLINES} P[j,i]=
    sum {(i,j) in POWERLINES} P[i,j] + LOADP[i];
...

data;
set CONSUMERS := LAKE MAIN ELM;
set GENERATORS := NORTH SOUTH;
set POWERLINES := (NORTH,SOUTH) (NORTH,LAKE)
          (SOUTH,LAKE)  (SOUTH,MAIN) (SOUTH,ELM)
          (LAKE,MAIN)   (MAIN,ELM);
param:   R   X :=
NORTH SOUTH 0.02 0.06
NORTH LAKE  0.08 0.24
SOUTH LAKE  0.06 0.24
...

MAIN  ELM   0.08 0.24;
param LOADP := NORTH 0; SOUTH 0.2, LAKE 0.45, MAIN 0.4, ELM 0.6;
...
```

by the constructions shown on page 207. Note the use of the keyword union to merge the nodes with and without power generation. Also observe the use of cross, which indicates all possible connections between the nodes, and within which indicates that the actual network is a subset of these. In fact the operator cross has higher priority than within so the brackets around the cross construction are not necessary. The user is using the convention that power flows *from* the first node *to* the second node (a negative value of flow is allowed and would indicate flow in the opposite sense). The program also shows the use of dependent parameters and variables. Thus the parameters ZSQ depend on R and X, and C and D both depend on R, X and ZSQ. It is necessary that the order in which these statements are given reflects these dependencies. The true variables in the problem (as shown here) are V, THETA and PG. Additional variables which depend on these variables, and also on the parameters, are PSI and P, as defined by the expressions which follow the '=' sign. The objective is to minimize the sum of generated power. Constraints include power balance constraints at consumer nodes and generator nodes, the latter including a term for power generation. Note the use of P[j,i] for power entering node i and P[i,j] for power exiting node i. The program also provides a useful illustration of how to supply data for network problems, and the use of the # sign for including comments. Note also the use of standard functions such as sin and cos in the expressions. The program shown is only part of the full model, which would include flow of reactive power, and upper and lower bounds on various of the variables.

## 7.3 Other Useful AMPL Features

The following problem is due to Bob Vanderbei (who gives many interesting AMPL programs: search for vanderbei princeton on Google and click on LOQO). A rocket starts at time $t = 0$, position $x(0) = 0$ and velocity $v(0) = 0$. It must reach position $x(T) = 100$, also with velocity $v(T) = 0$. We shall divide the total time $T$ into $n$ intervals of length $h$ and use finite difference approximations

$$v_i = \frac{x_{i+\frac{1}{2}} - x_{i-\frac{1}{2}}}{h} \quad \text{and} \quad a_i = \frac{v_{i+\frac{1}{2}} - v_{i-\frac{1}{2}}}{h}.$$

for velocity and acceleration. The maximum velocity is 5 units and the acceleration must lie within $\pm 1$ units. The aim is to minimize the total time $T = nh$. The AMPL program is

## A Rocket Problem

```
param n > 2;
set vrange = {0.5..n-0.5 by 1};
set arange = {1..n-1};
var x {0..n}; var v {vrange} <= 5;
var a {arange} <= 1, >= -1;
var h;
minimize T: n*h;
subject to
    xdiff {i in vrange}: x[i+0.5]-x[i-0.5]=h*v[i];
    vdiff {i in arange}: v[i+0.5]-v[i-0.5]=h*a[i];
    x0: x[0] = 0;   xn: x[n] = 100;
    v0: v[1.5] = 3*v[0.5];       # Implements v0 = 0
    vn: v[n-1.5] = 3*v[n-0.5];   # Implements vn = 0
```

The actual value of n must be supplied in the data section. An alternative implementation could be made in which v and a are expressed as dependent variables. Things to observe include the treatment of ranges and the use of by. Also note the use of both upper and lower bounds on the variables, and the beautifully concise form that AMPL permits.

AMPL can also be very descriptive when applied to finite element discretizations. I have used the following constructions in the context of a two dimensional p.d.e.

## 2-D Finite Element Constructions

```
set NODES;
set DIRICHLET within NODES;
set INTERIOR = NODES diff DIRICHLET;
set TRIANGLES within NODES cross NODES cross NODES;
set EDGES = setof {(i,j,k) in TRIANGLES} (i,j)
        union setof {(i,j,k) in TRIANGLES} (j,k)
        union setof {(i,j,k) in TRIANGLES} (k,i);
set SHAPE_FNS =
        setof {(i,j,k) in TRIANGLES} (i,j,k) union
        setof {(i,j,k) in TRIANGLES} (j,k,i) union
        setof {(i,j,k) in TRIANGLES} (k,i,j);
```

This illustrates the use of three suffix quantities (TRIANGLES), and the selection of all two suffix entities (EDGES) using setof. Shape functions are elementary functions defined on triangles taking the value 1 at one node and 0 at the others. Note also the use of `diff` for set difference, and the allowed use of underscore within an identifier.

AMPL contains many more useful constructions which we have not space to mention here. Purchasing a copy of the manual is essential! Worthy of mention however is the existence of `for` and `if then else` constructions. This can be very useful at the programming stage. An `if then else` construction is also allowed within a model but should be used with care, because it usually creates a nonsmooth function which many methods are not designed to handle. The same goes for `abs` and related nonsmooth functions. Another useful feature for creating loops at the programming stage is the `repeat` construction.

## 7.4  Accessing AMPL

Unfortunately the full AMPL system is a commercial product for which a licence fee must be paid. However there is a student version of AMPL available which is free, but restricted to no more than 300 variables and constraints. In this section we list the steps needed to install the student version of AMPL on a unix operating system, as follows

1. connect to `www.ampl.com`
2. follows the link to **download** the student edition, following the quick start instructions
3. choose the architecture
4. download `ampl.gz` and gunzip it
5. download one or more solvers (e.g. MINOS or SNOPT).

Ideally these files should be stored in a `\usr\bin` area with symbolic links that enable them to be called from other directories.

## 7.5  NEOS and Kestrel

NEOS (Network Enabled Optimization Server) is a system running at the Argonne National Laboratory that solves optimization problems submitted by anyone, through the medium of the internet. The simplest way to access NEOS is via email. The following template, using the solver SNOPT for example, should be sent to `neos@mcs.anl.gov`, after having included the model, data, etc., where indicated.

NEOS Template for AMPL

```
<document>
<category>nco</category>
<solver>SNOPT</solver>
<inputMethod>AMPL</inputMethod>

<model><![CDATA[
Insert Model Here
]]></model>

<data><![CDATA[
Insert Data Here
]]></data>

<commands><![CDATA[
Insert Programming Here
]]></commands>

<comments><![CDATA[
Insert Any Comments Here
]]></comments>

</document>
```

An alternative approach is to use the **Kestrel** interface to NEOS. This enables the remote solution of optimization problems within the AMPL and GAMS modelling languages. Quoting from the documentation, problem generation, including the run-time detection of syntax errors, occurs on the local machine using any available modelling language facilities. Solution takes place on a remote machine, with the result returned in the native modelling language format for further processing. To use Kestrel, the Kestrel interface must be downloaded at step 5 above, using the same process as for downloading the solvers. To initiate a solve with say SNOPT using Kestrel, the following protocol must be initiated when using AMPL on the local machine.

Accessing NEOS from the Kestrel interface

```
ampl: option solver kestrel;
ampl: option kestrel_options "solver=SNOPT";
```

An advantage of using NEOS from Kestrel (or by email as above) is that the restriction in size no longer applies. A disadvantage is that the response of the NEOS server can be slow at certain times of the day.

# References

1. E. M. L. Beale, *Numerical Methods* In: Nonlinear Programming, J. Abadie, ed., North-Holland, Amsterdam, 1967.
2. I. Bongartz, A. R. Conn, N. I. M. Gould and Ph. L. Toint, *CUTE: Constrained and Unconstrained Testing Environment* ACM Trans. Math. Software, 21, 1995, pp. 123–160.
3. R. H. Byrd, N. I. M. Gould, J. Nocedal and R. A. Waltz, *An Active-Set Algorithm for Nonlinear Programming Using Linear Programming and Equality Constrained Subproblems*, Math. Programming B, 100, 2004 pp. 27–48.
4. R. H. Byrd, J. Nocedal and R. B. Schnabel, *Representations of quasi-Newton matrices and their use in limited memory methods*, Math. Programming, 63, 1994, pp. 129–156.
5. C. M. Chin, *A new trust region based SLP filter algorithm which uses EQP active set strategy*, PhD thesis, Dept. of Mathematics, Univ. of Dundee, 2001.
6. C. M. Chin and R. Fletcher, *On the global convergence of an SLP-filter algorithm that takes EQP steps*, Math. Programming, 96, 2003, pp. 161–177.
7. R. M. Chamberlain, C. Lemarechal, H. C. Pedersen and M. J. D. Powell, *The watchdog technique for forcing convergence in algorithms forconstrained optimization*, In: Algorithms for Constrained Minimization of Smooth Nonlinear Functions, A. G. Buckley and J.-L. Goffin, eds., Math. Programming Studies, 16, 1982, pp. 1–17.
8. A. R. Conn, N. I. M. Gould and Ph. L. Toint, Trust Region Methods, MPS-SIAM Series on Optimization, SIAM Publications, Philadelphia, 2000.
9. R. Courant, *Variational methods for the solution of problems of equilibrium and vibration*, Bull. Amer. Math. Soc., 49, 1943, pp. 1–23.
10. J. E. Dennis and J. J. Moré, *A characterization of superlinear convergence and its application to quasi-Newton methods*, Math. Comp., 28, 1974, pp. 549–560.
11. R. Fletcher, *A model algorithm for composite nondifferentiable optimization problems*, In: Nondifferential and Variational Techniques in Optimization, D. C. Sorensen and R. J.-B. Wets eds., Math.Programming Studies, 17, 1982, pp. 67–76.
12. R. Fletcher, *Second order corrections for nondifferentiable optimization*, In: Numerical Analysis, Dundee 1981, G. A. Watson ed., Lecture Notes in Mathematics 912, Springer-Verlag, Berlin, 1982.
13. R. Fletcher, *Practical Methods of Optimization*, 1987, Wiley, Chichester.
14. R. Fletcher, *Dense Factors of Sparse Matrices*, In: Approximation Theory and Optimization, M. D. Buhmann and A. Iserles, eds, C.U.P., Cambridge, 1997.
15. R. Fletcher, *A New Low Rank Quasi-Newton Update Scheme for Nonlinear Programming*, In: System Modelling and Optimization, H. Futura, K. Marti and L. Pandolfi, eds., Springer IFIP series in Computer Science, 199, 2006, pp. 275–293, Springer, Boston.
16. R. Fletcher, N. I. M. Gould, S. Leyffer, Ph. L. Toint, and A. Wächter, *Global convergence of trust-region SQP-filter algorithms for general nonlinear programming*, SIAM J. Optimization, 13, 2002, pp. 635–659.
17. R. Fletcher, A. Grothey and S. Leyffer, *Computing sparse Hessian and Jacobian approximations with optimal hereditary properties*, In: Large-Scale Optimization with Applications, Part II: Optimal Design and Control, L. T. Biegler, T. F. Coleman, A. R. Conn and F. N. Santosa, Springer, 1997.
18. R. Fletcher and S. Leyffer, *Nonlinear programming without a penalty function*, Math. Programming, 91, 2002, pp. 239–270.
19. R. Fletcher and S. Leyffer, *Filter-type algorithms for solving systems of algebraic equations and inequalities*, In: G. di Pillo and A. Murli, eds, High Performance Algorithms and Software for Nonlinear Optimization, Kluwer, 2003.
20. R. Fletcher, S. Leyffer, and Ph. L. Toint, *On the global convergence of a filter-SQP algorithm*, SIAM J. Optimization, 13, 2002, pp. 44–59.
21. R. Fletcher, S. Leyffer, and Ph. L. Toint, *A Brief History of Filter Methods*, Preprint ANL/MCS-P1372-0906, Argonne National Laboratory, Mathematics and Computer Science Division, September 2006.

22. R. Fletcher and E. Sainz de la Maza, *Nonlinear programming and nonsmooth optimization by successive linear programming*, Math. Programming, 43, 1989, pp. 235–256.
23. R. Fourer, D. M. Gay and B. W. Kernighan, *AMPL: A Modeling Language for Mathematical Programming*, 2nd Edn., Duxbury Press, 2002.
24. K. R. Frisch, *The logarithmic potential method of convex programming*, Oslo Univ. Inst. of Economics Memorandum, May 1955.
25. P. E. Gill, W. Murray and M. A. Saunders, *SNOPT: An SQP Algorithm for Large-Scale Constrained Optimization*, SIAM Review, 47, 2005, pp. 99–131.
26. C. C. Gonzaga, E. Karas, and M. Vanti, *A globally convergent filter method for nonlinear programming*, SIAM J. Optimization, 14, 2003, pp. 646–669.
27. L. Grippo, F. Lampariello and S. Lucidi, *A nonmonotone line search technique for Newton's method*, SIAM J. Num. Anal., 23, pp. 707–716.
28. N. I. M. Gould and Ph. L. Toint, *Global Convergence of a Non-monotone Trust-Region SQP-Filter Algorithm for Nonlinear Programming*, In: Multiscale Optimization Methods and Applications, W. W. Hager, S. J. Huang, P. M. Pardalos and O. A. Prokopyev, eds., Springer Series on Nonconvex Optimization and Its Applications, Vol. 82, Springer Verlag, 2006.
29. S. P. Han, *A globally convergent method for nonlinear programming*, J. Opt. Theo. Applns., 22, 1976, pp. 297–309.
30. M. R. Hestenes, *Multiplier and gradient methods*, J. Opt. Theo. Applns, 4, 1969, pp. 303–320.
31. W. Karush, *Minima of functions of several variables with ineqalities as side conditions*, Master's Thesis, Dept. of Mathematics, Univ. of Chicago, 1939.
32. H. W. Kuhn and A. W. Tucker, *Nonlinear Programming*, In: Proceedings of the Second Berkeley Symposium on Mathematical Statistics and Probability, J. Neyman, ed., University of California Press, 1951.
33. O. L. Mangasarian and S. Fromowitz, *The Fritz John necessary optimality conditions in the presence of equality and inequality constraints* J. Math. Analysis and Applications, 17, 1967, pp. 37–47.
34. N. Maratos, *Exact penalty function algorithms for finite dimensional and control optimization problems*, Ph.D. Thesis, Univ. of London, 1978.
35. B. A. Murtagh and M. A. Saunders, *A projected Lagrangian algorithm and its implementation for sparse nonlinear constraints*, Math. Programming Studies, 16, 1982, pp. 84–117.
36. J. Nocedal, *Updating quasi-Newton matrices with limited storage*, Math. Comp., 35, 1980, pp. 773–782.
37. J. Nocedal and M. L. Overton, *Projected Hessian updating algorithms for nonlinearly constrained optimization*, SIAM J. Num. Anal., 22, 1985, pp. 821–850.
38. T. Pietrzykowski, *An exact potential method for constrained maxima*, SIAM J. Num. Anal., 6, 1969, pp. 217–238.
39. R. Polyak, *Modified barrier functions (theory and methods)*, Math. Programming, 54, 1992, pp. 177–222.
40. M. J. D. Powell, *A method for nonlinear constraints in minimization problems*, In: Optimization, R. Fletcher ed., Academic Press, London, 1969.
41. M. J. D. Powell, *A fast algorithm for nonlinearly constrained optimization calculations*, In: Numerical Analysis, Dundee 1977, G. A. Watson, ed., Lecture Notes in Mathematics 630, Springer Verlag, Berlin, 1978.
42. A. A. Ribeiro, E. W. Karas, and C. C. Gonzaga, *Global convergence of filter methods for nonlinear programming*, Technical report, Dept .of Mathematics, Federal University of Paraná, Brazil, 2006.
43. S. M. Robinson, *A quadratically convergent method for general nonlinear programming problems*, Math. Programming, 3, 1972, pp. 145–156.
44. R. T. Rockafellar, *Augmented Lagrange multiplier functions and duality in non-convex programming*, SIAM J. Control, 12, 1974, pp. 268–285.

45. Ph. L. Toint, *On sparse and symmetric updating subject to a linear equation*, Math. Comp., 31, 1977, pp. 954–961.
46. A. Wächter and L. T. Biegler, *Line search filter methods for nonlinear programming: Motivation and global convergence*, SIAM J. Optimization, 16, 2005, pp. 1–31.
47. A. Wächter and L. T. Biegler, *Line search filter methods for nonlinear programming: Local convergence*, SIAM J. Optimization, 16, 2005, pp. 32–48.
48. R. B. Wilson, *A simplicial algorithm for concave programming*, Ph.D. dissertation, Harvard Univ. Graduate School of Business Administration, 1960.

# Interior Point Methods for Nonlinear Optimization

Imre Pólik and Tamás Terlaky

# 1 Introduction

## 1.1 Historical Background

Interior-point methods (IPMs) are among the most efficient methods for solving linear, and also wide classes of other convex optimization problems. Since the path-breaking work of Karmarkar [48], much research was invested in IPMs. Many algorithmic variants were developed for Linear Optimization (LO). The new approach forced to reconsider all aspects of optimization problems. Not only the research on algorithms and complexity issues, but implementation strategies, duality theory and research on sensitivity analysis got also a new impulse. After more than a decade of turbulent research, the IPM community reached a good understanding of the basics of IPMs. Several books were published that summarize and explore different aspects of IPMs. The seminal work of Nesterov and Nemirovski [63] provides the most general framework for polynomial IPMs for convex optimization. Den Hertog [42] gives a thorough survey of primal and dual path-following IPMs for linear and structured convex optimization problems. Jansen [45] discusses primal-dual target following algorithms for linear optimization and complementarity problems. Wright [93] also concentrates on primal-dual IPMs, with special attention on infeasible IPMs, numerical issues and local, asymptotic convergence properties. The volume [80] contains 13 survey papers that cover almost all aspects of IPMs, their extensions and some applications. The book of Ye [96] is a rich source of polynomial IPMs not only for LO, but for convex optimization problems as well. It extends the IPM theory to derive bounds

I. Pólik and T. Terlaky (✉)
Department of Industrial and Systems Engineering, Lehigh University,
200 West Packer Avenue, 18015-1582, Bethlehem, PA, USA
e-mail: imre@polik.net; terlaky@lehigh.edu

G. Di Pillo et al. (eds.), *Nonlinear Optimization*, Lecture Notes in
Mathematics 1989, DOI 10.1007/978-3-642-11339-0_4,
© Springer-Verlag Berlin Heidelberg 2010

and approximations for classes of nonconvex optimization problems as well. Finally, Roos, Terlaky and Vial [72] present a thorough treatment of the IPM based theory – duality, complexity, sensitivity analysis – and wide classes of IPMs for LO.

Before going in a detailed discussion of our approach, some remarks are made on implementations of IPMs and on extensions and generalizations.

IPMs have also been implemented with great success for linear, conic and general nonlinear optimization. It is now a common sense that for large-scale, sparse, structured LO problems, IPMs are the method of choice and by today all leading commercial optimization software systems contain implementations of IPMs. The reader can find thorough discussions of implementation strategies in the following papers: [5, 53, 55, 94]. The books [72, 93, 96] also devote a chapter to that subject.

Some of the earlier mentioned books [42, 45, 63, 80, 96] discuss extensions of IPMs for classes of nonlinear problems. In recent years the majority of research is devoted to IPMs for nonlinear optimization, specifically for second order (SOCO) and semidefinite optimization (SDO). SDO has a wide range of interesting applications not only in such traditional areas as combinatorial optimization [1], but also in control, and different areas of engineering, more specifically structural [17] and electrical engineering [88]. For surveys on algorithmic and complexity issues the reader may consult [16, 18–20, 63, 64, 69, 75].

In the following sections we will build up the theory gradually, starting with linear optimization and generalizing through conic optimization to nonlinear optimization. We will demonstrate that the main idea behind the algorithms is similar but the details and most importantly the analysis of the algorithms are slightly different.

## 1.2 Notation and Preliminaries

After years of intensive research a deep understanding of IPMs is developed. There are easy to understand, simple variants of polynomial IPMs. The self-dual embedding strategy [47, 72, 97] provides an elegant solution for the initialization problem of IPMs. It is also possible to build up not only the complete duality theory of [72] of LO, but to perform sensitivity analysis [45, 46, 58, 72] on the basis of IPMs. We also demonstrate that IPMs not only converge to an optimal solution (if it exists), but after a finite number of iterations also allow a strongly polynomial rounding procedure [56, 72] to generate exact solutions. This all requires only the knowledge of elementary calculus and can be taught not only at a graduate, but at an advanced undergraduate level as well. Our aim is to present such an approach, based on the one presented in [72].

This chapter is structured as follows. First, in Section 2.1 we briefly review the general LO problem in canonical form and discuss how Goldman and Tucker's [32,85] self-dual and homogeneous model is derived. In Section 2.2 the Goldman-Tucker theorem, i.e., the existence of a strictly complementary solution for the skew-symmetric self-dual model will be proved. Here such basic IPM objects, as the interior solution, the central path, the Newton step, the analytic center of polytopes will be introduced. We will show that the central path converges to a strictly complementary solution, and that an exact strictly complementary solution for LO, or a certificate for infeasibility can be obtained after a finite number of iterations. Our theoretical development is summarized in Section 2.3. Finally, in Section 2.4 a general scheme of IPM algorithms is presented. This is the scheme that we refer back to in later sections. In Section 3 we extend the theory to conic (second order and semidefinite) optimization, discuss some applications and present a variant of the algorithm. Convex nonlinear optimization is discussed in Section 4 and a suitable interior point method is presented. Available software implementations are discussed in Section 5. Some current research directions and open problems are discussed in Section 6.

## 1.2.1 Notation

$\mathbb{R}^n_+$ denotes the set of nonnegative vectors in $\mathbb{R}^n$. Throughout, we use $\|\cdot\|_p$ ($p \in \{1, 2, \infty\}$) to denote the $p$-norm on $\mathbb{R}^n$, with $\|\cdot\|$ denoting the Euclidean norm $\|\cdot\|_2$. $I$ denotes the identity matrix, $e$ is used to denote the vector which has all its components equal to one. Given an $n$-dimensional vector $x$, we denote by $X$ the $n \times n$ diagonal matrix whose diagonal entries are the coordinates $x_j$ of $x$. If $x, s \in \mathbb{R}^n$ then $x^T s$ denotes the dot product of the two vectors. Further, $xs$, $x^\alpha$ for $\alpha \in \mathbb{R}$ and $\max\{x, y\}$ denotes the vectors resulting from coordinatewise operations. For any matrix $A \in \mathbb{R}^{m \times n}$, $A_j$ denotes the $j^{\text{th}}$ column of $A$. Furthermore,

$$\pi(A) := \prod_{j=1}^{n} \|A_j\|. \tag{1.1}$$

For any index set $J \subseteq \{1, 2, \ldots, n\}$, $|J|$ denotes the cardinality of $J$ and $A_J \in \mathbb{R}^{m \times |J|}$ the submatrix of $A$ whose columns are indexed by the elements in $J$. Moreover, if $K \subseteq \{1, 2, \ldots, m\}$, $A_{KJ} \in \mathbb{R}^{|K| \times |J|}$ is the submatrix of $A_J$ whose rows are indexed by the elements in $K$.

Vectors are assumed to be column vectors. The (vertical) concatenation of two vectors (or matrices of appropriate size) $u$ and $v$ is denoted by $(u; v)$, while the horizontal concatenation is $(u, v)$.

# 2  Interior Point Methods for Linear Optimization

This section is based on [81]. Here we build the theory of interior point methods for linear optimization including almost all the proofs. In later sections we refer back to these results.

## 2.1  The Linear Optimization Problem

We consider the general LO problem $(P)$ and its dual $(D)$ in canonical form:

$$\min\left\{c^T u : Au \geq b, \ u \geq 0\right\} \tag{P}$$

$$\max\left\{b^T v : A^T v \leq c, \ v \geq 0\right\}, \tag{D}$$

where $A$ is an $m \times k$ matrix, $b, v \in \mathbb{R}^m$ and $c, u \in \mathbb{R}^k$. It is well known that by using only elementary transformations, any given LO problem can easily be transformed into a "minimal" canonical form. These transformations can be summarized as follows:

- introduce slacks in order to get equations (if a variable has a lower and an upper bound, then one of these bounds is considered as an inequality constraint);
- shift the variables with lower or upper bound so that the respective bound becomes 0 and, if needed replace the variable by its negative;
- eliminate free variables;[1]
- use Gaussian elimination to transform the problem into a form where all equations have a singleton column (i.e., choose a basis and multiply the equations by the inverse basis) while dependent constraints are eliminated.

The weak duality theorem for the canonical LO problem is easily proved.

---

[1] Free variables can easily be eliminated one-by-one. If we assume that $x_1$ is a free variable and has a nonzero coefficient in a constraint, e.g., we have

$$\sum_{i=1}^{n} \alpha_i x_i = \beta$$

with $\alpha_1 \neq 0$, then we can express $x_1$ as

$$x_1 = \frac{\beta}{\alpha_1} - \sum_{i=1}^{n-1} \frac{\alpha_i}{\alpha_1} x_i. \tag{2.1}$$

Because $x_1$ has no lower or upper bounds, this expression for $x_1$ can be substituted into all the other constraints and in the objective function.

**Theorem 2.1 (Weak duality for linear optimization).** *Let us assume that $u \in \mathbb{R}^k$ and $v \in \mathbb{R}^m$ are feasible solutions for the primal problem $(P)$ and dual problem $(D)$, respectively. Then one has*

$$c^T u \geq b^T v$$

*where equality holds if and only if*

*(i) $u_i(c - A^T v)_i = 0$ for all $i = 1, \dots, k$ and*
*(ii) $v_j(Au - b)_j = 0$ for all $j = 1, \dots, m$.[2]*

*Proof.* Using primal and dual feasibility of $u$ and $v$ we may write

$$(c - A^T v)^T u \geq 0 \quad \text{and} \quad v^T(Au - b) \geq 0$$

with equality if and only if $(i)$, respectively $(ii)$ holds. Summing up these two inequalities we have the desired inequality

$$0 \leq (c - A^T v)^T u + v^T(Au - b) = c^T u - b^T v.$$

The theorem is proved.                                                                 □

One easily derives the following sufficient condition for optimality.

**Corollary 2.2.** *Let a primal and dual feasible solution $u \in \mathbb{R}^k$ and $v \in \mathbb{R}^m$ with $c^T u = b^T v$ be given. Then $u$ is an optimal solution of the primal problem $(P)$ and $v$ is an optimal solution of the dual problem $(D)$.*                     □

The Weak Duality Theorem provides a sufficient condition to check optimality of a feasible solution pair. However, it does not guarantee that, in case of feasibility, an optimal pair with zero duality gap always exists. This is the content of the so-called Strong Duality Theorem that we are going to prove in the next sections by using only simple calculus and basic concepts of IPMs.

As we are looking for optimal solutions of the LO problem with zero duality gap, we need to find a solution of the system formed by the primal and the dual feasibility constraints and by requiring that the dual objective is at least as large as the primal one. By the Weak Duality Theorem (Thm. 2.1) we know that any solution of this system is both primal and dual feasible with equal objective values. Thus, by Corollary 2.2, they are optimal. By introducing appropriate slack variables the following inequality system is derived.

$$
\begin{aligned}
Au - z = b, \quad & u \geq 0, \quad z \geq 0 \\
A^T v + w = c, \quad & v \geq 0, \quad w \geq 0 \\
b^T v - c^T u - \rho = 0, \quad & \rho \geq 0.
\end{aligned}
\tag{2.2}
$$

---

[2] These conditions are in general referred to as the *complementarity conditions*. Using the coordinatewise notation we may write $u(c - A^T v) = 0$ and $v(Au - b) = 0$. By the weak duality theorem complementarity and feasibility imply optimality.

By homogenizing, the *Goldman-Tucker model* [32, 85] is obtained.

$$
\begin{array}{llll}
Au - \tau b - z & = 0, & u \geq 0, & z \geq 0 \\
-A^T v & + \tau c & - w & = 0, & v \geq 0, & w \geq 0 \\
b^T v - c^T u & & -\rho = 0, & \tau \geq 0, & \rho \geq 0.
\end{array}
\tag{2.3}
$$

One easily verifies that if $(v, u, \tau, z, w, \rho)$ is a solution of the Goldman-Tucker system (2.3), then $\tau\rho > 0$ cannot hold. Indeed, if $\tau\rho$ were positive then the we would have

$$
0 < \tau\rho = \tau b^T v - \tau c^T u = u^T A^T v - z^T v - v^T A u - w^T u = -z^T v - w^T u \leq 0
$$

yielding a contradiction.

The homogeneous Goldman-Tucker system admits the trivial zero solution, but that has no value for our discussions. We are looking for some specific nontrivial solutions of this system. Clearly any solution with $\tau > 0$ gives a primal and dual optimal pair $(\frac{u}{\tau}, \frac{v}{\tau})$ with zero duality gap because $\rho$ must be zero if $\tau > 0$. On the other hand, any optimal pair $(u, v)$ with zero duality gap is a solution of the Goldman-Tucker system with $\tau = 1$ and $\rho = 0$.

Finally, if the Goldman-Tucker system admits a nontrivial feasible solution $(\bar{v}, \bar{u}, \bar{\tau}, \bar{z}, \bar{w}, \bar{\rho})$ with $\bar{\tau} = 0$ and $\bar{\rho} > 0$, then we may conclude that either $(P)$, or $(D)$, or both of them are infeasible. Indeed, $\bar{\tau} = 0$ implies that $A\bar{u} \geq 0$ and $A^T\bar{v} \leq 0$. Further, if $\bar{\rho} > 0$ then we have either $b^T\bar{v} > 0$, or $c^T\bar{u} < 0$, or both. If $b^T\bar{v} > 0$, then by assuming that there is a feasible solution $u \geq 0$ for $(P)$ we have

$$
0 < b^T\bar{v} \leq u^T A^T\bar{v} \leq 0
$$

which is a contradiction, thus if $b^T\bar{v} > 0$, then $(P)$ must be infeasible. Similarly, if $c^T\bar{u} < 0$, then by assuming that there is a dual feasible solution $v \geq 0$ for $(D)$ we have

$$
0 > c^T\bar{u} \geq v^T A\bar{u} \geq 0
$$

which is a contradiction, thus if $c^T\bar{u} > 0$, then $(D)$ must be infeasible.

Summarizing the results obtained so far, we have the following theorem.

**Theorem 2.3.** *Let a primal dual pair $(P)$ and $(D)$ of LO problems be given. The following statements hold for the solutions of the Goldman-Tucker system (2.3).*

1. *Any optimal pair $(u, v)$ of $(P)$ and $(D)$ with zero duality gap is a solution of the corresponding Goldman-Tucker system with $\tau = 1$.*
2. *If $(v, u, \tau, z, w, \rho)$ is a solution of the Goldman-Tucker system then either $\tau = 0$ or $\rho = 0$, i.e., $\tau\rho > 0$ cannot happen.*
3. *Any solution $(v, u, \tau, z, w, \rho)$ of the Goldman-Tucker system, where $\tau > 0$ and $\rho = 0$, gives a primal and dual optimal pair $(\frac{u}{\tau}, \frac{v}{\tau})$ with zero duality gap.*

4. *If the Goldman-Tucker system admits a feasible solution $(\bar{v}, \bar{u}, \bar{\tau}, \bar{z}, \bar{w}, \bar{p})$ with $\bar{\tau} = 0$ and $\bar{p} > 0$, then we may conclude that either $(P)$, or $(D)$, or both of them are infeasible.*

Our interior-point approach will lead us to a solution of the Goldman-Tucker system, where either $\tau > 0$ or $\rho > 0$, avoiding the undesired situation when $\tau = \rho = 0$.

Before proceeding, we simplify our notations. Observe that the Goldman-Tucker system can be written in the following compact form

$$Mx \geq 0, \qquad x \geq 0, \qquad s(x) = Mx, \tag{2.4}$$

where

$$x = \begin{pmatrix} v \\ u \\ \tau \end{pmatrix}, \qquad s(x) = \begin{pmatrix} z \\ w \\ \rho \end{pmatrix} \quad \text{and} \quad M = \begin{pmatrix} 0 & A & -b \\ -A^T & 0 & c \\ b^T & -c^T & 0 \end{pmatrix}$$

is a skew-symmetric matrix, i.e., $M^T = -M$. The Goldman-Tucker theorem [32, 72, 85] says that system (2.4) admits a strictly complementary solution. This theorem will be proved in the next section.

**Theorem 2.4 (Goldman, Tucker).** *System (2.4) has a strictly complementary feasible solution, i.e., a solution for which $x + s(x) > 0$.*

Observe that this theorem ensures that either case 3 or case 4 of Theorem 2.3 must occur when one solves the Goldman-Tucker system of LO. This is in fact the strong duality theorem of LO.

**Theorem 2.5.** *Let a primal and dual LO problem be given. Exactly one of the following statements hold:*

- *$(P)$ and $(D)$ are feasible and there are optimal solutions $u^*$ and $v^*$ such that $c^T u^* = b^T v^*$.*
- *Either problem $(P)$, or $(D)$, or both are infeasible.*

*Proof.* Theorem 2.4 implies that the Goldman-Tucker system of the LO problem admits a strictly complementary solution. Thus, in such a solution, either $\tau > 0$, and in that case item 3 of Theorem 2.3 implies the existence of an optimal pair with zero duality gap. On the other hand, when $\rho > 0$, item 4 of Theorem 2.3 proves that either $(P)$ or $(D)$ or both are infeasible.     □

Our next goal is to give an elementary constructive proof of Theorem 2.4. When this project is finished, we have the complete duality theory for LO.

## 2.2  The Skew-Symmetric Self-Dual Model

### 2.2.1  Basic Properties of the Skew-Symmetric Self-Dual Model

Following the approach in [72] we make our skew-symmetric model (2.4) a bit more general. Thus our prototype problem is

$$\min \ \{q^T x : Mx \geq -q, \ x \geq 0\}, \tag{SP}$$

where the matrix $M \in \mathbb{R}^{n \times n}$ is *skew-symmetric* and $q \in \mathbb{R}^n_+$. The set of feasible solutions of $(SP)$ is denoted by

$$SP := \{x \ : \ x \geq 0, \ Mx \geq -q \ \}.$$

By using the assumption that the coefficient matrix $M$ is skew-symmetric and the right-hand-side vector $-q$ is the negative of the objective coefficient vector, one easily verifies that the dual of (SP) is equivalent to (SP) itself, i.e., problem (SP) is *self-dual*. Due to the self-dual property the following result is trivial.

**Lemma 2.6.** *The optimal value of* (SP) *is zero and* (SP) *admits the zero vector $x = 0$ as a feasible and optimal solution.*

Given $(x, s(x))$, where $s(x) = Mx + q$ we may write

$$q^T x = x^T (s(x) - Mx) = x^T s(x) = e^T (x s(x)),$$

i.e., for any optimal solution $e^T (x s(x)) = 0$ implying that the vectors $x$ and $s(x)$ are complementary. For further use, the *optimal set* of (SP) is denoted by

$$SP^* := \{x \ : \ x \geq 0, \ s(x) \geq 0, \ x s(x) = 0\}.$$

A useful property of optimal solutions is given by the following lemma.

**Lemma 2.7.** *Let $x$ and $y$ be feasible for (SP). Then $x$ and $y$ are optimal if and only if*

$$xs(y) = ys(x) = xs(x) = ys(y) = 0. \tag{2.5}$$

*Proof.* Because $M$ is skew-symmetric we have $(x - y)^T M(x - y) = 0$, which implies that $(x - y)^T (s(x) - s(y)) = 0$. Hence $x^T s(y) + y^T s(x) = x^T s(x) + y^T s(y)$ and this vanishes if and only if $x$ and $y$ are optimal.  □

Thus, optimal solutions are complementary in the general sense, i.e., they are not only complementary w.r.t. their own slack vector, but complementary w.r.t. the slack vector for any other optimal solution as well.

All of the above results, including to find a trivial optimal solution were straightforward for (SP). The only nontrivial result that we need to prove is the existence of a strictly complementary solution.

First we prove the existence of a strictly complementary solution if the so-called interior-point condition holds.

**Assumption 2.8 (Interior-Point Condition (IPC))**
*There exists a point $x^0 \in SP$ such that*

$$(x^0, s(x^0)) > 0. \tag{2.6}$$

Before proceeding, we show that this condition can be assumed without loss of generality. If the reader is eager to know the proof of the existence of a strictly complementary solution for the self dual model (SP), he/she might temporarily skip the following subsection and return to it when all the results for the problem (SP) are derived under the IPC.

### 2.2.2 IPC for the Goldman-Tucker Model

Recall that (SP) is just the abstract model of the Goldman-Tucker problem (2.4) and our goal is to prove Theorem 2.4. In order to apply the results of the coming sections we need to modify problem (2.4) so that the resulting equivalent problem satisfies the IPC.

Self-dual embedding of (2.4) with IPC

Due to the second statement of Theorem 2.3, problem (2.4) cannot satisfy the IPC. However, because problem (2.4) is just a homogeneous feasibility problem, it can be transformed into an equivalent problem (SP) which satisfies the IPC. This happens by enlarging, i.e., embedding the problem and defining an appropriate nonnegative vector $q$.

Let us take $x = s(x) = e$. These vectors are positive, but they do not satisfy (2.4). Let us further define the error vector $r$ obtained this way by

$$r := e - Me, \quad \text{and let} \quad \lambda := n + 1.$$

Then we have

$$\begin{pmatrix} M & r \\ -r^T & 0 \end{pmatrix} \begin{pmatrix} e \\ 1 \end{pmatrix} + \begin{pmatrix} 0 \\ \lambda \end{pmatrix} = \begin{pmatrix} Me + r \\ -r^T e + \lambda \end{pmatrix} = \begin{pmatrix} e \\ 1 \end{pmatrix}. \tag{2.7}$$

Hence, the following problem

$$\min \left\{ \lambda \vartheta \ : \ -\begin{pmatrix} M & r \\ -r^T & 0 \end{pmatrix} \begin{pmatrix} x \\ \vartheta \end{pmatrix} + \begin{pmatrix} s \\ \nu \end{pmatrix} = \begin{pmatrix} 0 \\ \lambda \end{pmatrix}; \ \begin{pmatrix} x \\ \vartheta \end{pmatrix}, \begin{pmatrix} s \\ \nu \end{pmatrix} \geq 0 \right\} \quad (\overline{\text{SP}})$$

satisfies the IPC because for this problem the all-one vector is feasible. This problem is in the form of (SP), where

$$\overline{M} = \begin{pmatrix} M & r \\ -r^T & 0 \end{pmatrix}, \qquad \overline{x} = \begin{pmatrix} x \\ \vartheta \end{pmatrix} \qquad \text{and} \qquad \overline{q} = \begin{pmatrix} 0 \\ \lambda \end{pmatrix}.$$

We claim that finding a strictly complementary solution to (2.4) is equivalent to finding a strictly complementary optimal solution to problem $(\overline{SP})$. This claim is valid, because $(\overline{SP})$ satisfies the IPC and thus, as we will see, it admits a strictly complementary optimal solution. Because the objective function is just a constant multiple of $\vartheta$, this variable must be zero in any optimal solution, by Lemma 2.6. This observation implies the claimed result.


Conclusion

Every LO problem can be embedded in a self-dual problem $(\overline{SP})$ of the form (SP). This can be done in such a way that $\overline{x} = e$ is feasible for $(\overline{SP})$ and $\overline{s}(e) = e$. Having a strictly complementary solution of (SP) we either find an optimal solution of the embedded LO problem, or we can conclude that the LO problem does not have an optimal solution.

After this intermezzo, we return to the study of our prototype problem (SP) by assuming the IPC.


### 2.2.3 The Level Sets of (SP)

Let $x \in SP$ and $s = s(x)$ be a *feasible pair*. Due to self duality, the *duality gap* for this pair is twice the value

$$q^T x = x^T s,$$

however, for the sake of simplicity, the quantity $q^T x = x^T s$ itself will be referred to as *the duality gap*. First we show that the IPC implies the boundedness of the level sets.

**Lemma 2.9.** *Let the IPC be satisfied. Then, for each positive $K$, the set of all feasible pairs $(x, s)$ such that $x^T s \leq K$ is bounded.*

*Proof.* Let $(x^0, s^0)$ be an interior-point. Because the matrix $M$ is skew-symmetric, we may write

$$0 = (x - x^0)^T M (x - x^0) = (x - x^0)^T (s - s^0)$$
$$= x^T s + (x^0)^T s^0 - x^T s^0 - s^T x^0. \tag{2.8}$$

From here we get

$$x_j s_j^0 \leq x^T s^0 + s^T x^0 = x^T s + (x^0)^T s^0 \leq K + (x^0)^T s^0.$$

The proof is complete.                                                                    □

In particular, this lemma implies that the set of optimal solutions $SP^*$ is bounded as well.[3]

### 2.2.4  Central Path, Optimal Partition

First we define the central path [23, 27, 54, 74] of (SP).

**Definition 2.11.** Let the IPC be satisfied. The set of solutions

$$\{(x(\mu), s(x(\mu))) : Mx + q = s, \quad xs = \mu e, \quad x > 0 \quad \text{for some} \quad \mu > 0\} \quad (2.9)$$

is called the *central path* of (SP).

If no confusion is possible, instead of $s(x(\mu))$ the notation $s(\mu)$ will be used. Now we are ready to present our main theorem. This in fact establishes the existence of the central path. At this point our discussion deviates from the one presented in [72]. The proof presented here is more elementary because it does not make use of the logarithmic barrier function.

**Theorem 2.12.** *The next statements are equivalent.*

*i. (SP) satisfies the interior-point condition;*
*ii. For each $0 < \mu \in \mathbb{R}$ there exists $(x(\mu), s(\mu)) > 0$ such that*

$$Mx + q = s \qquad\qquad (2.10)$$
$$xs = \mu e.$$

*iii. For each $0 < w \in \mathbb{R}^n$ there exists $(x, s) > 0$ such that*

$$Mx + q = s \qquad\qquad (2.11)$$
$$xs = w.$$

---

[3] The following result shows that the IPC not only implies the boundedness of the level sets, but the converse is also true. We do not need this property in developing our main results, so this is presented without proof.

**Corollary 2.10.** *Let (SP) be feasible. Then the following statements are equivalent:*

*i. the interior-point condition is satisfied;*
*ii. the level sets of $x^T s$ are bounded;*
*iii. the optimal set $SP^*$ of (SP) is bounded.*

*Moreover, the solutions of these systems are unique.*

Before proving this highly important result we introduce the notion of optimal partition and present our main result. The partition $(B, N)$ of the index set $\{1, ..., n\}$ given by

$$B := \{i : x_i > 0, \text{ for some } x \in SP^*\}, \tag{2.12a}$$

$$N := \{i : s(x)_i > 0, \text{ for some } x \in SP^*\}, \tag{2.12b}$$

is called the *optimal partition*. By Lemma 2.7 the sets $B$ and $N$ are disjoint. Our main result says that the central path converges to a strictly complementary optimal solution, and this result proves that $B \cup N = \{1, ..., n\}$. When this result is established, the Goldman-Tucker theorem (Theorem 2.4) for the general LO problem is proved because we use the embedding method presented in Section 2.2.2.

**Theorem 2.13.** *If the IPC holds then there exists an optimal solution $x^*$ and $s^* = s(x^*)$ of problem* (SP) *such that $x_B^* > 0$, $s_N^* > 0$ and $x^* + s^* > 0$.*

First we prove Theorem 2.12.

*Proof.* We start the proof by demonstrating that the systems in $(ii)$ and $(iii)$ may have at most one solution. Because $(ii)$ is a special case of $(iii)$, it is sufficient to prove uniqueness for $(iii)$.

Let us assume to the contrary that for a certain $w > 0$ there are two vectors $(x, s) \neq (\overline{x}, \overline{s}) > 0$ solving $(iii)$. Then using the fact that matrix $M$ is skew-symmetric, we may write

$$0 = (x - \overline{x})^T M(x - \overline{x}) = (x - \overline{x})^T (s - \overline{s}) = \sum_{x_i \neq \overline{x}_i} (x - \overline{x})_i (s - \overline{s})_i. \tag{2.13}$$

Due to $xs = w = \overline{x}\,\overline{s}$ we have

$$x_i < \overline{x}_i \iff s_i > \overline{s}_i \tag{2.14a}$$

$$x_i > \overline{x}_i \iff s_i < \overline{s}_i. \tag{2.14b}$$

By considering these sign properties one easily verifies that the relation

$$0 = \sum_{x_i \neq \overline{x}_i} (x - \overline{x})_i (s - \overline{s})_i < 0 \tag{2.15}$$

should hold, but this is an obvious contradiction. As a result, we may conclude that if the systems in $(ii)$ and $(iii)$ admit a feasible solution, then such a solution is unique. $\qquad\square$

The Newton step

In proving the existence of a solution for the systems in $(ii)$ and $(iii)$ our main tool is a careful analysis of the Newton step when applied to the nonlinear systems in $(iii)$.[4]

Let a vector $(x, s) > 0$ with $s = Mx + q$ be given. For a particular $w > 0$ one wants to find the displacement $(\Delta x, \Delta s)$ that solves

$$M(x + \Delta x) + q = s + \Delta s \qquad (2.16)$$
$$(x + \Delta x)(s + \Delta s) = w.$$

This reduces to

$$M\Delta x = \Delta s \qquad (2.17)$$
$$x\Delta s + s\Delta x + \Delta x\Delta s = w - xs.$$

This equation system is still nonlinear. When we neglect the second order term $\Delta x\Delta s$ the *Newton equation*

$$M\Delta x = \Delta s \qquad (2.18)$$
$$x\Delta s + s\Delta x = w - xs$$

is obtained. This is a linear equation system and the reader easily verifies that the *Newton direction* $\Delta x$ is the solution of the nonsingular system of equations[5]

$$(M + X^{-1}S)\Delta x = x^{-1}w - s. \qquad (2.19)$$

When we perform a step in the Newton direction with step-length $\alpha$, for the new solutions $(x^+, s^+) = (x + \alpha\Delta x, s + \alpha\Delta s)$ we have

$$x^+ s^+ = (x + \alpha\Delta x)(s + \alpha\Delta s) = xs + \alpha(x\Delta s + s\Delta x) + \alpha^2\Delta x\Delta s \qquad (2.20)$$
$$= xs + \alpha(w - xs) + \alpha^2\Delta x\Delta s.$$

This relation clarifies that the local change of $xs$ is determined by the vector $w - xs$. Luckily this vector is known in advance when we apply a Newton step, thus for sufficiently small $\alpha$ we know precisely which coordinates of $xs$

---

[4] Observe that no preliminary knowledge on any variants of Newton's method is assumed. We just define and analyze the Newton step for our particular situation.

[5] Nonsingularity follows from the fact that the sum of a skew-symmetric, thus positive semi-definite, and a positive definite matrix is positive definite. Although it is not advised to use for numerical computations, the Newton direction can be expressed as $\Delta x = (M + X^{-1}S)^{-1} (x^{-1}w - s)$.

decrease locally (precisely those for which the related coordinate of $w - xs$ is negative) and which coordinate of $xs$ increase locally (precisely those for which the related coordinate of $w - xs$ is positive).

The equivalence of the three statements in Theorem 2.12.

Clearly $(ii)$ is a special case of $(iii)$ and the implication $(ii) \rightarrow (i)$ is trivial.

It only remains to be proved that $(i)$, i.e., the IPC, ensures that for each $w > 0$ the nonlinear system in $(iii)$ is solvable. To this end, let us assume that an $x^0 \in SP$ with $(x^0, s(x^0)) > 0$ is given. We use the notation $w^0 := x^0 s(x^0)$. The claim is proved in two steps.

**Step 1.** *For each $0 < \underline{w} < \overline{w} \in \mathbb{R}^n$ the following two sets are compact:*

$$L_{\overline{w}} := \{x \in SP : xs(x) \leq \overline{w}\} \text{ and}$$
$$U(\underline{w}, \overline{w}) := \{w : \underline{w} \leq w \leq \overline{w}, \ w = xs(x) \text{ for some } x \in L_{\overline{w}}\}.$$

Let us first prove that $L_{\overline{w}}$ is compact. For each $\overline{w} > 0$, the set $L_{\overline{w}}$ is obviously closed. By definition $L_{\overline{w}}$ is included in the level set $x^T s \leq e^T \overline{w}$, which by Lemma 2.9 is bounded, thus $L_{\overline{w}}$ is compact.

By definition the set $U(\underline{w}, \overline{w})$ is bounded. We only need to prove that it is closed. Let a convergent sequence $w^i \rightarrow \hat{w}$, $w^i \in U(\underline{w}, \overline{w})$, $i = 1, 2, \ldots$ be given. Then clearly $\underline{w} \leq \hat{w} \leq \overline{w}$ holds. Further, for each $i$ there exists $x^i \in L_{\overline{w}}$ such that $w^i = x^i s(x^i)$. Because the set $L_{\overline{w}}$ is compact, there is an $\hat{x} \in L_{\overline{w}}$ and a convergent subsequence $x^i \rightarrow \hat{x}$ (for ease of notation the subsequence is denoted again the same way). Then we have $\hat{x}s(\hat{x}) = \hat{w}$, proving that $U(\underline{w}, \overline{w})$ is closed, thus compact.

Observe that for each $w \in U(\underline{w}, \overline{w})$ by definition we have an $x \in SP$ with $w = xs(x)$. Due to $w > 0$ this relation implies that $x > 0$ and $s(x) > 0$.

**Step 2.** *For each $\hat{w} > 0$, the system $Mx + q = s$, $xs = \hat{w}$, $x > 0$ has a solution.*

If we have $\hat{w} = w^0 = x^0 s(x^0)$, then the claim is trivial. If $\hat{w} \neq w^0$ then we define $\overline{w} := \max\{\hat{w}, w^0\}$, $\overline{\eta} = \|\overline{w}\|_\infty + 1$, $\underline{w} := \min\{\hat{w}, w^0\}$ and $\underline{\eta} = \frac{1}{2} \min_i \underline{w}_i$. Then $\underline{\eta}e < \hat{w} < \overline{\eta}e$ and $\underline{\eta}e < w^0 < \overline{\eta}e$. Due to the last relation the set $\overline{U} := U(\underline{\eta}e, \overline{\eta}e)$ is nonempty and compact. We define the nonnegative function $d(w) : \overline{U} \rightarrow \mathbb{R}$ as

$$d(w) := \|w - \hat{w}\|_\infty.$$

The function $d(w)$ is continuous on the compact set $\overline{U}$, thus it attains its minimum

$$\tilde{w} := \arg\min_{w \in \overline{U}}\{d(w)\}.$$

If $d(\tilde{w}) = 0$, then $\tilde{w} = \hat{w} \Rightarrow \hat{w} \in \overline{U}$ and hence by the definition of $\overline{U}$ there is an $x \in SP$ satisfying $xs(x) = \hat{w}$ and the claim is proved.

If $d(\tilde{w}) > 0$ then we will show that a damped Newton step from $\tilde{w}$ towards $\hat{w}$ gives a point $w(a) \in \overline{U}$ such that $d(w(a)) < d(\tilde{w})$, contradicting the fact that $\tilde{w}$ minimizes $d(w)$. This situation is illustrated in Figure 1.

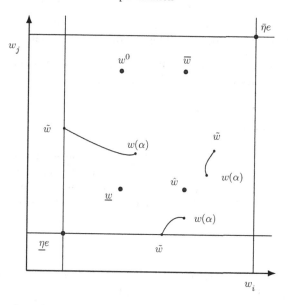

**Fig. 1** The situation when $\hat{w} \neq \tilde{w}$. A damped Newton step from $\tilde{w}$ to $\hat{w}$ is getting closer to $\hat{w}$. For illustration three possible different $\tilde{w}$ values are chosen.

The Newton step is well defined, because for the vector $\tilde{x} \in SP$ defining $\tilde{w}$ the relations $\tilde{x} > 0$ and $\tilde{s} = s(\tilde{x}) > 0$ hold. A damped Newton step from $\tilde{w}$ to $\hat{w}$ with sufficiently small $\alpha$ results in a point closer (measured by $d(\cdot) = \| \cdot \|_\infty$) to $\hat{w}$, because

$$w(\alpha) = x(\alpha)s(\alpha) := (\tilde{x} + \alpha\Delta x)(\tilde{s} + \alpha\Delta s) = \tilde{x}\tilde{s} + \alpha(\hat{w} - \tilde{x}\tilde{s}) + \alpha^2 \Delta x \Delta s$$
$$= \tilde{w} + \alpha(\hat{w} - \tilde{w}) + \alpha^2 \Delta x \Delta s. \tag{2.21}$$

This relation implies that

$$w(\mathbf{a}) - \hat{w} = (1 - \alpha)(\tilde{w} - \hat{w}) + \alpha^2 \Delta x \Delta s, \tag{2.22}$$

i.e., for $\alpha$ small enough[6] all nonzero coordinates of $|w(\mathbf{a}) - \hat{w}|$ are smaller than the respective coordinates of $|\tilde{w} - \hat{w}|$. Hence, $w(\mathbf{a})$ is getting closer to $\hat{w}$, closer than $\tilde{w}$. Due to $\underline{\eta}e < \hat{w} < \bar{\eta}e$ this result also implies that for the chosen

---

[6] The reader easily verifies that any value of $\alpha$ satisfying

$$\alpha < \min \left\{ \frac{\tilde{w}_i - \hat{w}_i}{\Delta x_i \Delta s_i} \; : \; (\tilde{w}_i - \hat{w}_i)(\Delta x_i \Delta s_i) > 0 \right\}$$

satisfies the requirement.

small $\mathbf{a}$ value the vector $w(\mathbf{a})$ stays in $\overline{U}$. Thus $\tilde{w} \neq \hat{w}$ cannot be a minimizer of $d(w)$, which is a contradiction. This completes the proof. □

Now we are ready to prove our main theorem, the existence of a strictly complementary solution, when the IPC holds.

Proof of Theorem 2.13.

Let $\mu_t \to 0$ $(t = 1, 2, \cdots)$ be a monotone decreasing sequence, hence for all $t$ we have $x(\mu_t) \in L_{\mu_1 e}$. Because $L_{\mu_1 e}$ is compact the sequence $x(\mu_t)$ has an accumulation point $x^*$ and without loss of generality we may assume that $x^* = \lim_{t \to \infty} x(\mu_t)$. Let $s^* := s(x^*)$. Clearly $x^*$ is optimal because

$$x^* s^* = \lim_{t \to \infty} x(\mu_t) s(x(\mu_t)) = \lim_{t \to \infty} \mu_t e = 0. \tag{2.23}$$

We still have to prove that $(x^*, s(x^*))$ is strictly complementary, i.e., $x^* + s^* > 0$. Let $\mathbf{B} = \{i : x_i^* > 0\}$ and $\mathbf{N} = \{i : s_i^* > 0\}$. Using the fact that $M$ is skew-symmetric, we have

$$0 = (x^* - x(\mu_t))^T (s^* - s(\mu_t)) = x(\mu_t)^T s(\mu_t) - x^{*T} s(\mu_t) - x(\mu_t)^T s^*, \tag{2.24}$$

which, by using that $x(\mu_t)_i s(\mu_t)_i = \mu_t$, can be rewritten as

$$\sum_{i \in \mathbf{B}} x_i^* s(\mu_t)_i + \sum_{i \in \mathbf{N}} s_i^* x(\mu_t)_i = n\mu_t, \tag{2.25a}$$

$$\sum_{i \in \mathbf{B}} \frac{x_i^*}{x(\mu_t)_i} + \sum_{i \in \mathbf{N}} \frac{s_i^*}{s(\mu_t)_i} = n. \tag{2.25b}$$

By taking the limit as $\mu_t$ goes to zero we obtain that

$$|\mathbf{B}| + |\mathbf{N}| = n,$$

i.e., $(\mathbf{B}, \mathbf{N})$ is a partition of the index set, hence $(x^*, s(x^*))$ is a strictly complementary solution. The proof of Theorem 2.13 is complete. □

As we mentioned earlier, this result is powerful enough to prove the strong duality theorem of LO in the strong form, including strict complementarity, i.e., the Goldman-Tucker Theorem (Thm. 2.4) for $SP$ and for $(P)$ and $(D)$.

Our next step is to prove that the accumulation point $x^*$ is unique.

## 2.2.5  Convergence to the Analytic Center

In this subsection we prove that the central path has only one accumulation point, i.e., it converges to a unique point, the so-called analytic center [74] of the optimal set $SP^*$.

**Definition 2.14.** Let $\bar{x} \in SP^*$, $\bar{s} = s(\bar{x})$ maximize the product

$$\prod_{i \in \mathbf{B}} x_i \prod_{i \in \mathbf{N}} s_i \qquad (2.26)$$

over $x \in SP^*$. Then $\bar{x}$ is called the *analytic center* of $SP^*$.

It is easily to verify that the analytic center is unique. Let us assume to the contrary that there are two different vectors $\bar{x} \neq \tilde{x}$ with $\bar{x}, \tilde{x} \in SP^*$ which satisfy the definition of analytic center, i.e.,

$$\vartheta^* = \prod_{i \in \mathbf{B}} \bar{x}_i \prod_{i \in \mathbf{N}} \bar{s}_i = \prod_{i \in \mathbf{B}} \tilde{x}_i \prod_{i \in \mathbf{N}} \tilde{s}_i = \max_{x \in SP^*} \prod_{i \in \mathbf{B}} x_i \prod_{i \in \mathbf{N}} s_i. \qquad (2.27)$$

Let us define $x^* = \frac{\bar{x} + \tilde{x}}{2}$. Then we have

$$\prod_{i \in \mathbf{B}} x_i^* \prod_{i \in \mathbf{N}} s_i^* = \prod_{i \in \mathbf{B}} \frac{1}{2}(\bar{x}_i + \tilde{x}_i) \prod_{i \in \mathbf{N}} (\bar{s}_i + \tilde{s}_i)$$

$$= \prod_{i \in \mathbf{B}} \frac{1}{2} \left( \sqrt{\frac{\bar{x}_i}{\tilde{x}_i}} + \sqrt{\frac{\tilde{x}_i}{\bar{x}_i}} \right) \prod_{i \in \mathbf{N}} \frac{1}{2} \left( \sqrt{\frac{\bar{s}_i}{\tilde{s}_i}} + \sqrt{\frac{\tilde{s}_i}{\bar{s}_i}} \right)$$

$$\sqrt{\prod_{i \in \mathbf{B}} \bar{x}_i \prod_{i \in \mathbf{N}} \bar{s}_i \prod_{i \in \mathbf{B}} \tilde{x}_i \prod_{i \in \mathbf{N}} \tilde{s}_i} > \prod_{i \in \mathbf{B}} \bar{x}_i \prod_{i \in \mathbf{N}} \bar{s}_i = \vartheta^*, \quad (2.28)$$

which shows that $\bar{x}$ is not the analytic center. Here the last inequality follows from the classical inequality $\mathbf{a} + \frac{1}{\mathbf{a}} \geq 2$ if $\mathbf{a} \in \mathbb{R}_+$ and strict inequality holds when $\mathbf{a} \neq 1$.

**Theorem 2.15.** *The limit point $x^*$ of the central path is the analytic center of $SP^*$.*

*Proof.* The same way as in the proof of Theorem 2.13 we derive

$$\sum_{i \in \mathbf{B}} \frac{\bar{x}_i}{x_i^*} + \sum_{i \in \mathbf{N}} \frac{\bar{s}_i}{s_i^*} = n. \qquad (2.29)$$

Now we apply the arithmetic-geometric mean inequality to derive

$$\left( \prod_{i \in \mathbf{B}} \frac{\bar{x}_i}{x_i^*} \prod_{i \in \mathbf{N}} \frac{\bar{s}_i}{s_i^*} \right)^{\frac{1}{n}} \leq \frac{1}{n} \left( \sum_{i \in \mathbf{B}} \frac{\bar{x}_i}{x_i^*} + \sum_{i \in \mathbf{N}} \frac{\bar{s}_i}{s_i^*} \right) = 1. \qquad (2.30)$$

Hence,

$$\prod_{i \in \mathbf{B}} \overline{x}_i \prod_{i \in \mathbf{N}} \overline{s}_i \leq \prod_{i \in \mathbf{B}} x_i^* \prod_{i \in \mathbf{N}} s_i^* \tag{2.31}$$

proving that $x^*$ is the analytic center of $SP^*$. The proof is complete.  □


### 2.2.6 Identifying the Optimal Partition

The condition number

In order to give bounds on the size of the variables along the central path we need to find a quantity that in some sense characterizes the set of optimal solutions. For an optimal solution $x \in SP^*$ we have

$$xs(x) = 0 \quad \text{and} \quad x + s(x) \geq 0.$$

Our next question is about the size of the nonzero coordinates of optimal solutions. Following the definitions in [72, 96] we define a condition number of the problem (SP) which characterizes the magnitude of the nonzero variables on the optimal set $SP^*$.

**Definition 2.16.** Let us define

$$\sigma^x := \min_{i \in \mathbf{B}} \max_{x \in SP^*} \{x_i\} \tag{2.32a}$$

$$\sigma^s := \min_{i \in \mathbf{N}} \max_{x \in SP^*} \{s(x)_i\}. \tag{2.32b}$$

Then the *condition number* of (SP) is defined as

$$\sigma = \min\{\sigma^x, \sigma^s\} = \min_{i} \max_{x \in SP^*} \{x_i + s(x)_i\}. \tag{2.33}$$

To determine the condition number $\sigma$ is in general more difficult than to solve the optimization problem itself. However, we can give an easily computable lower bound for $\sigma$. This bound depends only on the problem data.

**Lemma 2.17 (Lower bound for $\sigma$:).** *If $M$ and $q$ are integral[7] and all the columns of $M$ are nonzero, then*

$$\sigma \geq \frac{1}{\pi(M)}, \tag{2.34}$$

*where $\pi(M) = \prod_{i=1}^{n} \|M_i\|$.*

---

[7] If the problem data is rational, then by multiplying by the least common multiple of the denominators an equivalent LO problem with integer data is obtained.

*Proof.* The proof is based on Cramer's rule and on the estimation of determinants by using Hadamard's inequality.[8] Let $z = (x, s)$ be an optimal solution. Without loss of generality we may assume that the columns of the matrix $D = (-M, I)$ corresponding to the nonzero coordinates of $z = (x, s)$ are linearly independent. If they are not independent, then by using Gaussian elimination we can reduce the solution to get one with linearly independent columns. Let us denote this index set by $J$. Further, let the index set $K$ be such that $D_{KJ}$ is a nonsingular square submatrix of $D$. Such $K$ exists, because the columns in $D_J$ are linearly independent. Now we have $D_{KJ}z_J = q_K$, and hence, by Cramer's rule,

$$z_j = \frac{\det \left( D_{KJ}^{(j)} \right)}{\det \left( D_{KJ} \right)}, \quad \forall j \in J, \tag{2.35}$$

where $D_{KJ}^{(j)}$ denotes the matrix obtained when the $j$th column in $D_{KJ}$ is replaced by $q_K$. Assuming that $z_j > 0$ then, because the data is integral, the numerator in the quotient given above is at least one. Thus we obtain $z_j \geq \frac{1}{\det (D_{KJ})}$. By Hadamard's inequality the last determinant can be estimated by the product of the norm of its columns, which can further be bounded by the product of the norms of all the columns of the matrix $M$.

$\square$

The condition that none of the columns of the matrix $M$ is a zero vector is not restrictive. For the general problem (SP) a zero column $M_i$ would imply that $s_i = q_i$ for each feasible solution, thus the pair $(x_i, s_i)$ could be removed. More important is that for our embedding problem $(\overline{SP})$ none of the columns of the coefficient matrix

$$\begin{pmatrix} M & r \\ -r^T & 0 \end{pmatrix}$$

is zero. By definition we have $r = e - Me$ nonzero, because $e^T r = e^T e - e^T Me = n$. Moreover, if $M_i = 0$, then by using that matrix $M$ is skew-symmetric we have $r_i = 1$, thus the $i$th column of the coefficient matrix is again nonzero.

---

[8] Let $G$ be a nonsingular $n \times n$ matrix. Hadamard's inequality states that

$$\det (G) \leq \prod_{i=1}^{n} \|G_i\|$$

holds, see [37] for a reference.

The size of the variables along the central path

Now, by using the condition number $\sigma$, we are able to derive lower and upper bounds for the variables along the central path. Let $(B, N)$ be the optimal partition of the problem (SP).

**Lemma 2.18.** *For each positive $\mu$ one has*

$$x_i(\mu) \geq \frac{\sigma}{n} \quad i \in \mathbf{B}, \qquad x_i(\mu) \leq \frac{n\mu}{\sigma} \quad i \in \mathbf{N}, \tag{2.36a}$$

$$s_i(\mu) \leq \frac{n\mu}{\sigma} \quad i \in \mathbf{B}, \qquad s_i(\mu) \geq \frac{\sigma}{n} \quad i \in \mathbf{N}. \tag{2.36b}$$

*Proof.* Let $(x^*, s^*)$ be optimal, then by orthogonality we have

$$(x(\mu) - x^*)^T (s(\mu) - s^*) = 0,$$
$$x(\mu)^T s^* + s(\mu)^T x^* = n\mu,$$
$$x_i(\mu)s_i^* \leq x(\mu)^T s^* \leq n\mu, \quad 1 \leq i \leq n.$$

Since we can choose $(x^*, s^*)$ such that $s_i^* \geq \sigma$ and because $x_i(\mu)s_i(\mu) = \mu$, for $i \in \mathbf{N}$, we have

$$x_i(\mu) \leq \frac{n\mu}{s_i^*} \leq \frac{n\mu}{\sigma} \quad \text{and} \quad s_i(\mu) \geq \frac{\sigma}{n}, \ i \in \mathbf{N}.$$

The proofs of the other bounds are analogous.                                          $\square$


Identifying the optimal partition

The bounds presented in Lemma 2.18 make it possible to identify the optimal partition $(\mathbf{B}, \mathbf{N})$, when $\mu$ is sufficiently small. We just have to calculate the $\mu$ value that ensures that the coordinates going to zero are certainly smaller than the coordinates that converge to a positive number.

**Corollary 2.19.** *If we have a central solution $x(\mu) \in SP$ with*

$$\mu < \frac{\sigma^2}{n^2}, \tag{2.37}$$

*then the optimal partition $(\mathbf{B}, \mathbf{N})$ can be identified.*

The results of Lemma 2.18 and Corollary 2.19 can be generalized to the situation when a vector $(x, s)$ is not on, but just in a certain neighbourhood of the central path. In order to keep our discussion short, we do not go into those details. The interested reader is referred to [72].

## 2.2.7 Rounding to an Exact Solution

Our next goal is to find a strictly complementary solution. This could be done by moving along the central path as $\mu \to 0$. Here we show that we do not have to do that, we can stop at a sufficiently small $\mu > 0$, and round off the current "almost optimal" solution to a strictly complementary optimal one. We need some new notation. Let the optimal partition be denoted by $(\mathbf{B}, \mathbf{N})$, let $\omega := \|M\|_\infty = \max_{1 \le i \le n} \sum_{j=1}^n |M_{ij}|$ and $\pi := \pi(M) = \prod_{i=1}^n \|M_i\|$.

**Lemma 2.20.** *Let $M$ and $q$ be integral and all the columns of $M$ be nonzero. If $(x, s) := (x(\mu), s(x(\mu)))$ is a central solution with*

$$x^T s = n\mu < \frac{\sigma^2}{n^{\frac{3}{2}}(1+\omega)^2 \pi}, \quad \text{which certainly holds if} \quad n\mu \le \frac{1}{n^{\frac{3}{2}}(1+\omega)^2 \pi^3},$$

*then by a simple rounding procedure a strictly complementary optimal solution can be found in $\mathcal{O}(n^3)$ arithmetic operations.*

*Proof. Proof.* Let $x := x(\mu) > 0$ and $s := s(x) > 0$ be given. Because

$$\mu < \frac{\sigma^2}{n^{\frac{5}{2}}(1+\omega)^2 \pi} < \frac{\sigma^2}{n^2} \tag{2.38}$$

the optimal partition $(\mathbf{B}, \mathbf{N})$ is known. Let us simply set the small variables $x_{\mathbf{N}}$ and $s_{\mathbf{B}}$ to zero. Then we correct the created error and estimate the size of the correction.

For $(x, s)$ we have

$$M_{\mathbf{BB}} x_{\mathbf{B}} + M_{\mathbf{BN}} x_{\mathbf{N}} + q_{\mathbf{B}} = s_{\mathbf{B}}, \tag{2.39}$$

but by rounding $x_{\mathbf{N}}$ and $s_{\mathbf{B}}$ to zero the error $\hat{q}_{\mathbf{B}} = s_{\mathbf{B}} - M_{\mathbf{BN}} x_{\mathbf{N}}$ occurs. Similarly, we have

$$M_{\mathbf{NB}} x_{\mathbf{B}} + M_{\mathbf{NN}} x_{\mathbf{N}} + q_{\mathbf{N}} = s_{\mathbf{N}} \tag{2.40}$$

but by rounding $x_{\mathbf{N}}$ and $s_{\mathbf{B}}$ to zero the error $\hat{q}_{\mathbf{N}} = -M_{\mathbf{NN}} x_{\mathbf{N}}$ occurs.

Let us first estimate $\hat{q}_{\mathbf{B}}$ and $\hat{q}_{\mathbf{N}}$ by using the results of Lemma 2.18. For $\hat{q}_{\mathbf{B}}$ we have

$$\|\hat{q}_{\mathbf{B}}\| \le \sqrt{n}\|\hat{q}_{\mathbf{B}}\|_\infty = \sqrt{n}\|s_{\mathbf{B}} - M_{\mathbf{BN}} x_{\mathbf{N}}\|_\infty \le \sqrt{n}\|(I, -M_{\mathbf{BN}})\|_\infty \left\| \begin{matrix} s_{\mathbf{B}} \\ x_{\mathbf{N}} \end{matrix} \right\|_\infty$$

$$\le \sqrt{n}(1+\omega)\frac{n\mu}{\sigma} = \frac{n^{\frac{3}{2}}\mu(1+\omega)}{\sigma}. \tag{2.41}$$

We give a bound for the infinity norm of $\hat{q}_N$ as well:

$$\|\hat{q}_{\mathbf{N}}\|_\infty = \| - M_{\mathbf{NN}} x_{\mathbf{N}}\|_\infty \leq \|M_{\mathbf{NN}}\|_\infty \|x_{\mathbf{N}}\|_\infty \leq \omega \frac{n\mu}{\sigma}. \tag{2.42}$$

Now we are going to correct these errors by adjusting $x_{\mathbf{B}}$ and $s_{\mathbf{N}}$. Let us denote the correction by $\xi$ for $x_{\mathbf{B}}$ and by $\zeta$ for $s_{\mathbf{N}}$, further let $(\hat{x}, \hat{s})$ be given by $\hat{x}_{\mathbf{B}} := x_{\mathbf{B}} + \xi > 0$, $\hat{x}_{\mathbf{N}} = 0$, $\hat{s}_{\mathbf{B}} = 0$ and $\hat{s}_{\mathbf{N}} := s_{\mathbf{N}} + \zeta > 0$.

If we know the correction $\xi$ of $x_{\mathbf{B}}$, then from equation (2.40) the necessary correction $\zeta$ of $s_{\mathbf{N}}$ can easily be calculated. Equation (2.39) does not contain $s_N$, thus by solving the equation

$$M_{\mathbf{BB}}\xi = -\hat{q}_{\mathbf{B}} \tag{2.43}$$

the corrected value $\hat{x}_{\mathbf{B}} = x_{\mathbf{B}} + \xi$ can be obtained.

First we observe that the equation $M_{\mathbf{BB}}\xi = -\hat{q}_{\mathbf{B}}$ is solvable, because any optimal solution $x^*$ satisfies $M_{\mathbf{BB}} x_{\mathbf{B}}^* = -q_{\mathbf{B}}$, thus we may write $M_{\mathbf{BB}}(x_{\mathbf{B}}+\xi)$ $= M_{\mathbf{BB}} x_{\mathbf{B}}^* = -q_{\mathbf{B}}$, hence

$$M_{\mathbf{BB}}\xi = M_{\mathbf{BB}}(x_{\mathbf{B}}^* - x_{\mathbf{B}}) = -q_{\mathbf{B}} - s_{\mathbf{B}} + M_{\mathbf{BN}} x_{\mathbf{N}} + q_{\mathbf{B}} \tag{2.44}$$
$$= -s_{\mathbf{B}} + M_{\mathbf{BN}} x_{\mathbf{N}} = -\hat{q}_{\mathbf{B}}.$$

This equation system can be solved by Gaussian elimination. The size of $\xi$ obtained this way can be estimated by applying Cramer's rule and Hadamard's inequality, the same way as we have estimated $\sigma$ in Lemma 2.17. If $M_{\mathbf{BB}}$ is zero, then we have $q_{\mathbf{B}} = 0$ and $M_{\mathbf{BN}} x_{\mathbf{N}} = s_{\mathbf{B}}$, thus rounding $x_{\mathbf{N}}$ and $s_{\mathbf{B}}$ to zero does not produce any error here, hence we can choose $\xi = 0$. If $M_{\mathbf{BB}}$ is not the zero matrix, then let $\overline{M}_{\mathbf{BB}}$ be a maximal nonsingular square submatrix of $M_{\mathbf{BB}}$ and let $\bar{q}_{\mathbf{B}}$ be the corresponding part of $\hat{q}_{\mathbf{B}}$. By using the upper bounds on $x_{\mathbf{N}}$ and $s_{\mathbf{B}}$ by Lemma 2.18 we have

$$|\xi_i| = \frac{|\det (\overline{M}_{\mathbf{BB}}^{(i)})|}{|\det (\overline{M}_{\mathbf{BB}})|} \leq |\det (\overline{M}_{\mathbf{BB}}^{(i)})| \tag{2.45}$$
$$\leq \|\bar{q}_{\mathbf{B}}\| \, |\det (\overline{M}_{\mathbf{BB}})| \leq \frac{n^{\frac{3}{2}}\mu(1 + \omega)}{\sigma}\pi,$$

where (2.41) was used in the last estimation. This result, due to $\|x_{\mathbf{B}}\|_\infty \geq \frac{\sigma}{n}$, implies that $\hat{x}_{\mathbf{B}} = x_{\mathbf{B}} + \xi > 0$ certainly holds if $n\mu < \frac{\sigma^2}{n^{\frac{3}{2}}(1+\omega)\pi}$, and this is implied by the hypothesis of the theorem which was involving $(1 + \omega)^2$ instead of $(1 + \omega)$.

Finally, we simply correct $s_{\mathbf{N}}$ by using (2.40), i.e., we define $\zeta := \hat{q}_{\mathbf{N}} + M_{\mathbf{NB}}\xi$. We still must ensure that

$$\hat{s}_{\mathbf{N}} := s_{\mathbf{N}} + \hat{q}_{\mathbf{N}} + M_{\mathbf{NB}}\xi > 0. \tag{2.46}$$

Using again the bounds given in Lemma 2.17, the bound (2.42) and the estimate on $\xi$, one easily verifies that

$$\|\hat{q}_{\mathbf{N}} + M_{\mathbf{NB}}\xi\|_\infty \leq \|(I, M_{\mathbf{NB}})\|_\infty \left\| \begin{matrix} \hat{q}_{\mathbf{N}} \\ \xi \end{matrix} \right\|_\infty \tag{2.47}$$

$$\leq (1+w)\max\left\{ w\frac{n\mu}{\sigma}, \frac{n^{\frac{3}{2}}\mu(1+w)\pi}{\sigma}\right\} = \frac{n^{\frac{3}{2}}\mu(1+w)^2\pi}{\sigma}.$$

Thus, due to $\|s_{\mathbf{N}}\|_\infty \geq \frac{\sigma}{n}$, the vector $\hat{s}_{\mathbf{N}}$ is certainly positive if

$$\frac{\sigma}{n} > \frac{n^{\frac{3}{2}}\mu(1+w)^2\pi}{\sigma}. \tag{2.48}$$

This is exactly the first inequality given in the lemma. The second inequality follows by observing that $\pi\sigma \geq 1$, by Lemma 2.17.

The proof is completed by noting that the solution of an equation system by using Gaussian elimination, some matrix-vector multiplications and vector-vector summations, all with a dimension not exceeding $n$, are needed to perform our rounding procedure. Thus the computational complexity of our rounding procedure is at most $\mathcal{O}(n^3)$.                          □

Note that this rounding result can also be generalized to the situation when a vector $(x, s)$ is not on, but just in a certain neighbourhood of the central path. For details the reader is referred again to [72].[9]

## 2.3 Summary of the Theoretical Results

Let us return to our general LO problem in canonical form

$$\min\left\{c^T u \ : \ Au - z = b, \ u \geq 0, \ z \geq 0\right\} \tag{P}$$

$$\max\left\{b^T v \ : \ A^T v + w = c, \ v \geq 0, \ w \geq 0\right\}, \tag{D}$$

where the slack variables are already included in the problem formulation. In what follows we recapitulate the results obtained so far.

- In Section 2.1 we have seen that to solve the LO problem it is sufficient to find a strictly complementary solution to the Goldman-Tucker model

---

[9] This result makes clear that when one solves an LO problem by using an IPM, the iterative process can be stopped at a sufficiently small value of $\mu$. At that point a strictly complementary optimal solution can be identified easily.

$$
\begin{aligned}
Au \quad -\tau b -z \qquad\qquad &= 0 \\
-A^T v \qquad +\tau c \quad -w \quad &= 0 \\
b^T v \ -c^T u \qquad\qquad -\rho &= 0
\end{aligned}
$$

$$
v \geq 0,\ u \geq 0,\ \tau \geq 0,\ z \geq 0,\ w \geq 0,\ \rho \geq 0.
$$

- This *homogeneous* system always admits the zero solution, but we need a solution for which $\tau + \rho > 0$ holds.
- If $(u^*, z^*)$ is optimal for $(P)$ and $(v^*, w^*)$ for $(D)$ then $(v^*, u^*, 1, z^*, w^*, 0)$ is a solution for the Goldman-Tucker model with the requested property $\tau + \rho > 0$. See Theorem 2.3.
- Any solution of the Goldman-Tucker model $(v, u, \tau, z, w, \rho)$ with $\tau > 0$ yields an optimal solution pair (scale the variables $(u, z)$ and $(v, w)$ by $\frac{1}{\tau}$) for LO. See Theorem 2.3.
- Any solution of the Goldman-Tucker model $(u, z, v, w, \tau, \rho)$ with $\rho > 0$ provides a certificate of primal or dual infeasibility. See Theorem 2.3.
- If $\tau = 0$ in every solution $(v, u, \tau, z, w, \rho)$ then $(P)$ and $(D)$ have no optimal solutions with zero duality gap.
- The Goldman-Tucker model can be transformed into a skew-symmetric self-dual problem (SP) satisfying the IPC. See Section 2.2.2.
- If problem (SP) satisfies the IPC then
    - the central path exists (see Theorem 2.12);
    - the central path converges to a strictly complementary solution (see Theorem 2.13);
    - the limit point of the central path is the analytic center of the optimal set (see Theorem 2.15);
    - if the problem data is integral and a solution on the central path with a sufficiently small $\mu$ is given, then the optimal partition (see Corollary 2.19) and an exact strictly complementary optimal solution (see Lemma 2.20) can be found.

- These results give a constructive proof of Theorem 2.4.
- This way, as we have seen in Section 2.1, the Strong Duality theorem of linear optimization (Theorem 2.5) is proved.

The above summary shows that we have completed our project. The duality theory of LO is built up by using only elementary calculus and fundamental concepts of IPMs. In the following sections we follow this recipe to derive interior point methods for conic and general nonlinear optimization.

In the rest of this section a generic IP algorithm is presented.

## 2.4  A General Scheme of IP Algorithms for Linear Optimization

In this section a glimpse of the main elements of IPMs is given. We keep on working with our model problem (SP). In Sections 2.1 and 2.2.2 we have shown that a general LO problem can be transformed into a problem of the form (SP), and that problem satisfies the IPC. Some notes are due to the linear algebra involved. We know that the size of the resulting embedding problem (SP) is more than doubled comparing to the size of the original LO problem. Despite the size increase, the linear algebra of an IPM can be organized so that the computational cost of an iteration stays essentially the same.

Let us consider the problem (cf. page 223)

$$
\min \left\{ \lambda \vartheta \ : \ \begin{pmatrix} M & r \\ -r^T & 0 \end{pmatrix} \begin{pmatrix} x \\ \vartheta \end{pmatrix} + \begin{pmatrix} s \\ \nu \end{pmatrix} = \begin{pmatrix} 0 \\ \lambda \end{pmatrix}; \ \begin{pmatrix} x \\ \vartheta \end{pmatrix}, \begin{pmatrix} s \\ \nu \end{pmatrix} \geq 0 \right\}, \quad (\overline{\mathrm{SP}})
$$

where $r = e - Me$, $\lambda = n+1$ and the matrix $M$ is given by (2.4). This problem satisfies the IPC, because the all one vector $(x^0, \vartheta^0, s^0, \nu^0) = (e, 1, e, 1)$ is a feasible solution, moreover it is also on the central path by taking $\mu = 1$. In other words, it is a positive solution of the equation system

$$
\begin{pmatrix} M & r \\ -r^T & 0 \end{pmatrix} \begin{pmatrix} x \\ \vartheta \end{pmatrix} + \begin{pmatrix} s \\ \nu \end{pmatrix} = \begin{pmatrix} 0 \\ \lambda \end{pmatrix}; \ \begin{pmatrix} x \\ \vartheta \end{pmatrix}, \begin{pmatrix} s \\ \nu \end{pmatrix} \geq 0
$$
$$
\begin{pmatrix} x \\ \vartheta \end{pmatrix} \begin{pmatrix} s \\ \nu \end{pmatrix} = \begin{pmatrix} \mu e \\ \mu \end{pmatrix}, \qquad (2.49)
$$

which defines the central path of problem $(\overline{\mathrm{SP}})$. As we have seen, for each $\mu > 0$, this system has a unique solution. However, in general this solution cannot be calculated exactly. Therefore we are making Newton steps to get approximate solutions.

Newton step

Let us assume that a feasible interior-point $(x, \vartheta, s, \nu) > 0$ is given.[10] We want to find the solution of (2.49) for a given $\mu \geq 0$, in other words we want to determine the displacements $(\Delta x, \Delta \vartheta, \Delta s, \Delta \nu)$ so that

---

[10] Here we assume that all the linear equality constraints are satisfied. The resulting IPM is a feasible IPM. In the literature one can find infeasible IPMs [93] that do not assume that the linear equality constraints are satisfied.

$$\begin{pmatrix} M & r \\ -r^T & 0 \end{pmatrix}\begin{pmatrix} x+\Delta x \\ \vartheta+\Delta\vartheta \end{pmatrix}+\begin{pmatrix} s+\Delta s \\ \nu+\Delta\nu \end{pmatrix}=\begin{pmatrix} 0 \\ \lambda \end{pmatrix};$$

$$\begin{pmatrix} x+\Delta x \\ \vartheta+\Delta\vartheta \end{pmatrix},\ \begin{pmatrix} s+\Delta s \\ \nu+\Delta\nu \end{pmatrix}\geq 0; \tag{2.50}$$

$$\begin{pmatrix} x+\Delta x \\ \vartheta+\Delta\vartheta \end{pmatrix}\begin{pmatrix} s+\Delta s \\ \nu+\Delta\nu \end{pmatrix}=\begin{pmatrix} \mu e \\ \mu \end{pmatrix}.$$

By neglecting the second order terms $\Delta x\Delta s$ and $\Delta\vartheta\Delta\nu$, and the nonnegativity constraints, the Newton equation system is obtained (cf. page 227)

$$\begin{aligned} -M\Delta x\ -r\Delta\vartheta\ +\Delta s\ &\quad = 0 \\ r^T\Delta x\ &\quad +\Delta\nu = 0 \\ s\Delta x\ &+x\Delta s\ &= \mu e - xs \\ \nu\Delta\vartheta\ &+\vartheta\Delta\nu = \mu - \vartheta\nu. \end{aligned} \tag{2.51}$$

We start by making some observations. For any vector $(x,\vartheta,s,\nu)$ that satisfies the equality constraints of $(\overline{SP})$ we have

$$x^T s + \vartheta\nu = \vartheta\lambda. \tag{2.52}$$

Applying this to the solution obtained after making a Newton step we may write

$$(x+\Delta x)^T(s+\Delta s)+(\vartheta+\Delta\vartheta)^T(\nu+\Delta\nu)=(\vartheta+\Delta\vartheta)\lambda. \tag{2.53}$$

By rearranging the terms we have

$$(x^T s+\vartheta\nu)+(\Delta x^T\Delta s+\Delta\vartheta\Delta\nu)+(x^T\Delta s+s^T\Delta x+\vartheta\Delta\nu+\nu\Delta\vartheta)=\vartheta\lambda+\Delta\vartheta\lambda.$$

As we mentioned above, the first term in the left hand side sum equals to $\vartheta\lambda$, while from (2.51) we derive that the second sum is zero. From the last equations of (2.51) one easily derives that the third expression equals to $\mu(n+1)-x^T s-\vartheta\nu=\mu\lambda-\vartheta\lambda$. This way the equation $\mu\lambda-\vartheta\lambda=\Delta\vartheta\lambda$ is obtained, i.e., an explicit expression for $\Delta\vartheta$

$$\Delta\vartheta=\mu-\vartheta$$

is derived. This value can be substituted in the last equation of (2.51) to derive the solution

$$\Delta\nu=\frac{\mu}{\vartheta}-\nu-\frac{\nu(\mu-\vartheta)}{\vartheta},$$

i.e.,

$$\Delta\nu=\frac{\mu(1-\nu)}{\vartheta}.$$

On the other hand, $\Delta s$ can be expressed from the third equation of (2.51) as

$$\Delta s = \mu X^{-1}e - s - X^{-1}S\Delta x.$$

Finally, substituting all these values in the first equation of (2.51) we have

$$M\Delta x + X^{-1}S\Delta x = \mu X^{-1}e - s - (\mu - \vartheta)r,$$

i.e., $\Delta x$ is the unique solution of the positive definite system[11]

$$(M + X^{-1}S)\Delta x = \mu X^{-1}e - s - (\mu - \vartheta)r.$$

Having determined the displacements, we can make a (possibly damped) Newton step to update our current iterate:

$$x := x + \Delta x$$
$$\vartheta := \vartheta + \Delta\vartheta = \mu$$
$$s := s + \Delta s$$
$$\nu := \nu + \Delta\nu.$$

Proximity measures

We have seen that the central path is our guide to a strictly complementary solution. However, due to the nonlinearity of the equation system determining the central path, we cannot stay on the central path with our iterates, even if our initial interior-point was perfectly centred. For this reason we need some centrality, or with other words proximity, measure that enables us to control and keep our iterates in an appropriate neighbourhood of the central path. In general this measure depends on the current primal-dual iterate $x$ and $s$, and a value of $\mu$ on the central path. This measure quantifies how close the iterate is to the point corresponding to $\mu$ on the central path. We use $\delta(x, s, \mu)$ to denote this general proximity measure.

Let the vectors $\bar{x}$ and $\bar{s}$ be composed from $x$ and $\vartheta$, and from $s$ and $\nu$ respectively. Note that on the central path all the coordinates of the vector $\bar{x}\bar{s}$ are equal. This observation indicates that the proximity measure

---

[11] Observe that although the dimensions of problem $(\overline{\text{SP}})$ are larger than problem (SP), to determine the Newton step for both systems requires essentially the same computational effort. Note also, that the special structure of the matrix $M$ (see (2.4)) can be utilized when one solves this positive definite linear system. For details the reader is referred to [5, 72, 93, 97].

$$\delta_c(\overline{x}\,\overline{s}) := \frac{\max(\overline{x}\,\overline{s})}{\min(\overline{x}\,\overline{s})}, \tag{2.54}$$

where $\max(\overline{x}\,\overline{s})$ and $\min(\overline{x}\,\overline{s})$ denote the largest and smallest coordinate of the vector $\overline{x}\,\overline{s}$, is an appropriate measure of centrality. In the literature of IPMs various centrality measures were developed (see the books [42, 45, 72, 93, 97]). Here we present just another one, extensively used in [72]:

$$\delta_0(\overline{x}\,\overline{s}, \mu) := \frac{1}{2}\left\|\left(\frac{\overline{x}\,\overline{s}}{\mu}\right)^{\frac{1}{2}} - \left(\frac{\mu}{\overline{x}\,\overline{s}}\right)^{\frac{1}{2}}\right\|. \tag{2.55}$$

Both of these proximity measures allow us to design polynomial IPMs.

A generic interior point algorithm

Algorithm 1 gives a general framework for an interior point method.

---
**Algorithm 1** Generic Interior-Point Newton Algorithm
---
**Input:**
   A proximity parameter $\gamma$;
   an accuracy parameter $\mathbf{e} > 0$;
   a variable damping factor $\mathbf{a}$;
   update parameter $\theta$, $0 < \theta < 1$;
   $(\overline{x}^0, \overline{s}^0)$, $\mu^0 \leq 1$ s.t. $\mathbf{d}(\overline{x}^0\overline{s}^0, \mu^0) \leq \gamma$.
**begin**
   $\overline{x} := \overline{x}^0$; $\overline{s} := \overline{s}^0$; $\mu := \mu^0$;
   **while** $(n+1)\mu \geq \mathbf{e}$ **do**
   **begin**
      $\mu := (1-\theta)\mu$;
      **while** $\mathbf{d}(\overline{x}\,\overline{s}, \mu) \geq \gamma$ **do**
      **begin**
         $\overline{x} := \overline{x} + \alpha\Delta\overline{x}$;
         $\overline{s} := \overline{s} + \alpha\Delta\overline{s}$;
      **end**
   **end**
**end**
---

The following crucial issues remain:

- choose the proximity parameter $\gamma$,
- choose a proximity measure $\delta(x, s, \mu)$,
- choose an update scheme for $\mu$ and
- specify how to damp the Newton step when needed.

Our goal with the selection of these parameters is to be able to prove polynomial iteration complexity of the resulting algorithm.

Three sets of parameters are presented, which ensure that the resulting IPMs are polynomial. The proofs of complexity can, e.g., be found in [72]. Recall that $(\overline{SP})$ admits the all one vector as a perfectly centred initial solution with $\mu = 1$.

The first algorithm is a *primal-dual logarithmic barrier algorithm with full Newton steps*, studied e.g. in [72]. This IPM enjoys the best complexity known to date. Let us make the following choice:

- $\mathbf{d}(\overline{x}\,\overline{s}, \mu) := \delta_0(\overline{x}\,\overline{s}, \mu)$, this measure is zero on the central path;
- $\mu^0 := 1$;
- $\theta := \frac{1}{2\sqrt{n+1}}$;
- $\gamma = \frac{1}{\sqrt{2}}$;
- $(\Delta\overline{x}, \Delta\overline{s})$ is the solution of (2.51);
- $\alpha = 1$.

**Theorem 2.21 (Theorem II.52 in [72]).** *With the given parameter set the full Newton step algorithm requires not more than*

$$\left\lceil 2\sqrt{n+1}\log\frac{n+1}{e}\right\rceil$$

*iterations to produce a feasible solution* $(\overline{x}, \overline{s})$ *for* $(\overline{SP})$ *such that* $\delta_0(\overline{x}\,\overline{s}, \mu) \leq \gamma$ *and* $(n+1)\mu \leq \mathbf{e}$.

The second algorithm is a *large-update primal-dual logarithmic barrier algorithm*, studied also e.g. in [72]. Among our three algorithms, this is the most practical. Let us make the following choice:

- $\mathbf{d}(\overline{x}\,\overline{s}, \mu) := \delta_0(\overline{x}\,\overline{s}, \mu)$, this measure is zero on the central path;
- $\mu^0 := 1$;
- $0 < \theta < \frac{n+1}{n+1+\sqrt{n+1}}$;
- $\gamma = \frac{\sqrt{R}}{2\sqrt{1+\sqrt{R}}}$, where $R := \frac{\theta\sqrt{n+1}}{1-\theta}$;
- $(\Delta\overline{x}, \Delta\overline{s})$ is the solution of (2.51);
- a is the result of a line search, when along the search direction the primal-dual logarithmic barrier function

$$\frac{\overline{x}^T\overline{s}}{\mu} - (n+1) - \sum_{i=1}^{n+1}\log\frac{\overline{x}_i\overline{s}_i}{\mu}$$

is minimized.

**Theorem 2.22 (Theorem II.74 in [72]).** *With the given parameter set the large-update primal-dual logarithmic barrier algorithm requires not more than*

$$\left\lceil \frac{1}{\theta} \left[ 2 \left( 1 + \sqrt{\frac{\theta\sqrt{n+1}}{1-\theta}} \right) \right]^4 \log \frac{n+1}{e} \right\rceil$$

*iterations to produce a feasible solution* $(\overline{x},\overline{s})$ *for* $(\overline{SP})$ *such that* $\delta_0(\overline{x}\,\overline{s}, \mu) \le \tau$
*and* $(n+1)\mu \le \mathbf{e}$.

When we choose $\theta = \frac{1}{2}$, then the total complexity becomes $\mathcal{O}\big((n+1)\log\frac{n+1}{e}\big)$,
while the choice $\theta = \frac{K}{\sqrt{n+1}}$, with any fixed positive value $K$ gives a complexity
of $\mathcal{O}\big(\sqrt{n+1}\log\frac{n+1}{e}\big)$.

Other versions of this algorithm were studied in [66], where the analysis of
large-update methods was based purely on the use of the proximity $\delta_0(\overline{x}\,\overline{s}, \mu)$.

The last algorithm is the *Dikin step algorithm* studied in [72]. This is one
of the simplest IPMs, with an extremely elementary complexity analysis. The
price for simplicity is that the polynomial complexity result is not the best
possible. Let us make the following choices:

- $\mathbf{d}(\overline{x}\,\overline{s}, \mu) := \delta_c(\overline{x}\,\overline{s})$, this measure is always larger than or equal to 1;
- $\mu^0 := 0$, this implies that $\mu$ stays equal to zero, thus $\theta$ is irrelevant;
- $\gamma = 2$;
- $(\Delta\overline{x}, \Delta\overline{s})$ is the solution of (2.51) when the right-hand sides of the last two
  equations are replaced by $-\frac{x^2 s^2}{\|\overline{x}\,\overline{s}\|}$ and $-\frac{\vartheta\nu}{\|\overline{x}\,\overline{s}\|}$, respectively;
- $\alpha = \frac{1}{2\sqrt{n+1}}$.

**Theorem 2.23 (Theorem I.27 in [72]).** *With the given parameter set the*
*Dikin step algorithm requires not more than*

$$\left\lceil 2(n+1)\log\frac{n+1}{e} \right\rceil$$

*iterations to produce a feasible solution* $(\overline{x},\overline{s})$ *for* $(\overline{SP})$ *such that* $\delta_c(\overline{x}\,\overline{s}) \le 2$
*and* $(n+1)\mu \le \mathbf{e}$.

## 2.5  *The Barrier Approach*

In our approach so far we perturbed the optimality conditions for the primal
dual linear optimization problem to get the central path. In what follows
we show an alternative, sometimes more intuitive, sometimes more technical
route. Consider again the linear optimization problem in primal form:

$$\min c^T u$$
$$Au \ge b \qquad\qquad (P)$$
$$u \ge 0.$$

A standard convex optimization trick to treat inequalities is to add them to the objective function with a barrier term:

$$\min \ c^T u - \mu \sum_{i=1}^{n} \ln u_i - \mu \sum_{j=1}^{m} \ln (Au - b)_j, \qquad \text{(PBar)}$$

where $\mu > 0$. The function $-\ln t$ is a *barrier* function. In particular it goes to $\infty$ if $t$ goes to 0, and for normalization, it is 0 at 1. If $u_i$ is getting close to 0 then the modified objective function will converge to $\infty$. This way we received an unconstrained problem defined on the positive orthant, for which we can easily write the optimality conditions. The idea behind this method is to gradually reduce $\mu$ and at the same time try to solve the unconstrained problem approximately. If $\mu$ is decreased at the right rate then the algorithm will converge to the optimal solution of the original problem.

The first order necessary optimality conditions for system (PBar) are:

$$c_i - \mu \frac{1}{u_i} - \mu \sum_{j=1}^{m} \frac{A_{ji}}{(Au - b)_j} = 0, \ i = 1, \ldots, n. \qquad (2.56)$$

This equation yields the same central path equations that we obtained in Definition 2.11. An identical result can be derived starting from the dual for of the linear optimization problem.

A natural extension of this idea is to replace the $-\ln t$ function with another barrier function. Sometimes we can achieve better complexity results by doing so, see [63] (universal barrier), [9,10,87] (volumetric barrier), [66,67] (self-regular barrier) for details.

# 3  Interior Point Methods for Conic Optimization

## 3.1  Problem Description

Conic optimization is a natural generalization of linear optimization. As we will see, most of the results in Section 2.3 carry over to the conic case with some minor modifications and the structure and analysis of the algorithm will be similar to the linear case.

A general conic optimization problem in primal form can be stated as

$$\begin{aligned} \min \ & c^T x \\ & Ax = b \qquad \qquad \text{(PCon)} \\ & x \in \mathcal{K}, \end{aligned}$$

where $c, x \in \mathbb{R}^N$, $b \in \mathbb{R}^m$, $A \in \mathbb{R}^{m \times N}$ and $\mathcal{K} \subseteq \mathbb{R}^N$ is a cone. The standard Lagrange dual of this problem is

$$\max b^T y$$
$$A^T y + s = c \qquad\qquad\qquad \text{(DCon)}$$
$$s \in \mathcal{K}^*,$$

where $y \in \mathbb{R}^m$, $s \in \mathbb{R}^N$ and $\mathcal{K}^*$ is the dual cone of $\mathcal{K}$, namely $\mathcal{K}^* = \{s \in \mathbb{R}^N : s^T x \geq 0, \forall x \in \mathcal{K}\}$. The weak duality theorem follows without any further assumption:

**Theorem 3.1 (Weak duality for conic optimization).** *If $x$, $y$ and $s$ are feasible solutions of the problems* (PCon) *and* (DCon) *then*

$$s^T x = c^T x - b^T y \geq 0. \qquad\qquad (3.1)$$

*This quantity is the duality gap. Consequently, if the duality gap is 0 for some solutions $x$, $y$ and $s$, then they form an optimal solution.*

*Proof.* Let $x$, $y$ and $s$ be feasible solutions, then

$$c^T x = (A^T y + s)^T x = x^T A^T y + x^T s = b^T y + x^T s \geq b^T y, \qquad (3.2)$$

since $x \in \mathcal{K}$ and $s \in \mathcal{K}^*$ implies $x^T s \geq 0$. $\qquad\qquad\qquad\qquad\qquad$ $\square$

In order for this problem to be tractable we have to make some assumptions.

**Assumption 3.2** *Let us assume that $\mathcal{K}$ is a closed, convex, pointed (not containing a line) and solid (has nonempty interior) cone, and that it is self-dual, i.e., $\mathcal{K} = \mathcal{K}^*$.*

Cones in the focus of our study are called symmetric.

**Theorem 3.3 (Real symmetric cones).** *Any symmetric cone over the real numbers is a direct product of cones of the following type:*

*nonnegative orthant:* the set of nonnegative vectors, $\mathbb{R}^n_+$,
*Lorentz or quadratic cone:* the set $\mathbb{L}_{n+1} = \{(u_0, u) \in \mathbb{R}_+ \times \mathbb{R}^n : u_0 \geq \|u\|\}$,
   *and the*
*positive semidefinite cone:* the cone $\mathbb{PS}^{n \times n}$ of $n \times n$ real symmetric positive semidefinite matrices.

*The dimensions of the cones forming the product can be arbitrary.*

Let us assume further that the interior point condition is satisfied, i.e., there is a strictly feasible solution.[12] The strong duality theorem follows:

---
[12] This assumption is not needed if $\mathcal{K}$ is the linear cone, $\mathbb{R}^n_+$.

**Theorem 3.4 (Strong duality for conic optimization).** *If the primal problem* (PCon) *is strictly feasible, i.e., there exists an x for which $Ax = b$ and $x \in \text{int}(\mathcal{K})$, then the dual problem is solvable (the maximum is attained) and the optimal values of the primal and dual problems are the same.*

*If the dual problem* (DCon) *is strictly feasible, i.e., there exists a y for which $s = c - A^T y \in \text{int}(\mathcal{K})$, then the primal problem is solvable (the minimum is attained) and the optimal values of the primal and dual problems are the same.*

*If both problems are strictly feasible then both are solvable and the optimal values are the same.*

*Remark 3.5.* In conic optimization it can happen that one problem is infeasible but there is no certificate of infeasibility. Such problems are called weakly infeasible. Also, even if the duality gap is zero, the minimum or maximum might not be attained, meaning the problem is not solvable.

In what follows we treat the second order and the semidefinite cones separately. This simplification is necessary to keep the notation simple and to make the material more accessible. Interested readers can easily assemble the parts to get the whole picture.

First we introduce the following primal-dual second-order cone optimization problems:

$$
\begin{array}{ll}
\min \sum_{i=1}^{k} c^{i^T} x^i & \max b^T y \\[2ex]
\sum_{i=1}^{k} A^i x^i = b & A^{i^T} y + s^i = c^i, i = 1, \ldots, k \qquad \text{(SOCO)} \\[2ex]
x^i \in \mathbb{L}_{n_i}, i = 1, \ldots, k & s^i \in \mathbb{L}_{n_i}, i = 1, \ldots, k,
\end{array}
$$

where $x^i, s^i, c^i \in \mathbb{R}^{n_i}$, $b, y \in \mathbb{R}^m$ and $A^i \in \mathbb{R}^{m \times n_i}$, the number of cones is $k$ and the $i^{\text{th}}$ cone is of dimension $n_i$.

The semidefinite optimization problem requires slightly different notation:

$$
\begin{array}{ll}
\min \text{Tr}(CX) & \max b^T y \\[2ex]
\text{Tr}\left(A^{(i)} X\right) = b_i, i = 1, \ldots, m & \sum_{i=1}^{m} A^{(i)} y_i + S = C \qquad \text{(SDO)} \\[2ex]
X \in \mathbb{PS}^{n \times n} & S \in \mathbb{PS}^{n \times n},
\end{array}
$$

where $X, S, C, A^{(i)} \in \mathbb{R}^{n \times n}$, $b, y \in \mathbb{R}^m$. For symmetric matrices $U$ and $V$ the quantity $\text{Tr}(UV)$ is actually a scalar product defined on symmetric matrices, and is identical to the sum of the componentwise products of the matrix elements.

## 3.2 Applications of Conic Optimization

Let us briefly present three applications of conic optimization. For more details see [3,11,88,91] and the references therein.

### 3.2.1 Robust Linear Optimization

Consider the standard linear optimization problem:

$$\min c^T x \tag{3.3}$$
$$a_j^T x - b_j \geq 0, \ \forall j = 1, \ldots, m,$$

where the data $(a_j, b_j)$ is uncertain. The uncertainty is usually due to some noise, or implementation or measurement error, and thus it is modelled by Gaussian distribution. The level sets of the distribution are ellipsoids, so we assume that the data vectors $(a_j; b_j)$ come from an ellipsoid. The inequalities then have to be satisfied for all possible values of the data. More precisely, the set off all possible data values is

$$\left\{ \begin{pmatrix} a_j \\ -b_j \end{pmatrix} = \begin{pmatrix} a_j^0 \\ -b_j^0 \end{pmatrix} + Pu : u \in \mathbb{R}^m, \|u\| \leq 1 \right\}, \tag{3.4}$$

and the new, robust constraint is represented as the following set of infinitely many constraints

$$\left( \begin{pmatrix} a_j^0 \\ -b_j^0 \end{pmatrix} + Pu \right)^T \begin{pmatrix} x \\ 1 \end{pmatrix} \geq 0, \ \forall u : \|u\| \leq 1. \tag{3.5}$$

This constraint is equivalent to

$$\begin{pmatrix} a_j^0 \\ -b_j^0 \end{pmatrix}^T \begin{pmatrix} x \\ 1 \end{pmatrix} \geq \max_{\|u\| \leq 1} \left\{ -u^T P^T \begin{pmatrix} x \\ 1 \end{pmatrix} \right\}. \tag{3.6}$$

The maximum on right hand side is the maximum of a linear function over a sphere, so it can be computed explicitly. This gives a finite form of the robust constraint:

$$\left( a_j^0 \right)^T x - b_j^0 \geq \left\| P^T \begin{pmatrix} x \\ 1 \end{pmatrix} \right\|. \tag{3.7}$$

Introducing the linear equalities $z^j = \left(a_j^0\right)^T x - b_j^0$ and $z = P^T \begin{pmatrix} x \\ 1 \end{pmatrix}$ this constraint is a standard second order conic constraint. For more details on this approach see [11].

### 3.2.2 Eigenvalue Optimization

Given the $n \times n$ matrices $A^{(1)}, \ldots, A^{(m)}$ it is often required to find a nonnegative combination of them such that the smallest eigenvalue of the resulting matrix is maximal. The smallest eigenvalue function is not differentiable, thus we could not use it directly to solve the problem. Semidefinite optimization offers an efficient framework to solve these problems. The maximal smallest eigenvalue problem can be written as

$$\max \lambda$$

$$\sum_{i=1}^{m} A_i y_i - \lambda I \in \mathbb{PS}^{n \times n} \tag{3.8}$$

$$y_i \geq 0, \ i = 1, \ldots, m.$$

See [2, 63, 65] for more details.

### 3.2.3 Relaxing Binary Variables

A classical method to solve problems with binary variables is to apply a continuous relaxation. Given the binary variables $z_1, \ldots, z_n \in \{0, 1\}$ the most common solution is the linear relaxation $z_1, \ldots, z_n \in [0, 1]$. However, in many cases tighter relaxations can be obtained by introducing the new variables $x_i = (2z_i - 1)$ and relaxing the nonlinear nonconvex equalities $x_i^2 = 1$. Now consider the matrix $X = xx^T$. This matrix is symmetric, positive semidefinite, it has rank one and all the diagonal elements are 1. By relaxing the rank constraint we get a positive semidefinite relaxation of the original optimization problem. This technique was used extensively by Goemans and Williamson [29] to derive tight bounds for max-cut and satisfiability problems. For a survey of this area see [51] or the books [11, 40].

## 3.3 Initialization by Embedding

The key assumption for both the operation of an interior point method and the validity of the strong duality theorem is the existence of a strictly feasible solution of the primal-dual systems. Fortunately, the embedding technique we used for linear optimization generalizes to conic optimization [26]. Consider the following larger problem based on (PCon) and (DCon):

$$
\begin{array}{llllll}
\min(\bar{x}^T \bar{s} + 1)\theta \\
\quad Ax & -b\tau & +\bar{b}\theta & & = 0 \\
-A^T y & & +c\tau & -\bar{c}\theta & -s & = 0 \\
b^T y & -c^T x & & +\bar{z}\theta & & -\kappa = 0 \\
-\bar{b}^T y & +\bar{c}^T x & -\bar{z}^T \tau & & & = -\bar{x}^T \bar{s} - 1 \\
\quad x \in \mathcal{K}, \tau \geq 0 & & s \in \mathcal{K} & \kappa \geq 0,
\end{array}
\tag{HSD}
$$

where $\bar{x}, \bar{s} \in \operatorname{int}(\mathcal{K})$, $\bar{y} \in \mathbb{R}^m$ are arbitrary starting points, $\tau, \theta$ are scalars, $\bar{b} = b - A\bar{x}$, $\bar{c} = c - A^T\bar{y} - \bar{s}$ and $\bar{z} = c^T\bar{x} - b^T\bar{y} + 1$. This model has the following properties [19, 52].

**Theorem 3.6 (Properties of the HSD model).** *System* (HSD) *is self-dual and it has a strictly feasible starting point, namely* $(x, s, y, \tau, \theta, \kappa) = (\bar{x}, \bar{s}, \bar{y}, 1, 1, 0)$. *The optimal value of these problems is* $\theta = 0$, *and if* $\tau > 0$ *at optimality then* $(x/\tau, y/\tau, s/\tau)$ *is an optimal solution for the original primal-dual problem with equal objective values, i.e., the duality gap is zero. If* $\tau = 0$ *and* $\kappa > 0$, *then the problem is either unbounded, infeasible, or the duality gap at optimality is nonzero. If* $\tau = \kappa = 0$, *then either the problem is infeasible without a certificate (weakly infeasible) or the optimum is not attained.*

*Remark 3.7.* Due to strict complementarity, the $\tau = \kappa = 0$ case cannot happen in linear optimization. The duality theory of conic optimization is weaker, this leads to all those ill-behaved problems.

The importance of this model is that the resulting system is strictly feasible with a known interior point, thus it can be solved directly with interior point methods.

## 3.4 Conic Optimization as a Complementarity Problem

### 3.4.1 Second Order Conic Case

In order to be able to present the second order conic case we need to define some elements of the theory of Jordan algebras for our particular case. All the proofs, along with the general theory can be found in [22]. Here we include as much of the theory (without proofs) as needed for the discussion. Our main source here is [3].

Given two vectors $u, v \in \mathbb{R}^n$ we can define a special product on them, namely:

$$
u \circ v = (u^T v; u_1 v_{2:n} + v_1 u_{2:n}).
\tag{3.9}
$$

The most important properties of this bilinear product are summarized in the following theorem:

## Theorem 3.8 (Properties of o).

1. *Distributive law:* $u \circ (v + w) = u \circ v + u \circ w$.
2. *Commutative law:* $u \circ v = v \circ u$.
3. *The unit element is* $\iota = (1; 0)$, *i.e.,* $u \circ \iota = \iota \circ u = u$.
4. *Using the notation* $u^2 = u \circ u$ *we have* $u \circ (u^2 \circ v) = u^2 \circ (u \circ v)$.
5. *Power associativity:* $u^p = u \circ \cdots \circ u$ *is well-defined, regardless of the order of multiplication. In particular,* $u^p \circ u^q = u^{p+q}$.
6. *Associativity does not hold in general.*

The importance of this bilinear function lies in the fact that it can be used to generate the second order cone:

**Theorem 3.9.** *A vector $x$ is in a second order cone (i.e., $x_1 \geq \|x_{2:n}\|_2$) if and only if it can be written as the square of a vector under the multiplication $\circ$, i.e., $x = u \circ u$.*

Moreover, analogously to the spectral decomposition theorem of symmetric matrices, every vector $u \in \mathbb{R}^n$ can be written as

$$u = \lambda_1 c^{(1)} + \lambda_2 c^{(2)}, \qquad (3.10)$$

where $c^{(1)}$ and $c^{(2)}$ are on the boundary of the cone, and

$$c^{(1)^T} c^{(2)} = 0 \qquad (3.11a)$$

$$c^{(1)} \circ c^{(2)} = 0 \qquad (3.11b)$$

$$c^{(1)} \circ c^{(1)} = c^{(1)} \qquad (3.11c)$$

$$c^{(2)} \circ c^{(2)} = c^{(2)} \qquad (3.11d)$$

$$c^{(1)} + c^{(2)} = \iota \qquad (3.11e)$$

The vectors $c^{(1)}$ and $c^{(2)}$ are called the Jordan frame and they play the role of rank one matrices. The numbers $\lambda_1$ and $\lambda_2$ are called eigenvalues of $u$. They behave much the same way as eigenvalues of symmetric matrices, except that in our case there is an easy formula to compute them:

$$\lambda_{1,2}(u) = u_1 \pm \|u_{2:n}\|_2. \qquad (3.12)$$

This also shows that a vector is in the second order cone if and only if both of its eigenvalues are nonnegative.

The spectral decomposition enables us to compute functions over the vectors:

$$\|u\|_F = \sqrt{\lambda_1^2 + \lambda_2^2} = \sqrt{2}\,\|u\|_2, \qquad (3.13a)$$

$$\|u\|_2 = \max\left\{|\lambda_1|, |\lambda_2|\right\} = |u_1| + \|u_{2:n}\|_2, \tag{3.13b}$$

$$u^{-1} = \lambda_1^{-1} c^{(1)} + \lambda_2^{-1} c^{(2)}, \tag{3.13c}$$

$$u^{\frac{1}{2}} = \lambda_1^{\frac{1}{2}} c^{(1)} + \lambda_2^{\frac{1}{2}} c^{(2)}, \tag{3.13d}$$

where $u \circ u^{-1} = u^{-1} \circ u = \iota$ and $u^{\frac{1}{2}} \circ u^{\frac{1}{2}} = u$.

Since the mapping $v \mapsto u \circ v$ is linear, it can be represented with a matrix. Indeed, introducing the arrowhead matrix

$$\mathrm{Arr}\,(u) = \begin{pmatrix} u_1 & u_2 & \cdots & u_n \\ u_2 & u_1 & & \\ \vdots & & \ddots & \\ u_n & & & u_1 \end{pmatrix}, \tag{3.14}$$

we have $u \circ v = \mathrm{Arr}\,(u)\, v = \mathrm{Arr}\,(u)\, \mathrm{Arr}\,(v)\, \iota$. Another operator is the quadratic representation, which is defined as

$$Q_u = 2\,\mathrm{Arr}\,(u)^2 - \mathrm{Arr}\,(u^2), \tag{3.15}$$

thus $Q_u(v) = 2u \circ (u \circ v) - u^2 \circ v$ is a quadratic function[13] in $u$. This operator will play a crucial role in the construction of the Newton system.

Remember that second order cone optimization problems usually include several cones, i.e., $\mathcal{K} = \mathbb{L}_{n_1} \times \cdots \times \mathbb{L}_{n_k}$. For simplicity let us introduce the notation

$$\begin{aligned} A &= \left(A^1, \ldots, A^k\right), \\ x &= \left(x^1; \ldots; x^k\right), \\ s &= \left(s^1; \ldots; s^k\right), \\ c &= \left(c^1; \ldots; c^k\right). \end{aligned} \tag{3.16}$$

With this notation we can write

$$Ax = \sum_{i=1}^k A^i x^i, \tag{3.17}$$

$$A^T y = \left(A^{1^T} y; \ldots; A^{k^T} y\right).$$

Moreover, for a partitioned vector $u = (u^1; \ldots; u^k)$, $\mathrm{Arr}\,(u)$ and $Q_u$ are block diagonal matrices built from the blocks $\mathrm{Arr}\,(u^i)$ and $Q_{u^i}$, respectively.

---

[13] In fact, this operation is analogous to the mapping $V \mapsto UVU$ for symmetric matrices.

The optimality conditions for second order conic optimization are

$$Ax = b,\ x \in \mathcal{K}$$
$$A^T y + s = c,\ s \in \mathcal{K} \tag{3.18}$$
$$x \circ s = 0.$$

The first four conditions represent the primal and dual feasibility, while the last condition is called the complementarity condition. An equivalent form of the complementarity condition is $x^T s = 0$.

Now we perturb[14] the complementarity condition to get the central path:

$$Ax = b,\ x \in \mathcal{K} \tag{3.19}$$
$$A^T y + s = c,\ s \in \mathcal{K}$$
$$x^i \circ s^i = 2\mu \iota^i,\ i = 1, \ldots, k,$$

where $\iota^i = (1; 0; \ldots; 0) \in \mathbb{R}^{n_i}$. Finally, we apply the Newton method to this system to get the Newton step:

$$A\Delta x = 0 \tag{3.20}$$
$$A^T \Delta y + \Delta s = 0,$$
$$x^i \circ \Delta s^i + \Delta x^i \circ s^i = 2\mu \iota^i - x^i \circ s^i,\ i = 1, \ldots, k,$$

where $\Delta x = (\Delta x^1; \ldots; \Delta x^k)$ and $\Delta s = (\Delta s^1; \ldots; \Delta s^k)$. To solve this system we first rewrite it using the operator Arr ():

$$\begin{pmatrix} A & & \\ & A^T & I \\ & \mathrm{Arr}\,(s) & \mathrm{Arr}\,(x) \end{pmatrix} \begin{pmatrix} \Delta y \\ \Delta x \\ \Delta s \end{pmatrix} = \begin{pmatrix} 0 \\ 0 \\ 2\mu \iota - x \circ s \end{pmatrix}, \tag{3.21}$$

where $\iota = (\iota^1; \ldots; \iota^k)$. Eliminating $\Delta x$ and $\Delta s$ we get the so-called normal equation:

$$\left( A\,\mathrm{Arr}\,(s)^{-1}\,\mathrm{Arr}\,(x)\,A^T \right) \Delta y = -A\,\mathrm{Arr}\,(s)^{-1}\,(2\mu \iota - x \circ s). \tag{3.22}$$

The coefficient matrix is a $m \times m$. Unfortunately, not only this system is not symmetric, which is a disadvantage in practice, but in general it can be

---

[14] Our choice of perturbation might seem arbitrary but in fact this is the exact analog of what we did for linear optimization, since the vector $(1; 0)$ on the right hand side is the unit element for the multiplication $\circ$. See Section 3.6 to understand where the multiplier 2 comes from.

singular, even if $x$ and $s$ are in the interior of the cone $\mathcal{K}$. As an example[15] take $A = (0, \sqrt{3.69} + 0.7, 1)$, $\mathcal{K} = \left\{ x \in \mathbb{R}^3 : x_1 \geq \sqrt{x_2^2 + x_3^2} \right\}$. The points $x = (1; 0.8; 0.5)$ and $s = (1; 0.7; 0.7; )$ are strictly primal and dual feasible, but $A \operatorname{Arr}(s)^{-1} \operatorname{Arr}(x) A^T = 0$.

To prevent singularity and to get a symmetric system we rewrite the original optimization problem (SOCO) in an equivalent form. Let us fix a scaling vector $p \in \operatorname{int}(\mathcal{K})$ and consider the scaled problem[16]

$$
\begin{array}{ll}
\min \ (Q_{p^{-1}}c)^T (Q_p x) & \max \ b^T y \qquad \text{(SOCOscaled)} \\
(AQ_{p^{-1}}) (Q_p x) = b & (AQ_{p^{-1}})^T y + Q_{p^{-1}} s = Q_{p^{-1}} c \\
Q_p x \in \mathcal{K} & Q_{p^{-1}} s \in \mathcal{K}
\end{array}
$$

where $p^{-1}$ is defined by (3.13c), and $Q_p$ is given by (3.15). The exact form of $p$ will be specified later. This scaling has the following properties:

**Lemma 3.10.** *If $p \in \operatorname{int}(\mathcal{K})$, then*

1. *$Q_p$ and $Q_{p^{-1}}$ are inverses of each other, i.e., $Q_p Q_{p^{-1}} = I$.*
2. *The cone $\mathcal{K}$ is invariant, i.e., $Q_p(\mathcal{K}) = \mathcal{K}$.*
3. *Problems (SOCO) and (SOCOscaled) are equivalent.*

We can write the optimality conditions (3.18) for the scaled problem and perturb them to arrive at the central path for the symmetrized system. This defines a new Newton system:

$$
(AQ_{p^{-1}}) (Q_p \Delta x) = 0 \tag{3.23}
$$

$$
(AQ_{p^{-1}})^T \Delta y + Q_{p^{-1}} \Delta s = 0,
$$

$$
(Q_p x) \circ (Q_{p^{-1}} \Delta s) + (Q_p \Delta x) \circ (Q_{p^{-1}} s) = 2\mu\iota - (Q_p x) \circ (Q_{p^{-1}} s).
$$

Using Lemma 3.10 we can eliminate the scaling matrices from the first two equations, but not the third one. Although rather complicated, this system is still a linear system in the variables $\Delta x$, $\Delta y$ and $\Delta s$.

Before we can turn our attention to other elements of the algorithm we need to specify $p$. The most natural choice, i.e., $p = \iota$ is not viable as it does not provide a nonsingular Newton system. Another popular choice is the pair of primal-dual HKM directions, i.e.,

$$
p = s^{1/2} \text{ or } p = x^{1/2}, \tag{3.24}
$$

---

[15] See [67, S6.3.1].

[16] This scaling technique was originally developed for semidefinite optimization by Monteiro [57] and Zhang [99], and later generalized for second order cone optimization by Schmieta and Alizadeh [73].

in which case

$$Q_{p^{-1}}s = \iota \text{ or } Q_p x = \iota. \tag{3.25}$$

These directions are implemented as the default choice in the SOCO solver package SDPT3. Finally, probably the most studied and applied direction is the NT direction, defined as:

$$p = \left( Q_{x^{1/2}} \left( Q_{x^{1/2}} s \right)^{-1/2} \right)^{-1/2} = \left( Q_{s^{-1/2}} \left( Q_{s^{1/2}} x \right)^{1/2} \right)^{-1/2}. \tag{3.26}$$

This very complicated formula actually simplifies the variables, since

$$Q_p x = Q_{p^{-1}} s. \tag{3.27}$$

The NT scaling is implemented in SeDuMi and MOSEK and is also available in SDPT3.

We will now customize the generic IPM algorithm (see Algorithm 1 on page 242) for second order conic optimization. Let $\mu = \mu(x, s)$ be defined as

$$\mu(x, s) = \sum_{i=1}^{k} \frac{x^{i^T} s^i}{n_i}. \tag{3.28}$$

First let us define some centrality measures (see [3]). These measures are defined in terms of the scaled variable $w = (w_1; \ldots; w_k)$, where $w_i = Q_{x_i^{1/2}} s_i$.

$$\delta_F(x, s) := \| Q_{x^{1/2}} s - \mu \iota \|_F := \sqrt{\sum_{i=1}^{k} (\lambda_1(w_i) - \mu)^2 + (\lambda_2(w_i) - \mu)^2} \tag{3.29a}$$

$$\delta_\infty(x, s) := \| Q_{x^{1/2}} s - \mu \iota \|_2 := \max_{i=1,\ldots,k} \{ |\lambda_1(w_i) - \mu|, |\lambda_2(w_i) - \mu| \} \tag{3.29b}$$

$$\delta_\infty^-(x, s) := \| (Q_{x^{1/2}} s - \mu \iota)^- \|_\infty := \mu - \min_{i=1,\ldots,k} \{ \lambda_1(w_i), \lambda_2(w_2) \}, \tag{3.29c}$$

where the norms are special norms defined in (3.13) for the Jordan algebra. We can establish the following relations for these measures:

$$\delta_\infty^-(x, s) \le \delta_\infty(x, s) \le \delta_F(x, s). \tag{3.30}$$

The neighbourhoods are now defined as

$$\mathcal{N}(\gamma) := \{ (x, y, s) \text{ strictly feasible} : \delta(x, s) \le \gamma \mu(x, s) \}. \tag{3.31}$$

Choosing $\delta(x, s) = \delta_F(x, s)$ gives a narrow neighbourhood, while $\delta(x, s) = \delta_\infty^-(x, s)$ defines a wide one.

The results are summarized in the following theorem, taken from [3, 60].

**Theorem 3.11 (Short-step IPM for SOCO).** *Choose[17] $\gamma = 0.088$ and $\zeta = 0.06$. Assume that we have a starting point $(x^0, y^0, s^0) \in \mathcal{N}_F(\gamma)$. Compute the Newton step from the scaled Newton system (3.23). In every iteration, $\mu$ is decreased to $\left(1 - \frac{\zeta}{\sqrt{k}}\right)\mu$, i.e., $\theta = \frac{\zeta}{\sqrt{k}}$, and the stepsize is $\alpha = 1$. This algorithm finds an $\varepsilon$-optimal solution for the second order conic optimization problem (SOCO) with $k$ second order cones in at most*

$$\mathcal{O}\left(\sqrt{k}\log\frac{1}{\varepsilon}\right) \tag{3.33}$$

*iterations. The cost of one iteration depends on the sparsity structure of the coefficient matrix A. If all the data is dense then it is*

$$\mathcal{O}\left(m^3 + m^2 n + \sum_{i=1}^{k} n_i^2\right). \tag{3.34}$$

It might be surprising that the iteration complexity of the algorithm is independent of the dimensions of the cones. However, the cost of one iteration depends on the dimension of the cones.

Although this is essentially the best possible complexity result for second order cone optimization, this algorithm is not efficient enough in practice since $\theta$ is too small. Practical implementations use predictor-corrector schemes, see [67, 73, 77, 84] for more details.

Unlike the case of linear optimization, here we do not have a way to round an almost optimal interior solution to an optimal one, we have to live with approximate solutions.

### 3.4.2 Semidefinite Optimization

Interior point methods for semidefinite optimization have a very similar structure to the methods presented so far. We will apply the Newton method to the perturbed optimality conditions of semidefinite optimization.

---

[17] Any values $\gamma \in (0, 1/3)$ and $\zeta \in (0, 1)$ satisfying

$$\frac{4(\gamma^2 + \zeta^2)}{(1 - 3\gamma)^2}\left(1 - \frac{\zeta}{\sqrt{2n}}\right)^{-1} \leq \gamma \tag{3.32}$$

would work here.

The KKT optimality conditions for semidefinite optimization are:

$$\text{Tr}\left(A^{(i)}X\right) = b_i, i = 1, \ldots, m, \ X \in \mathbb{PS}^{n \times n}$$

$$\sum_{i=1}^{m} y_i A^{(i)} + S = C, \ S \in \mathbb{PS}^{n \times n} \tag{3.35}$$

$$XS = 0.$$

Again, the first four conditions ensure feasibility, while the last equation is the complementarity condition. The last equation can be written equivalently as $\text{Tr}\,(XS) = 0$. Now we perturb the complementarity condition, this way we arrive at the central path:

$$\text{Tr}\left(A^{(i)}X\right) = b_i, i = 1, \ldots, m, \ X \in \mathbb{PS}^{n \times n}$$

$$\sum_{i=1}^{m} y_i A^{(i)} + S = C, \ S \in \mathbb{PS}^{n \times n} \tag{3.36}$$

$$XS = \mu I,$$

where $I$ is the identity matrix. Now we try to apply the Newton method the same way we did for SOCO and LO, i.e., replace the variables with the updated ones and ignore the quadratic terms. This way we get:

$$\text{Tr}\left(A^{(i)}\Delta X\right) = 0, i = 1, \ldots, m$$

$$\sum_{i=1}^{m} \Delta y_i A^{(i)} + \Delta S = 0 \tag{3.37}$$

$$X\Delta S + \Delta X S = \mu I - XS.$$

We want to keep the iterates $X$ and $S$ symmetric and positive definite, thus we need $\Delta X$ and $\Delta S$ to be symmetric as well. However, solving (3.37) the displacement $\Delta X$ is typically not symmetric, simply due to the fact that the product of two symmetric matrices is not symmetric. Moreover, forcing the symmetry of $\Delta X$ by adding $\Delta X = \Delta X^T$ as a new constraint will make the problem overdetermined. Our first attempt at formulating the Newton system fails spectacularly.

Scaling techniques for semidefinite optimization

The solution to the problem we encountered at the end of the previous section is again to rewrite the optimality conditions (3.35) in an equivalent form and use that system to derive the central path. This technique is called scaling or symmetrization and there are many ways to rewrite the optimality conditions, see [82] for a thorough review. This symmetrization replaces $XS = \mu I$ in (3.36) with $\frac{1}{2}(MXS + SXM) = \mu M$, where $M$ might depend on $X$ and $S$, and can thus change from iteration to iteration. This choice defines the Monteiro-Zhang family of search directions. The new symmetrized central path equations are

$$\text{Tr}\left(A^{(i)}X\right) = b_i, i = 1, \ldots, m, \ X \in \mathbb{PS}^{n \times n}$$

$$\sum_{i=1}^{m} y_i A^{(i)} + S = C, \ S \in \mathbb{PS}^{n \times n} \tag{3.38}$$

$$MXS + SXM = \mu M,$$

and the Newton system is

$$\text{Tr}\left(A^{(i)}\Delta X\right) = 0, i = 1, \ldots, m$$

$$\sum_{i=1}^{m} \Delta y_i A^{(i)} + \Delta S = 0 \tag{3.39}$$

$$MX\Delta S + M\Delta XS + S\Delta XM + \Delta SXM = 2\mu I - MXS - SXM.$$

The solution matrices $\Delta X$ and $\Delta S$ of this system are symmetric, thus we can update the current iterates maintaining the symmetry of the matrices. Details on how to solve this system can be found in [77].

Some standard choices of the scaling matrix $M$ are (see [82] for more directions):

AHO scaling: The most natural choice, $M = I$. Unfortunately, the resulting system will have a solution only if $X$ and $S$ are in a small neighbourhood of the central path.

NT scaling: Probably the most popular choice,

$$M = S^{1/2}\left(S^{1/2}XS^{1/2}\right)^{-1/2} S^{1/2}. \tag{3.40}$$

This type of scaling has the strongest theoretical properties. Not surprisingly, most algorithmic variants use this scaling. It also facilitates the use of sparse linear algebra, see [77].

HKM scaling: In this case $M = S$ or $M = X^{-1}$. Typically, these scalings are somewhat faster to compute than the NT scaling, but certain large portions of the theory (such as [67]) are only developed for NT scaling.

Proximity measures

Let $\mu$ be defined as $\mu = \mu(X,S) := \frac{\text{Tr}(XS)}{n}$ for the rest of this section. Now we need to define some centrality measures similar to (2.55) and (3.29). The most popular choices for semidefinite optimization include

$$\delta_F(X,S) := \left\| X^{1/2}SX^{1/2} - \mu I \right\|_F = \sqrt{\sum_{i=1}^{n} \left(\lambda_i(X^{1/2}SX^{1/2}) - \mu\right)^2} \quad (3.41\text{a})$$

$$\delta_\infty(X,S) := \left\| X^{1/2}SX^{1/2} - \mu I \right\| = \max_i \left| \lambda_i(X^{1/2}SX^{1/2}) - \mu \right| \quad (3.41\text{b})$$

$$\delta_\infty^-(X,S) := \left\| \left(X^{1/2}SX^{1/2} - \mu I\right)^- \right\|_\infty$$

$$:= \max_i \left( \mu - \lambda_i(X^{1/2}SX^{1/2}) \right), \quad (3.41\text{c})$$

see [59] and the references therein for more details. For strictly feasible $X$ and $S$, these measures are zero only on the central path. Due to the properties of norms we have the following relationships:

$$\delta_\infty^-(X,S) \le \delta_\infty(X,S) \le \delta_F(X,S). \quad (3.42)$$

The neighbourhoods are defined as

$$\mathcal{N}(\gamma) := \{(X,y,S) \text{ strictly feasible} : \delta(X,S) \le \gamma\mu(X,S)\}. \quad (3.43)$$

Choosing $\delta(X,S) = \delta_F(X,S)$ gives a narrow neighbourhood, while $\delta(X,S) = \delta_\infty^-(X,S)$ defines a wide one.

A short-step interior point method

The following theorem, taken from [59], summarizes the details and the complexity of a short-step interior point algorithm for semidefinite optimization. Refer to Algorithm 1 on page 242 for the generic interior point algorithm.

**Theorem 3.12 (Short-step IPM for SDO).** *Choose*[18] $\gamma = 0.15$ *and* $\zeta = 0.13$. *Assume that we have a starting point* $(X^0, y^0, S^0) \in \mathcal{N}_F(\gamma)$. *We get the Newton step from (3.39). In every iteration, $\mu$ is decreased to* $\left(1 - \frac{\zeta}{\sqrt{n}}\right)\mu$,

---

[18] Any values $\gamma \in (0, 1/\sqrt{2})$ and $\zeta \in (0,1)$ satisfying

$$\frac{2(\gamma^2 + \zeta^2)}{(1 - \sqrt{2}\gamma)^2}\left(1 - \frac{\zeta}{\sqrt{n}}\right)^{-1} \le \gamma \quad (3.44)$$

would work here.

*i.e., $\theta = \frac{\varsigma}{\sqrt{n}}$, and the stepsize is $\alpha = 1$. This algorithm finds and $\varepsilon$-optimal solution for the semidefinite optimization problem* (SDO) *with an $n$ dimensional cone in at most*

$$\mathcal{O}\left(\sqrt{n}\log\frac{1}{\varepsilon}\right) \tag{3.45}$$

*iterations. If all the data matrices are dense[19] then the cost of one iteration is $\mathcal{O}\left(mn^3 + m^2n^2 + m^3\right)$.*

**Remark 3.13.** Depending on the magnitude of $m$ compared to $n$ any of the three terms of this expression can be dominant. The problem has $\mathcal{O}\left(n^2\right)$ variables, thus $m \leq n^2$. If $m$ is close to $n^2$ then the complexity of one iteration is $\mathcal{O}\left(n^6\right)$, while with a much smaller $m$ of order $\sqrt{n}$ the complexity is $\mathcal{O}\left(n^{3.5}\right)$.

Although this algorithmic variant is not very efficient in practice, this is still the best possible theoretical complexity result. Practical implementations usually use predictor-corrector schemes, see [77] for more details.

As we have already seen with second order conic optimization, it is not possible to obtain an exact solution to the problem. All we can get is an $\varepsilon$-optimal solution, see [68] for detailed complexity results.

## 3.5 Summary

To summarize the results about conic optimization let us go through our checklist from Section 2.3.

- We showed that the duality properties of conic optimization are slightly weaker than that of linear optimization, we need to assume strict feasibility (the interior point condition) for strong duality.
- We embedded the conic optimization problems (PCon) and (DCon) into a strictly feasible self-dual problem (HSD). From the optimal solutions of the self-dual model we can
  - derive optimal solutions for the original problem, or
  - decide primal or dual infeasibility, or
  - conclude that no optimal primal-dual solution pair exists with zero duality gap.
- If a strictly feasible solution exists (either in the original problem or in the self-dual model) then
  - the central path exists;
  - the central path converges to a maximally (not necessarily strictly) complementary solution;

---
[19] The complexity can be greatly reduced by exploiting the sparsity of the data, see [77] and the references therein.

– the limit point of the central path is not necessarily the analytic center of the optimal set (only if the problem has a strictly complementary solution).

- Due to the lack of a rounding scheme we cannot get exact optimal solutions from our algorithm and thus cannot use the algorithm to get exact solutions.

## 3.6  *Barrier Functions in Conic Optimization

Interior point methods for conic optimization can also be introduced through barrier functions in a similar fashion as we did in Section 2.5 for linear optimization. However, the barrier functions for conic optimization are more complicated and the discussion is a lot more technical, much less intuitive.

A suitable logarithmic barrier function for a second order cone is

$$\phi(x) = - \ln \left( x_1^2 - \|x_{2:n}\|_2^2 \right) = - \ln \lambda_1(x) - \ln \lambda_2(x), \qquad (3.46)$$

assuming that $x$ is in the interior of the second order cone. We can see that when the point $x$ is getting close to the boundary, then at least one of its eigenvalues is getting close to 0 and $\phi(x)$ is diverging to infinity. For the optimality conditions of this problem we will need the derivatives of the barrier function $\phi(x)$:

$$\nabla \phi(x) = -2 \frac{(x_1; -x_{2:n})^T}{x_1^2 - \|x_{2:n}\|_2^2} = -2 \left( x^{-1} \right)^T, \qquad (3.47)$$

where the inverse is taken in the Jordan algebra. The multiplier 2 appears due to the differentiation of a quadratic function, and it will also appear in the central path equations (3.19).

For the cone of positive semidefinite matrices we can use the barrier function

$$\phi(X) = - \ln \det \left( X \right) = - \sum_{i=1}^{n} \ln \lambda_i(X), \qquad (3.48)$$

which has the derivative

$$\nabla \phi(X) = - \left( X^{-1} \right)^T. \qquad (3.49)$$

Having these functions we can rewrite the conic optimization problem (PCon) as a linearly constrained problem

$$\min c^T x + \mu \phi(x)$$
$$Ax = b, \qquad \text{(PCon-Barrier)}$$

where $\mu \geq 0$. The KKT optimality conditions for this problem are the same systems as (3.19) and (3.36) defining the central path, thus the barrier approach again provides an alternative description of the central path. For more details on the barrier approach for conic optimization see, e.g., [4].

# 4  Interior Point Methods for Nonlinear Optimization

First we will solve the nonlinear optimization problem by converting it into a nonlinear complementarity problem. We will present an interior point algorithm for this problem, analyze its properties and discuss conditions for polynomial complexity. Then we present a direct approach of handling nonlinear inequality constraints using barrier functions and introduce the concept of self-concordant barrier functions.

## 4.1  Nonlinear Optimization as a Complementarity Problem

Let us consider the nonlinear optimization problem in the form

$$\min f(x) \qquad\qquad (\text{NLO})$$
$$g_j(x) \leq 0,\ j = 1,\dots,m$$
$$x \geq 0,$$

where $x \in \mathbb{R}^n$ and $f, g_j : \mathbb{R}^n \rightarrow \mathbb{R}$, are continuously differentiable convex functions. We will use the notation $g(x) = (g_1(x); \dots; g_m(x))$. The KKT optimality conditions for this problem are

$$\nabla f(x) + \sum_{i=1}^{m} \nabla g_j(x) y_j \geq 0$$
$$g_j(x) \leq 0$$
$$x, y \geq 0 \qquad\qquad (4.1)$$
$$\left( \nabla f(x) + \sum_{i=1}^{m} \nabla g_j(x) y_j \right)^T x = 0$$
$$g(x)^T y = 0.$$

Introducing

$$L(x, y) := f(x) + g(x)^T y \tag{4.2a}$$

$$F(\bar{x}) := \begin{pmatrix} \nabla_x L(x, y) \\ -g(x) \end{pmatrix} \tag{4.2b}$$

$$\bar{x} := \begin{pmatrix} x \\ y \end{pmatrix} \tag{4.2c}$$

we can write the nonlinear optimization problem as an equivalent nonlinear complementarity problem:

$$
\begin{aligned}
F(\bar{x}) - \bar{s} &= 0 \\
\bar{x}, \bar{s} &\geq 0 \\
\bar{x}\bar{s} &= 0.
\end{aligned}
\tag{4.3}
$$

## 4.2 Interior Point Methods for Nonlinear Complementarity Problems

In this section we derive an algorithm for this problem based on [70].

Let us now simplify the notation and focus on the nonlinear complementarity problem in the following form:

$$
\begin{aligned}
F(x) - s &= 0 \\
x, s &\geq 0 \\
xs &= 0,
\end{aligned}
\tag{NCP}
$$

where $x, s \in \mathbb{R}^n$, $F : \mathbb{R}^n \to \mathbb{R}^n$. After perturbing the third equation (the complementarity condition) we receive the equations for the central path. Note that the existence of the central path requires stronger assumptions than in the linear or conic case, see [25] and the references therein for details.

$$
\begin{aligned}
F(x) - s &= 0 \\
x, s &\geq 0 \\
xs &= \mu e,
\end{aligned}
\tag{4.4}
$$

where $\mu \geq 0$ and $e$ is the all one vector. We use the Newton method to solve this system, the corresponding equation for the Newton step is:

$$
\begin{aligned}
F'(x)\Delta x - \Delta s &= 0 \\
s\Delta x + x\Delta s &= \mu e - xs,
\end{aligned}
\tag{4.5}
$$

where $F'(x)$ is the Jacobian of $F(x)$. In general, the point $x + \Delta x$ is not feasible, i.e., $F(x + \Delta x) \geq 0$ and/or $x + \Delta x \geq 0$ is not satisfied, thus we will need to use a stepsize $\alpha > 0$ and consider a strictly feasible $x(\alpha) := x + \alpha \Delta x$ as the new (primal) iterate. The new dual iterate will be defined as $s(\alpha) = F(x + \alpha \Delta x)$. Note that unlike in linear and conic optimization, here $s(\alpha) \neq s + \alpha \Delta s$.

The algorithm is structured analogously to the generic structure of IPMs presented as Algorithm 1. All we need to do is to specify the details: the proximity measure $\delta(x, s)$, the choice of stepsize $\alpha$ and the update strategy of $\mu$.

## The proximity measure

There are several variants in existing implementations. The most important ones are

$$\delta_2(x, s) = \|xs - \mu e\|_2 \tag{4.6a}$$

$$\delta_\infty(x, s) = \|xs - \mu e\|_\infty \tag{4.6b}$$

$$\delta_\infty^-(x, s) = \left\|(xs - \mu e)^-\right\|_\infty := \max_i (\mu - x_i s_i), \tag{4.6c}$$

where $\mu = x^T s / n$. This enables us to define a neighbourhood of the central path:

$$\mathcal{N}(\gamma) = \{(x, s) \text{ strictly feasible} : \delta(x, s) \leq \gamma \mu\}, \tag{4.7}$$

where $\gamma \in (0, 1)$.

## Choosing the stepsize $\alpha$

For nonlinear optimization problems the stepsize is chosen using a line-search. We want to get a large step but stay away from the boundary of the feasible set. Let $\alpha_{\max}$ be the maximum feasible stepsize, i.e., the maximal value of $\alpha$ such that $x + \alpha \Delta x \geq 0$ and $F(x + \alpha \Delta x) \geq 0$.

We are looking for a stepsize $\alpha < \alpha_{\max}$ such that

• $(x(\alpha), s(\alpha))$ is inside the neighbourhood $\mathcal{N}(\gamma)$, and
• the complementarity gap $x(\alpha)^T F(x(\alpha))$ is minimized.

In some practical implementations $\alpha = 0.95\alpha_{\max}$ (or $\alpha = 0.99\alpha_{\max}$) is used as the stepsize, enhanced with a safeguarded backtracking strategy. The extra difficulty with general nonlinear optimization problems is that the line-search can get stuck in a local minimum, thus some globalization scheme is needed. Such ideas are implemented in the IPOPT solver [90].

Updating $\mu$

Usually we try to decrease $\mu$ at a superlinear rate, if possible. In short-step methods, $\mu$ is changed to $\mu\left(1 - \frac{\zeta}{\sqrt{n}}\right)$ after every iteration, i.e., $\theta = \frac{\zeta}{\sqrt{n}}$ in the general IPM framework on page 242, $\zeta$ is a constant depending on the neighbourhood parameter $\gamma$ and the smoothness of the mapping $F$. The smoothness is quantified with a Lipschitz constant $L$ in Assumption 4.1.

## 4.2.1  Complexity of IPM for NCP

Now assume that the Jacobian $F'(x)$ of $F(x)$ is a positive semidefinite matrix for all values of $x$. Then problem (NCP) is called a monotone nonlinear complementarity problem. If the original nonlinear optimization problem (NLO) is convex, then this always holds. To be able to prove polynomial convergence of IPMs for convex nonlinear problems we need to control the difference between $s(\alpha) = F(x(\alpha))$ and $s + \alpha \Delta s$. We assume a smoothness condition [8]:

**Assumption 4.1** *Consider the nonlinear complementarity problem* (NCP). *Assume that $F(x)$ satisfies the scaled Lipschitz property, i.e., for any $x > 0$, $h \in \mathbb{R}^n$, satisfying $|h_i/x_i| \leq \beta < 1$, there exists a constant $L(\beta) > 1$ such that*

$$\|x \cdot (F(x+h) - F(x) - F'(x)h)\|_1 \leq L(\beta)h^T F'(x)h. \qquad (4.8)$$

The complexity result is summarized in the following theorem:

**Theorem 4.2 (Complexity of short-step IPM for monotone NCP).**
*Assume that $F(x)$ is a monotone mapping satisfying the scaled Lipschitz property. The proximity measure is based on the 2-norm and assume that a strictly feasible starting point in $\mathcal{N}_2(\gamma)$ with $x^T s/n \leq 1$ is available.*

*The Newton step is computed from (4.5). If $\gamma$ and $\zeta$ are chosen properly, then $\alpha = 1$ is a valid stepsize, i.e., no line-search is necessary.*

*This algorithm yields an $\varepsilon$-complementary solution for* (NCP) *in at most $\mathcal{O}\left(\sqrt{n}L\log(1/\varepsilon)\right)$ iterations.*

Explicit forms of the constants and detailed proofs can be found in [8]. The cost of one iteration depends on the actual form of $F(x)$. It includes computing the Jacobian of $F$ at every iteration and solving an $n \times n$ linear system. When full Newton steps are not possible,[20] then finding $\alpha_{\max}$ and determining the stepsize $\alpha$ with a line-search are significant extra costs.

---

[20] This is the typical situation, as in practice we rarely have explicit information on the Lipschitz constant $L$.

## 4.3  Initialization by Embedding

Interior point methods require a strictly feasible starting point, but for nonlinear optimization problems even finding is feasible point is quite challenging. Moreover, if the original problem has nonlinear equality constraints which are modelled as two inequalities then the resulting system will not have an interior point solution. To remedy these problems we use a homogeneous embedding, similar to the ones presented in Section 2.2 and Section 3.3. Consider the following system [7, 8, 96]:

$$\nu F(x/\nu) - s = 0 \qquad\qquad \text{(NCP-H)}$$
$$x^T F(x/\nu) - \rho = 0$$
$$x, s, \nu, \rho \geq 0$$
$$xs = 0$$
$$\nu\rho = 0.$$

This is a nonlinear complementarity problem similar to (NCP). The properties of the homogenized system are summarized in the following theorem.

**Theorem 4.3.** *Consider the nonlinear complementarity problem (NCP) and its homogenized version (NCP-H). The following results hold:*

1. *The homogenized problem (NCP-H) is an (NCP).*
2. *If the original (NCP) is monotone then the homogenized (NCP) is monotone, too, thus we can use the algorithm presented in Section 4.2.*
3. *If the homogenized (NCP) has a solution $(x, s, \nu, \rho)$ with $\nu > 0$ then $(x/\nu, s/\nu)$ is a solution for the original system.*
4. *If $\nu = 0$ for all the solutions of (NCP-H) then the original system (NCP) does not have a solution.*

## 4.4  *The Barrier Method

An alternative way to introduce interior point methods for nonlinear optimization is to use the barrier technique already presented in Section 2.5 and Section 3.6. The basic idea is to place the nonlinear inequalities in the objective function inside a barrier function. Most barrier function are based on logarithmic functions.

The nonlinear optimization problem (NLO) can be rewritten as

$$\min\ f(x) - \mu \sum_{j=1}^{m} \ln(-g_j(x)) - \mu \sum_{i=1}^{n} \ln(x_i). \qquad (4.9)$$

If $x_i$ or $-g_j(x)$ gets close to 0, then the objective function grows to infinity. Our goal is to solve this barrier problem approximately for a given $\mu$, then decrease $\mu$ and resolve the problem. If $\mu$ is decreased at the right rate and the approximate solutions are good enough, then this method will converge to an optimal solution of the nonlinear optimization problem. See [63] for details on the barrier approach for nonlinear optimization.

# 5 Existing Software Implementations

After their early discovery in the 1950s, by the end of the 1960s IPMs were sidelined because their efficient implementation was quite problematic. As IPMs are based on Newton steps, they require significantly more memory than first order methods. Computers at the time had very limited memory. Furthermore, the Newton system is inherently becoming ill-conditioned as the iterates approach the optimal solution set. Double precision floating point arithmetic and regularization techniques were in their very early stage at that time. Solving large scale linear systems would have required sparse linear algebra routines, which were also unavailable. Most of these difficulties have been solved by now and so IPMs have become a standard choice in many branches of optimization.

In the following we give an overview of existing implementations of interior point methods. See Table 1 for a quick comparison their features. The web site of the solvers and the bibliographic references are listed in Table 2.

**Table 1** A comparison of existing implementations of interior point methods

| Solver | License | LO | SOCO | SDO | NLO |
|---|---|---|---|---|---|
| CLP barrier | open source | ✓ | QO | | |
| LIPSOL | open source | ✓ | | | |
| GLPK ipm | open source | ✓ | | | |
| HOPDM | commercial | ✓ | QO | | ✓ |
| MOSEK barrier | commercial | ✓ | ✓ | | ✓ |
| CPLEX barrier | commercial | ✓ | ✓[21] | | |
| XPRESS barrier | commercial | ✓ | QO | | |
| CSDP | open source | ✓ | | ✓ | |
| SDPA | open source | ✓ | | ✓ | |
| SDPT3 | open source | ✓ | ✓ | ✓ | |
| SeDuMi | open source | ✓ | ✓ | ✓ | |
| IPOPT | open source | ✓[22] | ✓[23] | | ✓ |
| KNITRO | commercial | ✓[22] | ✓[23] | | ✓ |
| LOQO | commercial | ✓[22] | ✓[23] | | ✓ |

[21] CPLEX solves second-order conic problems by treating them as special (nonconvex) quadratically constrained optimization problems.

[22] In theory all NLO solvers can solve linear optimization problems, but their efficiency and accuracy is worse than that of dedicated LO solvers.

[23] LOQO does solve second-order conic optimization problems but it uses a different approach. It handles the constraint $x_1 - \|x_{2:n}\|_2 \geq 0$ as a general nonlinear constraint, with

**Table 2** Availability of implementations of IPMs

| | |
|---|---|
| CLP | [24], http://www.coin-or.org/Clp |
| LIPSOL | [100], http://www.caam.rice.edu/~zhang/lipsol |
| GLPK | [28], http://www.gnu.org/software/glpk |
| HOPDM | [15], http://www.maths.ed.ac.uk/~gondzio/software/hopdm.html |
| MOSEK | [6], http://www.mosek.com |
| CPLEX | [12], http://www.ilog.com |
| XPRESS-MP | [41], http://www.dashoptimization.com |
| CSDP | [13], http://projects.coin-or.org/Csdp |
| SDPA | [95], http://homepage.mac.com/klabtitech/sdpa-homepage |
| SDPT3 | [86], http://www.math.nus.edu.sg/~mattohkc/sdpt3.html |
| SeDuMi | [76], http://sedumi.ie.lehigh.edu/ |
| IPOPT | [90], http://projects.coin-or.org/Ipopt |
| KNITRO | [14], http://www.ziena.com/knitro.htm |
| LOQO | [89], http://www.princeton.edu/~rvdb/loqo |

## 5.1 Linear Optimization

Interior point algorithms are the method of choice for large scale, sparse, degenerate linear optimization problems. Solvers using the simplex method are usually not competitive on those problems due to the large number of pivots needed to get to an optimal solution. However, interior point methods still do not have an efficient warm start strategy, something simplex based methods can do naturally, so their use for branch-and-bound type algorithms is limited.

IPMs have also been implemented in leading commercial packages, usually together with a simplex based solver. Comprehensive surveys of implementation strategies of IPMs can be found in, e.g., [5, 36]. For a review on the strengths and weaknesses of interior point methods versus variants of the simplex method see [43].

Linear optimization problems with up to a million variables can be solved routinely on a modern PC. On larger parallel architectures, linear and quadratic problems with billions of variables have been solved [34].

## 5.2 Conic Optimization

Interior point methods are practically the only choice for semidefinite optimization, most of the existing general purpose solvers fall into this category, only PENSDP[24] being a notable exception. Also, PENSDP is the only solver

---

some extra care taken due to the nondifferentiability of this form. In a similar way, other IPM based NLO solvers can solve SOCO problems in principle.

[24] [49], http://www.penopt.com/pensdp.html

that can handle nonlinear semidefinite problems and it is also the only commercial SDO solver (at least at the time this chapter is written).

The implementation of IPMs for conic optimization is more complicated than that for linear optimization, see [13, 77, 83] for more details.

Unfortunately, commercial modelling languages do not support SDO, thus limit its use in the commercial sector. Second order conic optimization is in a slightly better situation, since it is easily formulated, but there are only very few specialized solvers available. Only very few solvers can solve problems including both second order and semidefinite constraints, currently only SeDuMi and SDPT3. Both of these packages run under Matlab.

There are two open source modelling languages that support conic optimization: Yalmip[25] and CVX[26]. Both of these packages are written in Matlab.

## 5.3  Nonlinear Optimization

There are literally hundreds of solvers available for nonlinear optimization and only a small fraction of those use interior point methods. On the other hand, arguably, the most powerful, robust solvers are actually based on interior point methods, IPOPT, KNITRO and LOQO being the most successful ones. These are all general use nonlinear optimization solvers, they can handle nonconvex problems as well (yielding a locally optimal solution). Some codes have been specialized for optimization problems with complementarity constraints. The best known variant is IPOPT-C [71], an extension of IPOPT.

The implementation of these methods poses further challenges, see [90] for details.

## 6  Some Open Questions

Interior point algorithms have proved to be very successful methods for linear and nonlinear optimization, especially for large-scale problems. The "interior-point revolution" [92] has completely changed the field of optimization. By today, the fundamental theoretical questions regarding complexity and convergence of interior point methods have been addressed, see also [62] for a recent survey. Most importantly, we know that results about the iteration complexity of these methods cannot be improved further, see [21] for details on the worst-case complexity of interior point methods.

---

[25] [50], http://control.ee.ethz.ch/~joloef/yalmip.php
[26] [38, 39], http://www.stanford.edu/~boyd/cvx

## 6.1  Numerical Behaviour

Current research is focusing on efficient implementations of the methods. Due to the ill-conditioned nature of the Newton system in the core of IP methods, people are looking for ways to improve the numerical behaviour of the implementations. Some notable results are included in [31, 78, 79]. Most of these ideas are implemented in leading interior point solvers.

## 6.2  Rounding Procedures

Iterates of interior point methods stay inside the set of feasible solutions, while with a linear objective, the optimal solution is on the boundary of the feasible set. Rounding procedures try to jump from the last iterate of the IPM to an optimal solution on the boundary. This theory has been well-developed for linear optimization and linear complementarity problems [56, 72]. For conic optimization, the mere existence of such a method is an open question. In general we cannot expect to be able to get an exact optimal solution, but under special circumstances we might be able to get one.

## 6.3  Special Structures

Exploiting sparsity has always been one of the easiest ways to improve the performance of an optimization algorithm. With the availability of efficient sparse linear algebra libraries and matrix factorization routines, general (unstructured) sparsity seems to have been taken care of. On the other hand, sparse problems containing some dense parts pose a different challenge [30]. Moreover, even very sparse semidefinite optimization problems lead to a fully dense Newton system, which puts a limit on the size of the problems that can be solved.

There are several other special types of structures that cannot be fully exploited by current implementations of interior point methods. This limits the size of the problems that can be solved with IPMs. At the same time it offers a wide open area of further research.

## 6.4  Warmstarting

A serious deficiency of interior point methods is the lack of an efficient warm-starting scheme. The purpose of a warm-start scheme is to significantly reduce the number of iterations needed to reoptimize the problem after changes to the data (constraints are added or deleted, numbers are changed). Despite numerous attempts (see [33, 35, 98]), none of the methods are particularly successful.

If the change in the problem data is small enough then simplex based methods can very quickly find a new optimal solution. If the change is large (hundreds or thousands of new constraints are added) then interior point methods have a slight edge over first order methods.

## 6.5  Parallelization

With the general availability of inexpensive multiple core workstations and distributed computing environments, parallelization of optimization algorithms is more important than ever. Most developers are working on a parallelized version of their codes. Some success stories are reported in [13, 34, 44, 61].

**Acknowledgements** The authors were supported by the NSERC Discovery Grant #5-48923, the Canada Research Chair Program and a MITACS project. Research was done while both authors working at the School of Computational Engineering and Science, McMaster University, Hamilton, Ontario, Canada.

# References

1. F. Alizadeh. *Combinatorial optimization with interior point methods and semidefinite matrices*, Ph.D. thesis, University of Minnesota, Minneapolis, USA, 1991.
2. F. Alizadeh. *Interior point methods in semidefinite programming with applications to combinatorial optimization*, SIAM Journal on Optimization, 5(1),13–51, 1995.
3. F. Alizadeh and D. Goldfarb. *Second-order cone programming*, Mathematical Programming, Series B, 95,3–51, 2002.
4. F. Alizadeh and S. H. Schmieta. *Symmetric cones, potential reduction methods*, In H. Wolkowicz, R. Saigal, and L. Vandenberghe, editors, *Handbook of Semidefinite Programming: Theory, Algorithms, and Applications*, chapter 8, pages 195–233. Kluwer Academic Publishers, 2000.
5. E. D. Andersen, J. Gondzio, Cs. Mészáros, and X. Xu. *Implementation of interior point methods for large scale linear programming*, In T. Terlaky, editor, *Interior Point Methods of Mathematical Programming*, pages 189–252. Kluwer Academic Publishers, Dordrecht, The Netherlands, 1996.
6. E. D. Andersen, B. Jensen, R. Sandvik, and U. Worsøe. *The improvements in MOSEK version 5.*, Technical report 1-2007, MOSEK ApS, Fruebjergvej 3 Box 16, 2100 Copenhagen, Denmark, 2007.
7. E. D. Andersen and Y. Ye. *A computational study of the homogeneous algorithm for large-scale convex optimization*, Computational Optimization and Applications, 10(3), 243–280, 1998.
8. E. D. Andersen and Y. Ye. *On a homogeneous algorithm for the monotone complementarity problem*, Mathematical Programming, Series A, 84(2),375–399, 1999.

9. K. Anstreicher. *Large step volumetric potential reduction algorithms for linear programming*, Annals of Operations Research, 62, 521–538, 1996.

10. K. Anstreicher. *Volumetric path following algorithms for linear programming*, Mathematical Programming, 76, 245–263, 1997.

11. A. Ben-Tal and A. Nemirovski *Lectures on Modern Convex Optimization: Analysis, Algorithms, and Engineering Applications*, MPS-SIAM Series on Optimization. SIAM, Philadelphia, PA, 2001.

12. R. E. Bixby. *Solving real-world linear programs: A decade and more of progress*, Operations Research, 50(1), 3–15, 2002.

13. B. Borchers and J. G. Young. *Implementation of a primal-dual method for SDP on a shared memory parallel architecture*, Computational Optimization and Applications, 37(3), 355–369, 2007.

14. R. Byrd, J. Nocedal, and R. Waltz. *KNITRO: An integrated package for nonlinear optimization*, In G. Di Pillo and M. Roma, editors, *Large-Scale Nonlinear Optimization*, volume 83 of *Nonconvex Optimization and Its Applications*. Springer, 2006.

15. M. Colombo and J. Gondzio. *Further development of multiple centrality correctors for interior point methods*, Computational Optimization and Applications, 41(3), 277–305, 2008.

16. E. de Klerk, *Aspects of Semidefinite Programming: Interior Point Algorithms and Selected Applications*, Kluwer Academic Publichers, Dordrecht, The Netherlands, 2002.

17. E. de Klerk, C. Roos, and T. Terlaky. *Semi-definite problems in truss topology optimization*, Technical Report 95-128, Faculty of Technical Mathematics and Informatics, T.U. Delft, The Netherlands, 1995.

18. E. de Klerk, C. Roos, and T. Terlaky. *Initialization in semidefinite programming via a self-dual, skew-symmetric embedding*, OR Letters, 20, 213–221, 1997.

19. E. de Klerk, C. Roos, and T. Terlaky. *Infeasible-start semidefinite programming algorithms via self-dual embeddings*. In P. M. Pardalos and H. Wolkowicz, editors, *Topics in Semidefinite and Interior Point Methods,*, volume 18 of *Fields Institute Communications*, pages 215–236. AMS, Providence, RI, 1998.

20. E. de Klerk, C. Roos, and T. Terlaky. *On primal-dual path-following algorithms for semidefinite programming*, In F. Gianessi, S. Komlósi, and T. Rapcsák, editors, *New Trends in Mathematical Programming*, pages 137–157. Kluwer Academic Publishers, Dordrecht, The Netherlands, 1998.

21. A. Deza, E. Nematollahi, and T. Terlaky. *How good are interior point methods? Klee-Minty cubes tighten iteration-complexity bounds*, Mathematical Programming, 113, 1–14, 2008.

22. J. Faraut and A. Korányi *Analysis on Symmetric Cones*, Oxford Mathematical Monographs. Oxford University Press, 1994.

23. A. V. Fiacco and G. P. McCormick *Nonlinear Programming: Sequential Unconstrained Minimization Techniques*, John Wiley and Sons, New York, 1968. Reprint: Vol. 4. *SIAM Classics in Applied Mathematics*, SIAM Publications, Philadelphia, USA, 1990.

24. J. Forrest, D. de le Nuez, and R. Lougee-Heimer *CLP User Guide*, IBM Corporation, 2004.

25. A. Forsgren, Ph. E. Gill, and M. H. Wright. *Interior methods for nonlinear optimization*, SIAM Review, 44(4), 525–597, 2002.

26. R. M. Freund. *On the behavior of the homogeneous self-dual model for conic convex optimization*, Mathematical Programming, 106(3), 527–545, 2006.

27. R. Frisch. *The logarithmic potential method for solving linear programming problems*, Memorandum, University Institute of Economics, Oslo, Norway, 1955.

28. *GNU Linear Programming Kit Reference Manual, Version 4.28*, 2008.

29. M. X. Goemans and D. P. Williamson. *Improved approximation algorithms for maximum cut and satisfiability problems using semidefinite programming*, Journal of the ACM, 42, 1115–1145, 1995.

30. D. Goldfarb and K. Scheinberg. *A product-form cholesky factorization method for handling dense columns in interior point methods for linear programming*, Mathematical Programming, 99(1), 1–34, 2004.

31. D. Goldfarb and K. Scheinberg. *Product-form cholesky factorization in interior point methods for second-order cone programming*, Mathematical Programming, 103(1), 153–179, 2005.

32. A. J. Goldman and A. W. Tucker. *Theory of linear programming*, In H. W. Kuhn and A. W. Tucker, editors, *Linear Inequalities and Related Systems*, number 38 in Annals of Mathematical Studies, pages 53–97. Princeton University Press, Princeton, New Jersey, 1956.

33. J. Gondzio. *Warm start of the primal-dual method applied in the cutting plane scheme*, Mathematical Programming, 83(1), 125–143, 1998.

34. J. Gondzio and A. Grothey. *Direct solution of linear systems of size $10^9$ arising in optimization with interior point methods*, In R. Wyrzykowski, J. Dongarra, N. Meyer, and J. Wasniewski, editors, *Parallel Processing and Applied Mathematics*, number 3911 in Lecture Notes in Computer Science, pages 513–525. Springer-Verlag, Berlin, 2006.

35. J. Gondzio and A. Grothey. *A new unblocking technique to warmstart interior point methods based on sensitivity analysis*, SIAM Journal on Optimization, 3, 1184–1210, 2008.

36. J. Gondzio and T. Terlaky. *A computational view of interior point methods for linear programming*, In J. E. Beasley, editor, *Advances in Linear and Integer Programming*, pages 103–185. Oxford University Press, Oxford, 1996.

37. I. S. Gradshteyn and I. M. Ryzhik *Tables of Integrals, Series, and Products*, Academic Press, San Diego, CA, 6th edition, 2000.

38. M. Grant and S. Boyd. *CVX: Matlab software for disciplined convex programming*, Web page and software, http://stanford.edu/~boyd/cvx, 2008.

39. M. Grant and S. Boyd. *Graph implementations for nonsmooth convex programs*, In V. Blondel, S. Boyd, and H. Kimura, editors, *Recent Advances in Learning and Control (a tribute to M. Vidyasagar)*. Springer, 95–110, 2008.

40. M. Grötschel, L. Lovász, and A. Schrijver *Geometric Algorithms and Combinatorial Optimization*. Springer Verlag, 1988.

41. Ch. Guéret, Ch. Prins, and M. Sevaux *Applications of optimization with Xpress-MP*, Dash Optimization, 2002. Translated and revised by Susanne Heipcke.

42. D. den Hertog *Interior Point Approach to Linear, Quadratic and Convex Programming*, volume 277 of *Mathematics and its Applications*. Kluwer Academic Publishers, Dordrecht, The Netherlands, 1994.

43. T. Illés and T. Terlaky. *Pivot versus interior point methods: Pros and cons*, European Journal of Operational Research, 140(2), 170–190, 2002.

44. I. D. Ivanov and E. de Klerk. *Parallel implementation of a semidefinite programming solver based on CSDP in a distributed memory cluster*, Optimization Methods and Software, 25(3), 405–420, 2010.

45. B. Jansen, *Interior Point Techniques in Optimization. Complexity, Sensitivity and Algorithms*, volume 6 of *Applied Optimization*. Kluwer Academic Publishers, Dordrecht, The Netherlands, 1996.

46. B. Jansen, J. J. de Jong, C. Roos, and T. Terlaky. *Sensitivity analysis in linear programming: just be careful!*, European Journal of Operations Research, 101(1), 15–28, 1997.

47. B. Jansen, C. Roos, and T. Terlaky. *The theory of linear programming: Skew symmetric self-dual problems and the central path*, Optimization, 29, 225–233, 1994.

48. N. K. Karmarkar. *A new polynomial-time algorithm for linear programming*, Combinatorica, 4, 373–395, 1984.

49. M. Kočvara and M. Stingl *PENSDP Users Guide (Version 2.2)*, PENOPT GbR, 2006.

50. J. Löfberg. *YALMIP: a toolbox for modeling and optimization in MATLAB*, In *Proceedings of the 2004 IEEE International Symposium on Computer Aided Control Systems Design*, pages 284–289, 2004.

51. L. Lovász and A. Schrijver. *Cones of matrices and setfunctions, and 0-1 optimization*, SIAM Journal on Optimization, 1(2), 166–190, 1991.

52. Z.-Q. Luo, J. F. Sturm, and S. Zhang. *Conic linear programming and self-dual embedding*, Optimization Methods and Software, 14, 169–218, 2000.

53. K. A. McShane, C. L. Monma, and D. F. Shanno. *An implementation of a primal-dual interior point method for linear programming*, ORSA Journal on Computing, 1, 70–83, 1989.

54. N. Megiddo. *Pathways to the optimal set in linear programming*, In N. Megiddo, editor, *Progress in Mathematical Programming: Interior Point and Related Methods*, pages 131–158. Springer Verlag, New York, 1989. Identical version in: *Proceedings of the 6th Mathematical Programming Symposium of Japan, Nagoya, Japan*, pages 1–35, 1986.

55. S. Mehrotra. *On the implementation of a (primal-dual) interior point method*, SIAM Journal on Optimization, 2(4), 575–601, 1992.

56. S. Mehrotra and Y. Ye. *On finding the optimal facet of linear programs*, Mathematical Programming, 62, 497–515, 1993.

57. R. D. C. Monteiro. *Primal-dual path-following algorithms for semidefinite programming*, SIAM Journal on Optimization, 7, 663–678, 1997.

58. R. D. C. Monteiro and S. Mehrotra. *A general parametric analysis approach and its implications to sensitivity analysis in interior point methods*, Mathematical Programming, 72, 65–82, 1996.

59. R. D. C. Monteiro and M. J. Todd. *Path-following methods*. In H. Wolkowicz, R. Saigal, and L. Vandenberghe, editors, *Handbook of Semidefinite Programming: Theory, Algorithms, and Applications*, chapter 10, pages 267–306. Kluwer Academic Publishers, 2000.

60. R. D. C. Monteiro and T. Tsuchiya. *Polynomial convergence of primal-dual algorithms for the second-order cone program based on the MZ-family of directions*, Mathematical Programming, 88(1), 61–83, 2000.

61. K. Nakata, M. Yamashita, K. Fujisawa, and M. Kojima. *A parallel primal-dual interior-point method for semidefinite programs using positive definite matrix completion*, Parallel Computing, 32, 24–43, 2006.

62. A. S. Nemirovski and M. J. Todd. *Interior-point methods for optimization*, Acta Numerica, 17, 191–234, 2008.

63. Y. E. Nesterov and A. Nemirovski *Interior-Point Polynomial Algorithms in Convex Programming*, volume 13 of *SIAM Studies in Applied Mathematics*. SIAM Publications, Philadelphia, PA, 1994.

64. Y. E. Nesterov and M. J. Todd. *Self-scaled barriers and interior-point methods for convex programming*, Mathematics of Operations Research, 22(1), 1–42, 1997.

65. M. L. Overton. *On minimizing the maximum eigenvalue of a symmetric matrix*, SIAM Journal on Matrix Analysis and Applications, 9(2), 256–268, 1988.

66. J. Peng, C. Roos, and T. Terlaky. *New complexity analysis of the primal-dual Newton method for linear optimization*, Annals of Operations Research, 99, 23–39, 2000.

67. J. Peng, C. Roos, and T. Terlaky *Self-Regularity: A New Paradigm for Primal-Dual Interior-Point Algorithms*. Princeton University Press, Princeton, NJ, 2002.

68. L. Porkoláb and L. Khachiyan. *On the complexity of semidefinite programs*, Journal of Global Optimization, 10(4), 351–365, 1997.

69. F. Potra and R. Sheng. *On homogeneous interior-point algorithms for semidefinite programming*, Optimization Methods and Software, 9(3), 161–184, 1998.

70. F. Potra and Y. Ye. *Interior-point methods for nonlinear complementarity problems*, Journal of Optimization Theory and Applications, 68, 617–642, 1996.

71. A. U. Raghunathan and L. T. Biegler. *An interior point method for mathematical programs with complementarity constraints (MPCCs)*, SIAM Journal on Optimization, 15(3), 720–750, 2005.
72. C. Roos, T. Terlaky, and J.-Ph. Vial *Theory and Algorithms for Linear Optimization. An Interior Approach*, Springer, New York, USA, 2nd edition, 2006.
73. S. H. Schmieta and F. Alizadeh. *Associative and Jordan algebras, and polynomial time interior-point algorithms*, Mathematics of Operations Research, 26(3), 543–564, 2001.
74. Gy. Sonnevend. *An "analytic center" for polyhedrons and new classes of global algorithms for linear (smooth, convex) programming*, In A. Prékopa, J. Szelezsán, and B. Strazicky, editors, *System Modeling and Optimization: Proceedings of the 12th IFIP-Conference held in Budapest, Hungary, September 1985*, volume 84 of *Lecture Notes in Control and Information Sciences*, pages 866–876. Springer Verlag, Berlin, West Germany, 1986.
75. J. F. Sturm. *Primal-dual interior point approach to semidefinite programming*, In J. B. G. Frenk, C. Roos, T. Terlaky, and S. Zhang, editors, *High Performance Optimization*. Kluwer Academic Publishers, Dordrecht, The Netherlands, 1999.
76. J. F. Sturm. *Using SeDuMi 1.02, a Matlab toolbox for optimization over symmetric cones*, Optimization Methods and Software, 11-12, 625–653, 1999.
77. J. F. Sturm. *Implementation of interior point methods for mixed semidefinite and second order cone optimization problems*, Optimization Methods and Software, 17(6), 1105–1154, 2002.
78. J. F. Sturm. *Avoiding numerical cancellation in the interior point method for solving semidefinite programs*, Mathematical Programming, 95(2), 219–247, 2003.
79. J. F. Sturm and S. Zhang. *An interior point method, based on rank-1 updates, for linear programming*, Mathematical Programming, 81, 77–87, 1998.
80. T. Terlaky, editor *Interior Point Methods of Mathematical Programming*, volume 5 of *Applied Optimization*, Kluwer Academic Publishers, Dordrecht, The Netherlands, 1996.
81. T. Terlaky. *An easy way to teach interior-point methods*, European Journal of Operational Research, 130(1), 1–19, 2001.
82. M. J. Todd. *A study of search directions in primal-dual interior-point methods for semidefinite programming*, Optimization Methods and Software, 11, 1–46, 1999.
83. K. C. Toh, R. H. Tütüncü, and M. J. Todd. *On the implementation of SDPT3 (version 3.1) – a Matlab software package for semidefinite-quadratic-linear programming*, In *Proceedings of the IEEE Conference on Computer-Aided Control System Design*, 2004.
84. T. Tsuchiya. *A convergence analysis of the scaling-invariant primal-dual path-following algorithms for second-order cone programming*, Optimization Methods and Software, 11, 141–182, 1999.
85. A. W. Tucker. *Dual systems of homogeneous linear relations*, In H. W. Kuhn and A. W. Tucker, editors, *Linear Inequalities and Related Systems*, number 38 in Annals of Mathematical Studies, pages 3–18. Princeton University Press, Princeton, New Jersey, 1956.
86. R. H. Tütüncü, K. C. Toh, and M. J. Todd. *Solving semidefinite-quadratic-linear programs using SDPT3*, Mathematical Programming Series B, 95, 189–217, 2003.
87. P. M. Vaidya. *A new algorithm for minimizing convex functions over convex sets*, Mathematical Programming, 73(3), 291–341, 1996.
88. L. Vandenberghe and S. Boyd. *Semidefinite programming*, SIAM Review, 38, 49–95, 1996.
89. R. J. Vanderbei *LOQO User's Guide – Version 4.05*, Princeton University, School of Engineering and Applied Science, Department of Operations Research and Financial Engineering, Princeton, New Jersey, 2006.

90. A. Wächter and L. T. Biegler. *On the implementation of a primal-dual interior point filter line search algorithm for large-scale nonlinear programming,* Mathematical Programming, 106(1), 25–57, 2006.

91. H. Wolkowicz, R. Saigal, and L. Vandenberghe, editors *Handbook of Semidefinite Programming: Theory, Algorithms, and Applications,* Kluwer Academic Publishers, 2000.

92. M. H. Wright. *The interior-point revolution in optimization: History, recent developments, and lasting consequences,* Bulletin (New Series) of the American Mathematical Society, 42(1), 39–56, 2004.

93. S. J. Wright. *Primal-Dual Interior-Point Methods,* SIAM, Philadelphia, USA, 1997.

94. X. Xu, P. Hung, and Y. Ye. *A simplified homogeneous and self-dual linear programming algorithm and its implementation,* Annals of Operations Research, 62, 151–172, 1996.

95. M. Yamashita, K. Fujisawa, and M. Kojima. *Implementation and evaluation of SDPA 6.0 (SemiDefinite Programming Algorithm 6.0),* Optimization Methods and Software, 18, 491–505, 2003.

96. Y. Ye *Interior-Point Algorithms: Theory and Analysis,* Wiley-Interscience Series in Discrete Mathematics and Optimization. John Wiley and Sons, New York, 1997.

97. Y. Ye, M. J. Todd, and S. Mizuno. *An $O(\sqrt{n}L)$-iteration homogeneous and self-dual linear programming algorithm,* Mathematics of Operations Research, 19, 53–67, 1994.

98. E. A. Yildirim and S. J. Wright. *Warm start strategies in interior-point methods for linear programming,* SIAM Journal on Optimization, 12(3), 782–810, 2002.

99. Y. Zhang. *On extending primal-dual interior-point algorithms from linear programming to semidefinite programming,* SIAM Journal of Optimization, 8, 356–386, 1998.

100. Y. Zhang. *Solving large-scale linear programs by interior-point methods under the MATLAB environment,* Optimization Methods Software, 10, 1–31, 1998.

# List of Participants

1. Benedetti Irene, Italy,
   `benedetti@math.unifi.it`
2. Bogani Claudio, Italy,
   `bogani@math.unifi.it`
3. Bomze Immanuel M., Austria, (**lecturer**)
   `immanuel.bomze@univie.ac.at`
4. Bosio Sandro, Italy,
   `sandro.bosio@polimi.it`
5. Cassioli Andrea, Italy,
   `cassioli@hotmail.com`
6. Di Pillo Gianni, Italy, (**editor**)
   `dipillo@dis.uniroma1.it`
7. Donato Maria Bernadette, Italy,
   `bdonato@dipmat.unime.it`
8. Demyanov Vladimir F., Russia, (**lecturer**)
   `vfd@ad9503.spb.edu`
9. Fagundez Fabio, Brazil,
   `fabio.fagundez@gmail.com`
10. Fedele Mariagrazia, Italy,
    `m.fedele@unifg.it`
11. Fletcher Roger, UK, (**lecturer**)
    `fletcher@maths.dundee.ac.uk`
12. Flores Salvador, France,
    `flores@cict.fr`
13. Jamali Abdur Rakib Muhammad Jalal Udd, Italy,
    `jamali@di.unito.it`
14. Khalaf Walaa, Italy,
    `walaa@deis.unical.it`
15. Kinzebulatov Damir, Russia,
    `knzbltv@udm.net`
16. Locatelli Marco, Italy,
    `locatell@di.unito.it`

17. Lorenz Thomas, Germany,
    `thomas.lorenz@iwr.uni-heidelberg.de`
18. Maggioni Francesca, Italy,
    `francesca.maggioni@unibg.it`
19. Marina Basova, Russia,
    `basova_marina@mail.ru`
20. Milasi Monica, Italy,
    `monica@dipmat.unime.it`
21. Palagi Laura, Italy,
    `palagi@dis.uniroma1.it`
22. Panicucci Barbara, Italy,
    `panicucc@mail.dm.unipi.it`
23. Passacantando Mauro, Italy,
    `passacantando@dma.unipi.it`
24. Piccialli Veronica, Italy,
    `Veronica.Piccialli@dis.uniroma1.it`
25. Poderskiy Ilya, Russia,
    `ilven@yandex.ru`
26. Porcelli Margherita, Italy,
    `margherita.porcelli@math.unifi.it`
27. Rinaldi Francesco, Italy,
    `frinaldi@iasi.cnr.it`
28. Risi Arnaldo, Italy,
    `risi@iasi.cnr.it`
29. Salvi Francesca, Italy,
    `francesca.salvi@unifi.it`
30. Schoen Fabio, Italy, (**editor**)
    `fabio.schoen@unifi.it`
31. Sciacca Eva, Italy,
    `sciacca@dmi.unict.it`
32. Sciandrone Marco, Italy,
    `sciandro@dsi.unifi.it`
33. Stevanato Elisa, Italy,
    `elisa.stevanato@pd.infn.it`
34. Terlaky Tamás, USA, (**lecturer**)
    `terlaky@lehigh.edu`
35. Toninelli Roberta, Italy,
    `roberta.toninelli@libero.it`
36. Tosques Mario, Italy,
    `mario.tosques@unipr.it`
37. Totz Bettina, Austria,
    `bettina.totz@cmx.at`

38. Zakharova Anastasia, Russia,
    `nastjka@list.ru`
39. Zecca Pietro, Italy,
    `zecca@unifi.it`
40. Zilinskas Julius, Lithuania,
    `julius.zilinskas@mii.lt`

# LIST OF C.I.M.E. SEMINARS

## Published by C.I.M.E

# Published by Ed. Cremonese, Firenze

1966
39. Calculus of variations
40. Economia matematica
41. Classi caratteristiche e questioni connesse
42. Some aspects of diffusion theory

1967
43. Modern questions of celestial mechanics
44. Numerical analysis of partial differential equations
45. Geometry of homogeneous bounded domains

1968
46. Controllability and observability
47. Pseudo-differential operators
48. Aspects of mathematical logic

1969
49. Potential theory
50. Non-linear continuum theories in mechanics and physics and their applications
51. Questions of algebraic varieties

1970
52. Relativistic fluid dynamics
53. Theory of group representations and Fourier analysis
54. Functional equations and inequalities
55. Problems in non-linear analysis

1971
56. Stereodynamics
57. Constructive aspects of functional analysis (2 vol.)
58. Categories and commutative algebra

1972
59. Non-linear mechanics
60. Finite geometric structures and their applications
61. Geometric measure theory and minimal surfaces

1973
62. Complex analysis
63. New variational techniques in mathematical physics
64. Spectral analysis

1974
65. Stability problems
66. Singularities of analytic spaces
67. Eigenvalues of non linear problems

1975
68. Theoretical computer sciences
69. Model theory and applications
70. Differential operators and manifolds

# Published by Ed. Liguori, Napoli

1976
71. Statistical Mechanics
72. Hyperbolicity
73. Differential topology

1977
74. Materials with memory
75. Pseudodifferential operators with applications
76. Algebraic surfaces

# Published by Ed. Liguori, Napoli & Birkhäuser

1978
77. Stochastic differential equations
78. Dynamical systems

1979
79. Recursion theory and computational complexity
80. Mathematics of biology

1980
81. Wave propagation
82. Harmonic analysis and group representations
83. Matroid theory and its applications

# Published by Springer-Verlag

# Lecture Notes in Mathematics

For information about earlier volumes
please contact your bookseller or Springer
LNM Online archive: springerlink.com

Vol. 1846: H. Ammari, H. Kang, Reconstruction of Small Inhomogeneities from Boundary Measurements (2004)

Vol. 1847: T.R. Bielecki, T. Björk, M. Jeanblanc, M. Rutkowski, J.A. Scheinkman, W. Xiong, Paris-Princeton Lectures on Mathematical Finance 2003 (2004)

Vol. 1848: M. Abate, J. E. Fornaess, X. Huang, J. P. Rosay, A. Tumanov, Real Methods in Complex and CR Geometry, Martina Franca, Italy 2002. Editors: D. Zaitsev, G. Zampieri (2004)

Vol. 1849: Martin L. Brown, Heegner Modules and Elliptic Curves (2004)

Vol. 1850: V. D. Milman, G. Schechtman (Eds.), Geometric Aspects of Functional Analysis. Israel Seminar 2002-2003 (2004)

Vol. 1851: O. Catoni, Statistical Learning Theory and Stochastic Optimization (2004)

Vol. 1852: A.S. Kechris, B.D. Miller, Topics in Orbit Equivalence (2004)

Vol. 1853: Ch. Favre, M. Jonsson, The Valuative Tree (2004)

Vol. 1854: O. Saeki, Topology of Singular Fibers of Differential Maps (2004)

Vol. 1855: G. Da Prato, P.C. Kunstmann, I. Lasiecka, A. Lunardi, R. Schnaubelt, L. Weis, Functional Analytic Methods for Evolution Equations. Editors: M. Iannelli, R. Nagel, S. Piazzera (2004)

Vol. 1856: K. Back, T.R. Bielecki, C. Hipp, S. Peng, W. Schachermayer, Stochastic Methods in Finance, Bressanone/Brixen, Italy, 2003. Editors: M. Fritelli, W. Runggaldier (2004)

Vol. 1857: M. Émery, M. Ledoux, M. Yor (Eds.), Séminaire de Probabilités XXXVIII (2005)

Vol. 1858: A.S. Cherny, H.-J. Engelbert, Singular Stochastic Differential Equations (2005)

Vol. 1859: E. Letellier, Fourier Transforms of Invariant Functions on Finite Reductive Lie Algebras (2005)

Vol. 1860: A. Borisyuk, G.B. Ermentrout, A. Friedman, D. Terman, Tutorials in Mathematical Biosciences I. Mathematical Neurosciences (2005)

Vol. 1861: G. Benettin, J. Henrard, S. Kuksin, Hamiltonian Dynamics – Theory and Applications, Cetraro, Italy, 1999. Editor: A. Giorgilli (2005)

Vol. 1862: B. Helffer, F. Nier, Hypoelliptic Estimates and Spectral Theory for Fokker-Planck Operators and Witten Laplacians (2005)

Vol. 1863: H. Führ, Abstract Harmonic Analysis of Continuous Wavelet Transforms (2005)

Vol. 1864: K. Efstathiou, Metamorphoses of Hamiltonian Systems with Symmetries (2005)

Vol. 1865: D. Applebaum, B.V. R. Bhat, J. Kustermans, J. M. Lindsay, Quantum Independent Increment Processes I. From Classical Probability to Quantum Stochastic Calculus. Editors: M. Schürmann, U. Franz (2005)

Vol. 1866: O.E. Barndorff-Nielsen, U. Franz, R. Gohm, B. Kümmerer, S. Thorbjønsen, Quantum Independent Increment Processes II. Structure of Quantum Lévy Processes, Classical Probability, and Physics. Editors: M. Schürmann, U. Franz, (2005)

Vol. 1867: J. Sneyd (Ed.), Tutorials in Mathematical Biosciences II. Mathematical Modeling of Calcium Dynamics and Signal Transduction. (2005)

Vol. 1868: J. Jorgenson, S. Lang, $Pos_n(R)$ and Eisenstein Series. (2005)

Vol. 1869: A. Dembo, T. Funaki, Lectures on Probability Theory and Statistics. Ecole d'Eté de Probabilités de Saint-Flour XXXIII-2003. Editor: J. Picard (2005)

Vol. 1870: V.I. Gurariy, W. Lusky, Geometry of Müntz Spaces and Related Questions. (2005)

Vol. 1871: P. Constantin, G. Gallavotti, A.V. Kazhikhov, Y. Meyer, S. Ukai, Mathematical Foundation of Turbulent Viscous Flows, Martina Franca, Italy, 2003. Editors: M. Cannone, T. Miyakawa (2006)

Vol. 1872: A. Friedman (Ed.), Tutorials in Mathematical Biosciences III. Cell Cycle, Proliferation, and Cancer (2006)

Vol. 1873: R. Mansuy, M. Yor, Random Times and Enlargements of Filtrations in a Brownian Setting (2006)

Vol. 1874: M. Yor, M. Émery (Eds.), In Memoriam Paul-André Meyer - Séminaire de Probabilités XXXIX (2006)

Vol. 1875: J. Pitman, Combinatorial Stochastic Processes. Ecole d'Eté de Probabilités de Saint-Flour XXXII-2002. Editor: J. Picard (2006)

Vol. 1876: H. Herrlich, Axiom of Choice (2006)

Vol. 1877: J. Steuding, Value Distributions of $L$-Functions (2007)

Vol. 1878: R. Cerf, The Wulff Crystal in Ising and Percolation Models, Ecole d'Eté de Probabilités de Saint-Flour XXXIV-2004. Editor: Jean Picard (2006)

Vol. 1879: G. Slade, The Lace Expansion and its Applications, Ecole d'Eté de Probabilités de Saint-Flour XXXIV-2004. Editor: Jean Picard (2006)

Vol. 1880: S. Attal, A. Joye, C.-A. Pillet, Open Quantum Systems I, The Hamiltonian Approach (2006)

Vol. 1881: S. Attal, A. Joye, C.-A. Pillet, Open Quantum Systems II, The Markovian Approach (2006)

Vol. 1882: S. Attal, A. Joye, C.-A. Pillet, Open Quantum Systems III, Recent Developments (2006)

Vol. 1883: W. Van Assche, F. Marcellàn (Eds.), Orthogonal Polynomials and Special Functions, Computation and Application (2006)

Vol. 1884: N. Hayashi, E.I. Kaikina, P.I. Naumkin, I.A. Shishmarev, Asymptotics for Dissipative Nonlinear Equations (2006)

Vol. 1885: A. Telcs, The Art of Random Walks (2006)

Vol. 1886: S. Takamura, Splitting Deformations of Degenerations of Complex Curves (2006)

Vol. 1887: K. Habermann, L. Habermann, Introduction to Symplectic Dirac Operators (2006)

Vol. 1888: J. van der Hoeven, Transseries and Real Differential Algebra (2006)

Vol. 1889: G. Osipenko, Dynamical Systems, Graphs, and Algorithms (2006)

Vol. 1890: M. Bunge, J. Funk, Singular Coverings of Toposes (2006)

Vol. 1891: J.B. Friedlander, D.R. Heath-Brown, H. Iwaniec, J. Kaczorowski, Analytic Number Theory, Cetraro, Italy, 2002. Editors: A. Perelli, C. Viola (2006)

Vol. 1892: A. Baddeley, I. Bárány, R. Schneider, W. Weil, Stochastic Geometry, Martina Franca, Italy, 2004. Editor: W. Weil (2007)

Vol. 1893: H. Hanßmann, Local and Semi-Local Bifurcations in Hamiltonian Dynamical Systems, Results and Examples (2007)

Vol. 1894: C.W. Groetsch, Stable Approximate Evaluation of Unbounded Operators (2007)

Vol. 1895: L. Molnár, Selected Preserver Problems on Algebraic Structures of Linear Operators and on Function Spaces (2007)

Vol. 1896: P. Massart, Concentration Inequalities and Model Selection, Ecole d'Été de Probabilités de Saint-Flour XXXIII-2003. Editor: J. Picard (2007)

## Recent Reprints and New Editions

# LECTURE NOTES IN MATHEMATICS

Edited by J.-M. Morel, F. Takens, B. Teissier, P.K. Maini

**Editorial Policy** (for Multi-Author Publications: Summer Schools/Intensive Courses)

1. Lecture Notes aim to report new developments in all areas of mathematics and their applications - quickly, informally and at a high level. Mathematical texts analysing new developments in modelling and numerical simulation are welcome. Manuscripts should be reasonably self-contained and rounded off. Thus they may, and often will, present not only results of the author but also related work by other people. They should provide sufficient motivation, examples and applications. There should also be an introduction making the text comprehensible to a wider audience. This clearly distinguishes Lecture Notes from journal articles or technical reports which normally are very concise. Articles intended for a journal but too long to be accepted by most journals, usually do not have this "lecture notes" character.

2. In general SUMMER SCHOOLS and other similar INTENSIVE COURSES are held to present mathematical topics that are close to the frontiers of recent research to an audience at the beginning or intermediate graduate level, who may want to continue with this area of work, for a thesis or later. This makes demands on the didactic aspects of the presentation. Because the subjects of such schools are advanced, there often exists no textbook, and so ideally, the publication resulting from such a school could be a first approximation to such a textbook. Usually several authors are involved in the writing, so it is not always simple to obtain a unified approach to the presentation.

   For prospective publication in LNM, the resulting manuscript should not be just a collection of course notes, each of which has been developed by an individual author with little or no co-ordination with the others, and with little or no common concept. The subject matter should dictate the structure of the book, and the authorship of each part or chapter should take secondary importance. Of course the choice of authors is crucial to the quality of the material at the school and in the book, and the intention here is not to belittle their impact, but simply to say that the book should be planned to be written by these authors jointly, and not just assembled as a result of what these authors happen to submit.

   This represents considerable preparatory work (as it is imperative to ensure that the authors know these criteria before they invest work on a manuscript), and also considerable editing work afterwards, to get the book into final shape. Still it is the form that holds the most promise of a successful book that will be used by its intended audience, rather than yet another volume of proceedings for the library shelf.

3. Manuscripts should be submitted either online at www.editorialmanager.com/lnm/ to Springer's mathematics editorial, or to one of the series editors. Volume editors are expected to arrange for the refereeing, to the usual scientific standards, of the individual contributions. If the resulting reports can be forwarded to us (series editors or Springer) this is very helpful. If no reports are forwarded or if other questions remain unclear in respect of homogeneity etc, the series editors may wish to consult external referees for an overall evaluation of the volume. A final decision to publish can be made only on the basis of the complete manuscript; however a preliminary decision can be based on a pre-final or incomplete manuscript. The strict minimum amount of material that will be considered should include a detailed outline describing the planned contents of each chapter.

   Volume editors and authors should be aware that incomplete or insufficiently close to final manuscripts almost always result in longer evaluation times. They should also be aware that parallel submission of their manuscript to another publisher while under consideration for LNM will in general lead to immediate rejection.

4. Manuscripts should in general be submitted in English. Final manuscripts should contain at least 100 pages of mathematical text and should always include

   - a general table of contents;
   - an informative introduction, with adequate motivation and perhaps some historical remarks: it should be accessible to a reader not intimately familiar with the topic treated;
   - a global subject index: as a rule this is genuinely helpful for the reader.

   Lecture Notes volumes are, as a rule, printed digitally from the authors' files. We strongly recommend that all contributions in a volume be written in the same LaTeX version, preferably LaTeX2e. To ensure best results, authors are asked to use the LaTeX2e style files available from Springer's web-server at

   ftp://ftp.springer.de/pub/tex/latex/svmonot1/ (for monographs) and
   ftp://ftp.springer.de/pub/tex/latex/svmultt1/ (for summer schools/tutorials).

   Additional technical instructions are available on request from: lnm@springer.com.

5. Careful preparation of the manuscripts will help keep production time short besides ensuring satisfactory appearance of the finished book in print and online. After acceptance of the manuscript authors will be asked to prepare the final LaTeX source files and also the corresponding dvi-, pdf- or zipped ps-file. The LaTeX source files are essential for producing the full-text online version of the book. For the existing online volumes of LNM see: http://www.springerlink.com/openurl.asp?genre=journal&issn=0075-8434.

   The actual production of a Lecture Notes volume takes approximately 12 weeks.

6. Volume editors receive a total of 50 free copies of their volume to be shared with the authors, but no royalties. They and the authors are entitled to a discount of 33.3% on the price of Springer books purchased for their personal use, if ordering directly from Springer.

7. Commitment to publish is made by letter of intent rather than by signing a formal contract. Springer-Verlag secures the copyright for each volume. Authors are free to reuse material contained in their LNM volumes in later publications: a brief written (or e-mail) request for formal permission is sufficient.

**Addresses:**

Professor J.-M. Morel, CMLA,
École Normale Supérieure de Cachan,
61 Avenue du Président Wilson,
94235 Cachan Cedex, France
E-mail: Jean-Michel.Morel@cmla.ens-cachan.fr

Professor F. Takens, Mathematisch Instituut,
Rijksuniversiteit Groningen, Postbus 800,
9700 AV Groningen, The Netherlands
E-mail: F.Takens@rug.nl

Professor B. Teissier,
Institut Mathématique de Jussieu,
UMR 7586 du CNRS,
Équipe "Géométrie et Dynamique",
175 rue du Chevaleret,
75013 Paris, France
E-mail: teissier@math.jussieu.fr

*For the "Mathematical Biosciences Subseries" of LNM:*

Professor P.K. Maini, Center for Mathematical Biology,
Mathematical Institute, 24-29 St Giles,
Oxford OX1 3LP, UK
E-mail: maini@maths.ox.ac.uk

Springer, Mathematics Editorial I, Tiergartenstr. 17,
69121 Heidelberg, Germany,
Tel.: +49 (6221) 487-8259
Fax: +49 (6221) 4876-8259
E-mail: lnm@springer.com